Elliptic Curves

2ND EDITION

Elliptic Curves

JAMES S MILNE

University of Michigan, USA

World Scientific

NEW JERSEY • LONDON • SINGAPORE • BEIJING • SHANGHAI • HONG KONG • TAIPEI • CHENNAI • TOKYO

Published by

World Scientific Publishing Co. Pte. Ltd.

5 Toh Tuck Link, Singapore 596224

USA office: 27 Warren Street, Suite 401-402, Hackensack, NJ 07601

UK office: 57 Shelton Street, Covent Garden, London WC2H 9HE

Library of Congress Cataloging-in-Publication Data
Names: Milne, J. S., 1942– author.
Title: Elliptic curves / James S. Milne, University of Michigan.
Description: Second edition. | New Jersey : World Scientific Publishing Co., [2020] |
Includes bibliographical references and index.
Identifiers: LCCN 2020027465 | ISBN 9789811221835 (hardcover) | ISBN 9789811221842
(ebook for institutions) | ISBN 9789811221859 (ebook for individuals)
Subjects: LCSH: Curves, Elliptic.
Classification: LCC QA567.2.E44 M55 2020 | DDC 516.3/52--dc23
LC record available at https://lccn.loc.gov/2020027465

British Library Cataloguing-in-Publication Data
A catalogue record for this book is available from the British Library.

First edition published as: Milne, J. S., Elliptic curves. BookSurge Publishers, Charleston, SC, 2006.
viii+238 pp. ISBN: 1-4196-5257-5

For any available supplementary material, please visit
https://www.worldscientific.com/worldscibooks/10.1142/11870#t=suppl

Desk Editor: Liu Yumeng

To the memory of my advisor, John T. Tate 1925–2019.

Preface

In early 1996, I taught a course on elliptic curves. Since this was not long after Wiles had proved Fermat's Last Theorem and I promised to explain some of the ideas underlying his proof, the course attracted an unusually large and diverse audience. As a result, I attempted to make the course accessible to all students with a knowledge only of the standard first-year graduate courses.

When it was over, I collected the notes that I had handed out during the course into a single file, made a few corrections, and posted them on the Web, where they were downloaded tens of thousands of times.

In 2006, I rewrote the notes and made them available as a paperback (the first edition of this work). For this second edition, I have rewritten and updated the notes once again, while retaining most of the numbering from the first edition.

Beyond its intrinsic interest, the study of elliptic curves makes an excellent introduction to some of the deeper aspects of current research in number theory. In the book, I have attempted to place the theory of elliptic curves in this wider context, as well as in its historical context.

In reviewing the theory of elliptic curves, I have been struck by how much of it originated with calculations: those of Ramanujan, which suggested Hecke operators and the Ramanujan conjecture; those of Sato, which suggested the Sato–Tate conjecture in its general forms; those of Selmer, which suggested the Cassels–Tate duality theorem; those of Birch and Stephens, which suggested the Gross–Zagier formula; and, of course, those of Birch and Swinnerton-Dyer, which suggested their conjecture and its generalizations.

I wish to thank Rochelle Kronzek of World Scientific Publishers for encouraging me to prepare this second edition.

Contents

Introduction

An elliptic curve over a field k is a nonsingular projective curve of genus 1 with a distinguished point. When the characteristic of k is not 2 or 3, it can be realized as a plane projective curve

$$Y^2Z = X^3 + aXZ^2 + bZ^3, \qquad 4a^3 + 27b^2 \neq 0.$$

The distinguished point is $(0\colon 1\colon 0)$. For example, the following pictures show the real points (except the point at infinity) of two elliptic curves:

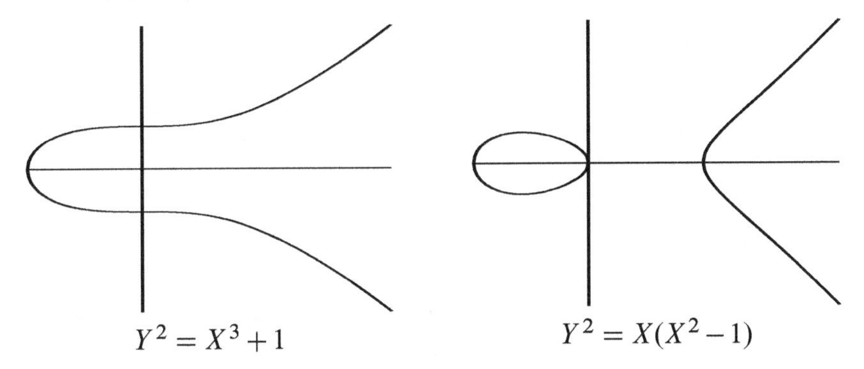

$Y^2 = X^3 + 1$ $\qquad$ $Y^2 = X(X^2 - 1)$

Although the problem of computing the points on an elliptic curve E with rational numbers as coordinates has fascinated mathematicians since the time of the ancient Greeks, it was not until 1922 that it was proved that it is possible to construct all the points starting from a finite number by drawing chords and tangents. This is the famous theorem of Mordell, which shows more precisely that the abelian group $E(\mathbb{Q})$ of rational points on E is finitely generated. There is a simple algorithm for computing the torsion subgroup of $E(\mathbb{Q})$, but there is still no *proven* algorithm for computing the rank. In one of the earliest applications of computers to mathematics, Birch and Swinnerton-Dyer discovered experimentally a relation between the rank and the numbers N_p of the points on the curve read modulo the different prime numbers p. The problem of proving this relation (the conjecture of Birch and Swinnerton-Dyer) is one of the most important in mathematics. Chapter IV of the book proves Mordell's theorem and explains the conjecture of Birch and Swinnerton-Dyer.

In 1955, Taniyama noted that it was plausible that the N_p attached to a given elliptic curve always arise in a simple way from a modular form (in modern terminology, that the elliptic curve is modular). In 1985 Frey observed that this did not appear to be true for the elliptic curve attached to a nontrivial solution of the Fermat equation $X^n + Y^n = Z^n$, $n > 2$. His observation prompted Serre to revisit some old conjectures implying this, and Ribet proved enough of his conjectures to deduce that Frey's observation is correct: the elliptic curve attached to a nontrivial solution of the Fermat equation is not modular. Finally, in 1994 Wiles (assisted by Taylor) proved that every elliptic curve in a large class is modular, thereby proving Fermat's Last Theorem. Chapter V of the book is devoted to explaining these results.

The first three chapters of the book develop the basic theory of elliptic curves.

Elliptic curves have been used to shed light on some important problems that, at first sight, appear to have nothing to do with them. I mention three such problems.

Fast factorization of integers

There is an algorithm for factoring integers that uses elliptic curves and is in many respects better than previous algorithms. People have been factoring integers for millennia, but recently the topic has become of practical significance: given an integer n that is the product $n = pq$ of two (large) primes p and q, there is a code for which everyone who knows n can encode a message, but only those who know p and q can decode it. The security of the code depends on no unauthorized person being able to factor n. See Koblitz 1994, VI, 4, or Silverman and Tate 1992, IV, 4.

Lattices and the sphere packing problem

The sphere packing problem is that of finding an arrangement of n-dimensional unit balls in Euclidean space that covers as much of the space as possible without overlaps. The arrangement is called a lattice packing if the centres of the spheres are the points of a lattice in n-space.

The best packing in the plane is a lattice packing.

Elliptic curves have been used to find lattice packings in many dimensions that are denser than any previously known. See Chapter IV, §11.

Congruent numbers

A natural number n is said to be ***congruent*** if it is the area of a right-angle triangle whose sides have rational length. For example, 6 is congruent because it is the area of a $(3,4,5)$ triangle. Such triangles occur in the writings of Diophantus, and the problem of determining whether a given natural number n is congruent is stated already in an Arabic manuscript from the tenth century (Dickson 1966). Fibonacci showed that 5 and 6 are congruent, Fermat that $1,2,3$, are not congruent, and Euler proved that 7 is congruent. However, an elementary argument shows that n is congruent if and only if the elliptic curve

$$Y^2 = X^3 - n^2 X$$

has a point (x,y) with $x,y \in \mathbb{Q}$ and $y \neq 0$, and it is only with recent advances in the study of elliptic curves that significant progress has been made on the ancient problem. See p. 220.

Why are they called elliptic curves?

Computing the length of an arc of an ellipse leads to integrals that cannot be evaluated in terms of elementary functions. These integrals were called "elliptic", and the name was then carried over to the curves on which they naturally lie (see p. 275). For a friendly explanation of how elliptic integrals led to elliptic curves, and why ellipses are not elliptic curves (and never will be), see Rice and Brown 2012.

PREREQUISITES

A knowledge of the basic algebra, analysis, and topology usually taught in advanced undergraduate or first-year graduate courses. Some knowledge of algebraic geometry and algebraic number theory will be useful but not essential.

NOTATION

We use the following standard notation: $\mathbb{N}$ is the set of natural numbers $\{0, 1, 2, \ldots\}$, $\mathbb{Z}$ the ring of integers, $\mathbb{Q}$ the field of rational numbers, $\mathbb{R}$ the field of real numbers, $\mathbb{C}$ the field of complex numbers, and $\mathbb{F}_p$ the field with p elements. A number field is a finite extension of $\mathbb{Q}$.

Throughout the book, k is a field and k^{al} is an algebraic closure of k. All rings are commutative with 1, and homomorphisms of rings are required to map 1 to 1. For a ring A, $A^\times$ is the group of units in A:

$$A^\times = \{a \in A \mid \text{there exists a } b \in A \text{ such that } ab = 1\}.$$

For an abelian group X,

$$X_n = \{x \in X \mid nx = 0\}.$$

For a finite set S, $\#S$ or occasionally $[S]$ denotes the number of elements of S. For an element a of a set with an equivalence relation, we sometimes use $[a]$ to denote the equivalence class of a. The symbol log always means the natural logarithm.

$X \stackrel{\text{def}}{=} Y$ X is defined to be Y, or equals Y by definition;
$X \subset Y$ X is a subset of Y (not necessarily proper);
$X \approx Y$ X and Y are isomorphic;
$X \simeq Y$ X and Y are canonically isomorphic, or there is a given isomorphism from one to the other.

REFERENCES

In addition to references listed at the end, I refer to the following of my online notes, available at `https://www.jmilne.org/math/`.

ANT Algebraic Number Theory, v3.07.

CA A Primer of Commutative Algebra, v4.03.

FT Fields and Galois Theory, v4.61.

ACKNOWLEDGEMENTS

I thank the following for providing corrections and comments for earlier versions of this work: Alan Bain, Leen Bleijenga, Keith Conrad, Jean Cougnard, Rankeya Datta, Mark Faucette, Jochen Gerhard, Enis Kaya, Michael Müller, Stefan Müller, Holger Partsch, PENG Bo, Jasper Scholten, Nicholas Wilson, Dmitriy Zanin, and others.

Chapter I

Algebraic Curves

Thus we see that many interesting curves can be defined as the zero sets of equations $f(x, y) = 0$, where $f(x, y)$ is a polynomial in x and y. Such curves were called algebraic curves *by Leibniz.*

Brieskorn and Knörrer 1986, p. 78.

In this chapter, we develop some of the basic theory of algebraic curves, concentrating on plane curves. In Section 3 we describe the group structure on a cubic curve.

1 Plane algebraic curves; Bezout's theorem

We introduce plane algebraic curves, and develop the intersection theory of curves in the plane.

Polynomial rings

We shall need to use that polynomial rings over fields are unique factorization domains. This is proved by induction starting from the fact that a polynomial ring $A[X]$ is a unique factorization domain if A is (CA, 4.9). As the units in $k[X_1, \ldots, X_n]$ are just the nonzero constants, every element f of the ring can be written as a product

$$f = f_1^{m_1} \cdots f_r^{m_r}$$

of powers of irreducible polynomials f_i with no f_i a constant multiple of another, and the factorization is unique up to replacing the f_i with constant multiples. The ***repeated factors*** of f are those f_i with $m_i > 1$. A polynomial f is irreducible if and only if the principal ideal (f) is prime (CA, 4.1).

Affine plane curves

The ***affine plane*** over k is $\mathbb{A}^2(k) = k \times k$. A nonconstant polynomial $f \in k[X,Y]$, assumed to have no repeated factor in $k^{\mathrm{al}}[X,Y]$, defines an ***affine plane curve*** C_f over k whose points with coordinates in any field K containing k are the zeros of f in K^2,

$$C_f(K) = \{(x,y) \in K^2 \mid f(x,y) = 0\}.$$

For $c \in k^{\times}$, the curves C_f and C_{cf} have the same points in every field K, and will not be distinguished.

For a curve over k, a point with coordinates in k is often called a ***rational point*** on the curve.

The curve C_f is said to be ***irreducible*** if f is irreducible in $k[X,Y]$, and it is said be ***geometrically irreducible*** if f remains irreducible in $k^{\mathrm{al}}[X,Y]$. For a general curve C_f, we can write $f = f_1 f_2 \cdots f_r$ with the f_i distinct irreducible polynomials in $k[X,Y]$, and then

$$C_f(K) = C_{f_1}(K) \cup \cdots \cup C_{f_r}(K), \quad \text{all } K \supset k.$$

The curves C_{f_i} are called the ***irreducible components*** of C_f (or just ***components***).

The notation $C\colon f = 0$ means that C is the curve C_f and the notation $C\colon f_1 = f_2$ means that C is the curve $C_{f_1 - f_2}$.

EXAMPLE 1.1 The curve $X^2 - 2Y^2 = 0$ is irreducible over $\mathbb{Q}$, but becomes the pair of lines $(X - \sqrt{2}Y)(X + \sqrt{2}Y) = 0$ over $\mathbb{Q}[\sqrt{2}]$. Therefore, it is not geometrically irreducible.

Recall that in characteristic $p \neq 0$, the binomial theorem says that

$$(X+Y)^p = X^p + Y^p.$$

EXAMPLE 1.2 If the field k is not perfect, then it has characteristic $p \neq 0$ for some prime p and there exists an a in k that is not a pth power in k. The element a becomes an pth power in k^{al}, say, $a = \alpha^p$. The polynomial $f = X^p + aY^p$ is irreducible in $k[X,Y]$, but it becomes a pth power

$f = (X + \alpha Y)^p$ in $k^{\mathrm{al}}[X,Y]$, and so it does not define an affine plane curve in our sense. This problem occurs only for non perfect k. When k is perfect, a polynomial with no repeated factor in $k[X,Y]$ will not acquire a repeated factor in $k^{\mathrm{al}}[X,Y]$.

We define the partial derivatives of polynomials by the obvious formulas,

$$\frac{\partial}{\partial X}\left(\sum\nolimits_{i,j} a_{ij}X^iY^j\right) = \sum\nolimits_{i,j} i a_{ij}X^{i-1}Y^j.$$

When we write $f(X,Y)$ as a polynomial in $X-a$ and $Y-b$, we find that

$$f(X,Y) = f(a,b) + \left(\frac{\partial f}{\partial X}\right)_{(a,b)}(X-a) + \left(\frac{\partial f}{\partial Y}\right)_{(a,b)}(Y-b)$$
$$+ \text{ terms of degree } \geq 2 \text{ in } X-a,\, Y-b.$$

We call this the ***Taylor expansion*** of f.

Let $P = (a,b) \in C_f(K)$, some $K \supset k$. If at least one of the partial derivatives $\frac{\partial f}{\partial X}, \frac{\partial f}{\partial Y}$ is nonzero at P, then P is said to be ***nonsingular***, and the ***tangent line*** to C at P is defined to be

$$\left(\frac{\partial f}{\partial X}\right)_P (X-a) + \left(\frac{\partial f}{\partial Y}\right)_P (Y-b) = 0.$$

A curve C is said to be ***nonsingular*** if all the points in $C(k^{\mathrm{al}})$ are nonsingular, and otherwise it is said to be ***singular***.

ASIDE 1.3 Let $f(x,y)$ be a real-valued function on $\mathbb{R}^2$. The condition that at least one of $\frac{\partial f}{\partial X}, \frac{\partial f}{\partial Y}$ is nonzero at P means that the curve has a local parameterization near P. The gradient $\nabla f \stackrel{\text{def}}{=} \left(\frac{\partial f}{\partial X}, \frac{\partial f}{\partial Y}\right)$ of f is a vector field on $\mathbb{R}^2$ whose value $(\nabla f)_P$ at a point $P = (a,b) \in \mathbb{R}^2$ is normal to any level curve $f(x,y) = c$ through P, and so

$$(\nabla f)_P \cdot (x-a, y-b) = 0$$

is the tangent line to the level curve. This explains the above definition.

REMARK 1.4 (a) If C is nonsingular, then every point in $C(K)$ for K a field containing k is nonsingular, because otherwise the polynomials f, $\frac{\partial f}{\partial X}$, $\frac{\partial f}{\partial Y}$ would generate a *proper* ideal of $k[X,Y]$, and so would have a common zero in k^{al} by the Nullstellensatz.[1]

[1]This says that polynomials $f_1,\ldots,f_r$ in $k[X_1,\ldots,X_n]$ have a common zero in k^{al} unless the ideal $(f_1,\ldots,f_r)$ they generate equals $k[X_1,\ldots,X_n]$. See CA, 13.8.

(b) A point on the intersection of two irreducible components of a curve is always singular. Consider, for example, a curve C_f with $f = f_1 f_2$ and $f_1(0,0) = 0 = f_2(0,0)$. Then

$$\frac{\partial f}{\partial X}(0,0) = \left(f_1 \frac{\partial f_2}{\partial X} + \frac{\partial f_1}{\partial X} f_2\right)(0,0) = 0$$
$$\frac{\partial f}{\partial Y}(0,0) = \left(f_1 \frac{\partial f_2}{\partial Y} + \frac{\partial f_1}{\partial Y} f_2\right)(0,0) = 0,$$

and so $P = (0,0)$ is singular on C_f.

(c) If C_f and C_g are affine plane curves over k with no common component, then $C_f(k^{\mathrm{al}}) \cap C_g(k^{\mathrm{al}})$ is finite. The theory of resultants (I, 1.22, 1.24) shows that there exist polynomials $a(X,Y)$, $b(X,Y)$, and $r(X)$ with coefficients in k such that

$$a(X,Y)F(X,Y) + b(X,Y)G(X,Y) = r(X)$$

and $r(x_0) = 0$ if and only if $F(x_0,Y)$ and $G(x_0,Y)$ have a common zero. The polynomial $r(X)$ has only finitely many roots, and for each root x_0, $F(x_0,Y)$ and $G(x_0,Y)$ have only finitely many common zeros.

(d) Let $f = f(X,Y) \in k[X,Y]$ be a nonconstant polynomial without repeated factors. If $\frac{\partial f}{\partial X}$ is the zero polynomial, then either X does not occur in f or else $\mathrm{char}(k) = p \neq 0$ and $f \in k[X^p, Y]$. If $\frac{\partial f}{\partial X}$ and $\frac{\partial f}{\partial Y}$ are both zero, then $\mathrm{char}(k) = p \neq 0$ and $f \in k[X^p, Y^p]$, say,

$$f(X,Y) = \sum\nolimits_{i,j} a_{ij} X^{ip} Y^{jp}.$$

If k is perfect, then

$$\sum\nolimits_{i,j} a_{ij} X^{ip} Y^{jp} = \left(\sum\nolimits_{i,j} a_{ij}^{1/p} X^i Y^j\right)^p,$$

and so f is a pth power in $k[X,Y]$, contradicting the hypothesis on f. Thus, if k is perfect, then $\frac{\partial f}{\partial X}$ and $\frac{\partial f}{\partial Y}$ are not both the zero polynomial, and there are only finitely many singular points in $C_f(k^{\mathrm{al}})$.

EXAMPLE 1.5 Consider the curve

$$C: \quad Y^2 = X^3 + aX + b.$$

At a singular point (x,y) of C,

$$2Y = 0, \quad 3X^2 + a = 0, \quad Y^2 = X^3 + aX + b.$$

If $\mathrm{char}(k) \neq 2$, then $y = 0$ and x is a common root of $X^3 + aX + b$ and its derivative, i.e., it is a double root of $X^3 + aX + b$; it follows that

$$\begin{aligned} C \text{ is nonsingular} &\iff X^3 + aX + b \text{ has no multiple root (in } k^{\mathrm{al}}) \\ &\iff \text{its discriminant}^2\, \Delta = 4a^3 + 27b^2 \text{ is nonzero.} \end{aligned}$$

If $\mathrm{char}(k) = 2$, then C always has a singular point in k^{al}, namely, (α, β), where $3\alpha^2 + a = 0$ and $\beta^2 = \alpha^3 + a\alpha + b$.

Let C_f be an affine curve over k, and let $P = (a, b) \in C_f(K)$. We can write f as a polynomial in $X - a$ and $Y - b$ with coefficients in K, say,

$$f(X,Y) = f_1(X-a, Y-b) + \cdots + f_n(X-a, Y-b), \tag{1}$$

where f_i is homogeneous of degree i in $X - a$ and $Y - b$. The point P is nonsingular if and only if $f_1 \neq 0$, in which case the tangent line to C_f at P has equation $f_1 = 0$ (compare (1) with the Taylor expansion of f).

Let $P = (a, b)$ be a singular point on C_f, so that

$$f(X,Y) = f_m(X-a, Y-b) + \text{ terms of higher degree,}$$

with $f_m \neq 0$, $m \geq 2$. Then m is called the ***multiplicity*** of P on C, denoted $m_P(C)$. If $m = 2$ (resp. 3), then P is called a ***double*** (resp. ***triple) point.*** Over k^{al},

$$f_m(X,Y) = \prod L_i^{r_i},$$

where each L_i is a polynomial $c_i(X - a) + d_i(Y - b)$ of degree one with coefficients in k^{al} and the lines $L_i = 0$ are distinct. They are called the ***tangent lines*** to C_f at P, and r_i is the ***multiplicity*** of $L_i = 0$. We say that P is an ***ordinary singularity*** if the tangent lines all have multiplicity 1. An ordinary double point is called a ***node.***

EXAMPLE 1.6 The curve $Y^2 = X^3 + aX^2$ has a singularity at $(0,0)$. If $a \neq 0$, it is a node, and the tangent lines at $(0,0)$ are $Y = \pm\sqrt{a}X$. They are defined over k if and only if a is a square in k. If $a = 0$, the singularity is a cusp (see 1.12 below).

ASIDE 1.7 (a) In the definition of an affine plane curve, we require f to have no repeated factors over k^{al} so that our curves are geometrically reduced. Allowing

[2] According to the usual definition, this is actually the negative of the discriminant.

repeated factors would allow the curve to have multiple components, not just multiple points, and it would not be determined by its points in k^{al}.

(b) (For the experts.) Essentially, we have defined an affine plane curve to be a geometrically reduced closed subscheme of $\mathbb{A}^2_k$ of dimension 1. Such a scheme corresponds to an ideal in $k[X,Y]$ of height one, which is principal because $k[X,Y]$ is a unique factorization domain. The polynomial generating the ideal is uniquely determined by the scheme up to multiplication by a nonzero constant.

Intersection numbers

We wish to define the intersection number of curves C_f and C_g at a point P of $C_f \cap C_g$. For $P = (0,0)$, the next proposition shows that there is exactly one reasonable way of doing this.

Let $\mathcal{F}(k)$ be the set of pairs of polynomials $f, g \in k[X,Y]$ having no common factor h in $k[X,Y]$ with $h(0,0) = 0$.

PROPOSITION 1.8 *There is a unique map $I: \mathcal{F}(k) \to \mathbb{N}$ such that*

(a) $I(X,Y) = 1$;

(b) $I(f,g) = I(g,f)$ *all* $(f,g) \in \mathcal{F}(k)$;

(c) $I(f,gh) = I(f,g) + I(f,h)$ *all* $(f,g), (f,h) \in \mathcal{F}(k)$;

(d) $I(f, g+hf) = I(f,g)$ *all* $(f,g) \in \mathcal{F}(k)$, $h \in k[X,Y]$;

(e) $I(f,g) = 0$ *if* $g(0,0) \neq 0$.

PROOF. We first prove the uniqueness. Let $f, g \in \mathcal{F}(k)$. The theory of resultants (see 1.24 below) provides us with polynomials $a(X,Y)$ and $b(X,Y)$ such that $af + bg = r \in k[X]$ and $\deg_Y(a) < \deg_Y(g)$, $\deg_Y(b) < \deg_Y(f)$. If $\deg_Y(f) \leq \deg_Y(g)$, then we write

$$I(f,g) \overset{(c)}{=} I(f,gb) - I(f,b) \overset{(d)}{=} I(f,r) - I(f,b),$$

and otherwise, by a similar argument, we write

$$I(f,g) = I(r,g) - I(a,g).$$

Continue in this fashion until Y is eliminated from one of the polynomials, say, from g, so that $g = g(X) \in k[X]$. Write $g(X) = X^m g_0(X)$, where $g_0(0) \neq 0$. Then

$$I(f,g) \overset{(c)}{=} mI(f,X) + I(f,g_0) \overset{(e)}{=} mI(f,X).$$

After subtracting a multiple of X from $f(X,Y)$, we can assume that it is a polynomial in Y alone. Write $f(Y) = Y^n f_0(Y)$, where $f_0(0) \neq 0$. Then

$$I(f,X) \overset{(c)}{=} nI(Y,X) + I(f_0,X) \overset{(a,b,e)}{=} n.$$

This completes the proof of uniqueness.

For the existence, let

$$k[X,Y]_{(0,0)} = \{h_1/h_2 \mid h_1,h_2 \in k[X,Y],\, h_2(0,0) \neq 0\}$$

(local ring at the maximal ideal (X,Y)). If $f,g \in \mathcal{F}(k)$, then the quotient ring $k[X,Y]_{(0,0)}/(f,g)$ is finite dimensional as a k-vector space, and when we set $I(f,g)$ equal to its dimension we obtain a map with the required properties (Fulton 1969, §3.3; in the first and second editions, this is Chap. 3, §3). □

REMARK 1.9 Let K be a field containing k. The theory of resultants shows that $\mathcal{F}(k)$ is contained in $\mathcal{F}(K)$ (see 1.25). It follows that, for $f,g \in \mathcal{F}(k)$, $I(f,g)$ is the same whether we regard f and g as polynomials with coefficients in k or K. In fact, it is possible to show directly that $\dim_k k[X,Y]_{(0,0)}/(f,g)$ does not change when we pass from k to K.

EXAMPLE 1.10 Applying 1.8d, we find that

$$I(Y^2 - X^2(X+1), X) = I(Y^2, X) = 2.$$

Although the Y-axis is not tangent to the curve $Y^2 = X^2(X+1)$, the intersection number is > 1 because the origin is singular on the curve.

The argument in the proof of the proposition is a practical algorithm for computing $I(f,g)$, but if the polynomials are monic when regarded as polynomials in Y, the following method is faster. If $\deg_Y(g) \geq \deg_Y(f)$, then we can divide f into g (as polynomials in Y) and obtain

$$g = fh + r, \quad \deg_Y r < \deg_Y f \text{ or } r = 0.$$

By property (d),

$$I(f,g) = I(f,r).$$

Continue in this fashion until one of the polynomials has degree 1 in Y, and apply the following lemma.

LEMMA 1.11 *If $f(0)=0$, then $I(Y-f(X),g(X,Y))=m$, where X^m is the power of X dividing $g(X,f(X))$.*

PROOF. Divide $Y-f(X)$ into $g(X,Y)$ (as polynomials in Y) to obtain

$$g(X,Y)=(Y-f(X))h(X,Y)+g(X,f(X)),$$

from which it follows that

$$I(Y-f(X),g(X,Y))=I(Y-f(X),g(X,f(X))=mI(Y-f(X),X).$$

Finally, since we are assuming $f(0)=0$, $f(X)=Xh(X)$, and so

$$I(Y-f(X),X)=I(Y,X)=1. \qquad \square$$

Consider two curves C_f and C_g in $\mathbb{A}^2(k)$, and let

$$P=(a,b)\in C_f(K)\cap C_g(K), \quad K\supset k.$$

We say that P is an ***isolated point*** of $C_f\cap C_g$ if C_f and C_g do not have a common component passing through P. Then f and g have no common factor h with $h(a,b)=0$, and so we can define the ***intersection number*** of C_f and C_g at P to be

$$I(P,C_f\cap C_g)\stackrel{\text{def}}{=}I(f(X+a,Y+b),g(X+a,Y+b)).$$

For example, if $P=(0,0)$, then $I(P,C_f\cap C_g)=I(f,g)$. From the definition of I, we find that

$$I(P,C_f\cap C_g)=\dim_K K[X,Y]_{(a,b)}/(f,g)$$

where $K[X,Y]_{(a,b)}$ is the local ring

$$K[X,Y]_{(a,b)}=\{h_1/h_2 \mid h_1,h_2\in K[X,Y],\ h_2(a,b)\neq 0\}$$

at the point (a,b).

If C_f and C_g are affine plane curves over k with no common component, then

$$\sum_{P\in C_f(k^{\mathrm{al}})\cap C_g(k^{\mathrm{al}})} I(P,C_f\cap C_g)=\dim_k k[X,Y]/(f,g)$$

(Fulton 1969, §3.3 (9)).

REMARK 1.12 Let P be a double point on a curve C, and suppose that C has only one tangent line L at P. Then $I(P, L \cap C) \geq 3$ and, when equality holds, we call P a ***cusp***. For example, let C be the curve $Y^2 = X^3$ and let $P = (0,0)$. Then $L : Y = 0$ is the only tangent line at P and

$$I(P, L \cap C) \stackrel{\text{def}}{=} I(Y^2 - X^3, Y) \stackrel{1.8\text{d}}{=} I(X^3, Y) = 3,$$

and so P is a cusp (see the picture in 1.14).

REMARK 1.13 As one would hope, $I(P, C \cap D) = 1$ if and only if P is nonsingular on both C and D and the tangent lines to C and D at P are distinct. More generally,

$$I(P, C \cap D) \geq m_P(C) \cdot m_P(D),$$

with equality if and only if C and D have no tangent line in common at P (Fulton 1969, §3.3 (5)).

ASIDE 1.14 Intuitively, the intersection number of two curves at a point is the actual number of intersection points after one of the curves has been moved slightly. For example, the intersection number at $(0,0)$ of the Y-axis with the curve $Y^2 = X^3$ should be 2 because, after the Y-axis has been moved slightly, the single point of intersection becomes two points — see the picture at right. Similar, the intersection number at $(0,0)$ of the Y-axis with the curves $Y^2 = X^2(X+1)$ (see 1.10) and $Y^2 = X(X^2-1)$ (see p. 1) should be 2. Of course, this picture is complicated by the fact that the intersection points may become visible only when we use complex numbers.

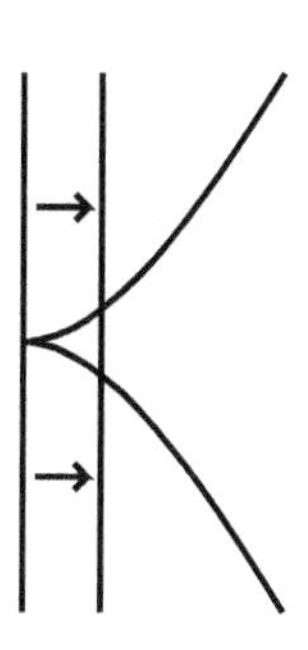

Projective plane curves

The ***projective plane*** over k is

$$\mathbb{P}^2(k) = \{(x, y, z) \in k^3 \mid (x, y, z) \neq (0,0,0)\} / \sim,$$

where $(x, y, z) \sim (x', y', z')$ if and only if there exists a $c \neq 0$ in k such that $(x', y', z') = (cx, cy, cz)$. We write $(x{:}y{:}z)$ for the equivalence class of (x, y, z) — the colon is meant to suggest that only the ratios matter. Let $P \in \mathbb{P}^2(k)$; the triples (x, y, z) representing P lie on a single line $L(P)$ through the origin in k^3, and $P \mapsto L(P)$ is a bijection from $\mathbb{P}^2(k)$ to the set of all such lines. Projective n-space $\mathbb{P}^n(k)$, $n \in \mathbb{N}$, can be defined similarly.

Let $U_2(k) = \{(x\colon y\colon z) \mid z \neq 0\}$ and $L_\infty(k) = \{(x\colon y\colon z) \mid z = 0\}$. Then

$$(x, y) \mapsto (x\colon y\colon 1) : \mathbb{A}^2(k) \to U_2(k)$$
$$(x\colon y) \mapsto (x\colon y\colon 0) : \mathbb{P}^1(k) \to L_\infty(k)$$

are bijections. Moreover, $\mathbb{P}^2(k)$ is the disjoint union

$$\mathbb{P}^2(k) = U_2(k) \sqcup L_\infty(k)$$

of the "affine plane" U_2 with the "line at infinity" L_∞. The line

$$aX + bY + cZ = 0 \quad (a \text{ and } b \text{ not both zero})$$

meets L_∞ at the point $(-b\colon a\colon 0)$, which depends only on the slope of the affine line $aX + bY + c = 0$. We can think of $\mathbb{P}^2$ as being the affine plane U_2 with exactly one point added for each family of parallel lines.

A nonconstant homogeneous polynomial $F \in k[X, Y, Z]$, assumed to have no repeated factor in k^{al}, defines a ***projective plane curve*** C_F over k whose points in any field $K \supset k$ are the zeros of F in $\mathbb{P}^2(K)$,

$$C_F(K) = \{(x\colon y\colon z) \in \mathbb{P}^2(k) \mid F(x, y, z) = 0\}.$$

Note that, because F is homogeneous,

$$F(cx, cy, cz) = c^{\deg F} F(x, y, z), \quad c \in k^\times,$$

and so, although it does not make sense to speak of the value of F at a point P of $\mathbb{P}^2$, it does make sense to say whether or not F is zero at P. Again, we do not distinguish the curves C_F and C_{cF} with $c \in k^\times$. A plane projective curve is (uniquely) a union of irreducible projective plane curves (those defined by irreducible polynomials). The degree of F is called the ***degree*** of the curve C_F.

EXAMPLE 1.15 The curve

$$Y^2 Z = X^3 + aXZ^2 + bZ^3$$

intersects the line at infinity at the point $(0\colon 1\colon 0)$, i.e., at the same point as all the vertical lines do. This is plausible geometrically, because, as $x \to \infty$, the slope of the tangent line to the real affine curve

$$Y^2 = X^3 + aX + b$$

tends to ∞ (see the pictures p. 1).

Let $U_1(k) = \{(x\colon y\colon z) \mid y \neq 0\}$ and $U_0(k) = \{(x\colon y\colon z) \mid x \neq 0\}$. Then U_1 and U_0 are again, in a natural way, affine planes,

$$(x,z) \leftrightarrow (x\colon 1\colon z)\colon \mathbb{A}^2(k) \leftrightarrow U_1(k),$$
$$(y,z) \leftrightarrow (1\colon y\colon z)\colon \mathbb{A}^2(k) \leftrightarrow U_0(k).$$

Since at least one of x, y, or z is nonzero,

$$\mathbb{P}^2 = U_0 \cup U_1 \cup U_2.$$

A projective plane curve $C = C_F$ is the union of three affine plane curves,

$$C = C_0 \cup C_1 \cup C_2, \quad C_i = C \cap U_i.$$

When we identify each U_i with $\mathbb{A}^2$ in the natural way, C_0, C_1, and C_2 become identified with the affine curves defined by the polynomials $F(1,Y,Z)$, $F(X,1,Z)$, and $F(X,Y,1)$ respectively.

EXAMPLE 1.16 The curve

$$C: \quad Y^2Z = X^3 + aXZ^2 + bZ^3$$

is covered by two affine curves, namely,

$$C_2 : Y^2 = X^3 + aX + b \quad \text{and} \quad C_1 : Z = X^3 + aXZ^2 + bZ^3.$$

The notions of tangent line, multiplicity, intersection number, etc. can be carried over to projective curves by using that each point P of a projective curve C lies on at least one of the affine curves C_i. If $P \in C(k)$ is nonsingular as a point of one affine curve C_i, then it is nonsingular as a point of all C_i on which it lies. Similar statements hold for the other notions.

EXERCISE 1.17 Let P be a point on a projective plane curve $C = C_F$. Show that P is singular if and only if

$$\left(\frac{\partial F}{\partial X}\right)_P = \left(\frac{\partial F}{\partial Y}\right)_P = \left(\frac{\partial F}{\partial Z}\right)_P = 0.$$

When P is nonsingular, show that the projective line

$$L: \quad \left(\frac{\partial F}{\partial X}\right)_P X + \left(\frac{\partial F}{\partial Y}\right)_P Y \quad + \left(\frac{\partial F}{\partial Z}\right)_P Z = 0$$

has the property that $L \cap U_i$ is the tangent line at P to the affine curve C_i for all i such that P lies in U_i.

Bezout's theorem

THEOREM 1.18 (BEZOUT) *Let C and D be projective plane curves over k of degrees m and n respectively having no common component. Then C and D intersect over k^{al} in exactly mn points counted with multiplicity, i.e.,*

$$\sum_{P \in C(k^{\mathrm{al}}) \cap D(k^{\mathrm{al}})} I(P, C \cap D) = mn.$$

PROOF. For an elementary proof, see Fulton 1969, §5.3. □

Let C and D be projective plane curves over k of degrees m and n respectively. If there are mn points in $C(k^{\mathrm{al}}) \cap D(k^{\mathrm{al}})$, then Bezout's theorem and 1.13 show that the points must be nonsingular on both C and D; if there are more than mn points, then the curves must have a common component.

ASIDE 1.19 Over $\mathbb{C}$, the map

$$F, G \mapsto \sum_{P \in C_F(\mathbb{C}) \cap C_G(\mathbb{C})} I(P, C_F \cap C_G)$$

is continuous in the coefficients of F and G for the discrete topology on $\mathbb{N}$. Thus, the total intersection number is constant on continuous families of homogeneous polynomials, which allows us, in proving Bezout's theorem, to take F and G to be X^m and Y^n. Then $P = (0{:}0{:}0)$ is the only point on both curves, and

$$I(P, C_{X^m} \cap C_{Y^n}) = I(X^m, Y^n) = mn,$$

which proves the theorem. A similar argument works over any field once one has shown that the total intersection number is constant on *algebraic families* (cf. Shafarevich 1994, III, 2.2).

EXAMPLE 1.20 According to Bezout's theorem, a curve of degree m will meet the line at infinity in exactly m points counted with multiplicity. Our favourite curve

$$C: \quad Y^2Z = X^3 + aXZ^2 + bZ^3$$

meets L_∞ at a single point $P = (0{:}1{:}0)$, but (see 1.8)

$$\begin{aligned} I(P, L_\infty \cap C) &= I(Z, Z - X^3 - aXZ - bZ^3) \\ &= I(Z, X^3) \\ &= 3. \end{aligned}$$

In general, a nonsingular point P on a curve C is called a ***point of inflection*** (or ***flex***) if the intersection multiplicity of the tangent line and C at P is ≥ 3.[3] If the intersection multiplicity equals 3, then the point of inflection is said to be ***ordinary***. This is always the case for cubic curves.

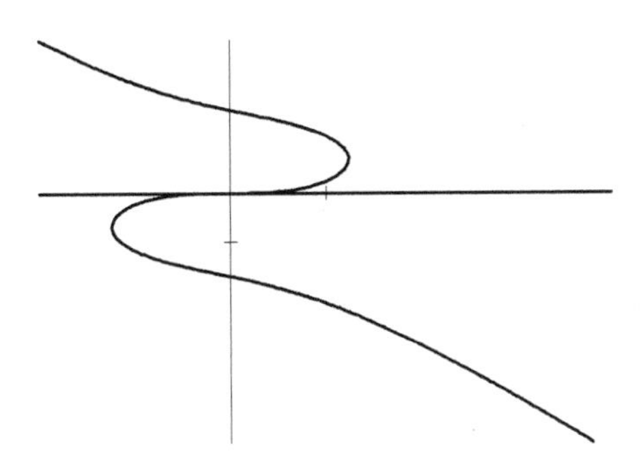

$$Z = X^3 + XZ^2 + Z^3$$

REMARK 1.21 According to Bezout's theorem, any two irreducible components of a projective plane curve have a common point (possibly in a finite extension of k), which will be singular by 1.4b. Therefore, nonsingular projective plane curves are geometrically irreducible.

Appendix: resultants

Let $f(X) = s_0X^m + s_1X^{m-1} + \cdots + s_m$ and $g(X) = t_0X^n + t_1X^{n-1} + \cdots + t_n$ be polynomials with coefficients in k. The ***resultant*** $\operatorname{Res}(f, g)$ of f and g is defined to be the determinant

$$\begin{vmatrix} s_0 & s_1 & \ldots & s_m & & \\ & s_0 & \ldots & & s_m & \\ & & \ldots & & & \ldots \\ t_0 & t_1 & \ldots & t_n & & \\ & t_0 & \ldots & & t_n & \\ & & \ldots & & & \ldots \end{vmatrix} \quad \begin{matrix} \\ n \text{ rows} \\ \\ \\ m \text{ rows} \\ \\ \end{matrix}$$

[3]Sometimes the intersection multiplicity is also required to be odd.

There are n rows of s_i and m rows of t_j, so that the matrix is $(m+n)\times(m+n)$; all blank spaces are to be filled with zeros. The resultant is a polynomial in the coefficients of f and g.

PROPOSITION 1.22 *The resultant* $\mathrm{Res}(f,g)=0$ *if and only if*

(a) *both* s_0 *and* t_0 *are zero; or*

(b) *the two polynomials have a common root in* k.

PROOF. If (a) holds, then the first column of the determinant is zero, and so certainly $\mathrm{Res}(f,g)=0$. Suppose that a is a common root of f and g, so that there exist polynomials f_1 and g_1 in $k[X]$ of degrees $m-1$ and $n-1$ respectively such that

$$f(X)=(X-a)f_1(X), \qquad g(X)=(X-a)g_1(X).$$

From these equalities we find that

$$f(X)g_1(X)-g(X)f_1(X)=0. \tag{2}$$

On equating the coefficients of $X^{m+n-1},\ldots,X,1$ in (2) to zero, we find that the coefficients of f_1 and g_1 are the solutions of a system of $m+n$ linear equations in $m+n$ unknowns. The matrix of coefficients of the system is the transpose of the matrix

$$\begin{pmatrix} s_0 & s_1 & \ldots & s_m & & \\ & s_0 & \ldots & & s_m & \\ & & \ldots & & & \ldots \\ t_0 & t_1 & \ldots & t_n & & \\ & t_0 & \ldots & & t_n & \\ & & \ldots & & & \ldots \end{pmatrix} \tag{3}$$

The existence of the solution shows that this matrix has determinant zero, which implies that $\mathrm{Res}(f,g)=0$.

Conversely, suppose that $\mathrm{Res}(f,g)=0$ but neither s_0 nor t_0 is zero. Because the above matrix has determinant zero, we can solve the linear equations to find polynomials f_1 and g_1 satisfying (2). A root a of f must also be a root of f_1 or of g. If the former, cancel $X-a$ from the left hand side of (2), and consider a root of $f_1/(X-a)$. As $\deg f_1<\deg f$, this argument eventually leads to a root of f that is not a root of f_1, and so must be a root of g. □

Note that polynomials $f, g \in k[X]$ have a common root in k if and only if they have a common root in k^{al}, because Euclid's algorithm shows that the greatest common divisor of two polynomials does not change when the field is extended (FT, 2.10).

EXAMPLE 1.23 For a monic polynomial $f(X)$ of degree m, $\mathrm{Res}(f, f') = (-1)^{m(m-1)/2}\mathrm{disc}(f)$.

Let $\mathbf{c}_1, \ldots, \mathbf{c}_{m+n}$ be the columns of the matrix (3). Then

$$\begin{pmatrix} X^{m-1} f(X) \\ X^{m-2} f \\ \vdots \\ f(X) \\ X^{n-1} g(X) \\ \vdots \\ g(X) \end{pmatrix} = X^{m+n-1}\mathbf{c}_0 + \cdots + 1\mathbf{c}_{m+n},$$

and so

$$\mathrm{Res}(f, g) \stackrel{\text{def}}{=} \det(\mathbf{c}_0, \ldots, \mathbf{c}_{m+n}) = \det(\mathbf{c}_0, \cdots, \mathbf{c}_{m+n-1}, \mathbf{c}),$$

where $\mathbf{c}$ is the vector on the left of the above equation. On expanding out this last determinant, we find that

$$\mathrm{Res}(f, g) = a(X) f(X) + b(X) g(X),$$

where $a(X)$ and $b(X)$ are polynomials of degrees $\leq n-1$ and $\leq m-1$ respectively.

If $f(X)$ and $g(X)$ have coefficients in an integral domain R, for example, $\mathbb{Z}$ or $k[Y]$, then $\mathrm{Res}(f, g) \in R$, and the polynomials $a(X)$ and $b(X)$ have coefficients in R.

PROPOSITION 1.24 *Let $f(X,Y), g(X,Y) \in k[X,Y]$, and let $r(X) \in k[X]$ be the resultant of f and g regarded as polynomials in Y with coefficients in $k[X]$.*

(a) *There exist $a(X,Y), b(X,Y) \in k[X,Y]$ such that*

$$af + bg = r(X) \in k[X]$$

and $\deg_Y(a) < \deg_Y(g)$, $\deg_Y(b) < \deg_Y(f)$.

(b) *The polynomial* $r = 0$ *if and only if* f *and* g *have a common factor in* $k[X,Y]$.

PROOF. (a) Immediate from the above discussion.

(b) We know that $r = 0$ if and only if f and g have a common factor in $k(X)[Y]$, but by Gauss's lemma (CA, 4.5), this is equivalent to their having a common factor in $k[X,Y]$. □

COROLLARY 1.25 *If the polynomials* f *and* g *have no common factor in* $k[X,Y]$, *then they have no common factor in* $K[X,Y]$ *for any field* $K \supset k$.

PROOF. Since the resultant of f and g is the same whether we work over k or K, this follows from (b) of the proposition. □

The resultant of homogeneous polynomials

$$F(X,Y) = s_0 X^m + s_1 X^{m-1} Y + \cdots + s_m Y^m$$
$$G(X,Y) = t_0 X^n + t_1 X^{n-1} Y + \cdots + t_n Y^n$$

is defined as for inhomogeneous polynomials.

PROPOSITION 1.26 *The resultant* $\mathrm{Res}(F,G) = 0$ *if and only if* F *and* G *have a nontrivial zero in* $\mathbb{P}^1(k^{\mathrm{al}})$.

PROOF. The nontrivial zeros of $F(X,Y)$ in $\mathbb{P}^1(k^{\mathrm{al}})$ are of the form

(a) $(a{:}1)$ with a a root of $F(X,1)$, or

(b) $(1{:}0)$ in the case that $s_0 = 0$.

Since a similar statement is true for $G(X,Y)$, this proposition is a restatement of Proposition 1.22. □

Computer algebra programs can find resultants. For example, entering "`polresultant`$((x+a)^5,(x+b)^5,x)$" into Pari/gp gives the answer $(-a+b)^{25}$, which means that the polynomials have a common root if and only if $a = b$, and this can happen in 25 ways.

ASIDE 1.27 There is a geometric interpretation of the last proposition. Take k to be algebraically closed, and regard the coefficients of F and G as indeterminates. Let V be the subset of $\mathbb{A}^{m+n+2} \times \mathbb{P}^1$ where both $F(s_0,\ldots,s_m;X,Y)$ and $G(t_0,\ldots,t_n;X,Y)$ vanish. The proposition says that the projection of V on $\mathbb{A}^{m+n+2}$ is the set where $\mathrm{Res}(F,G)$, regarded as a polynomial in the s_i and t_i, vanishes. In other words, the proposition tells us that the projection of the Zariski-closed set V is the Zariski-closed set defined by the resultant of F and G.

Elimination theory does this more generally. Given polynomials $P_i(T_1,\ldots,T_m;X_0,\ldots,X_n)$, homogeneous in the X_i, it provides an algorithm for finding polynomials $R_j(T_1,\ldots,T_m)$ such that the $P_i(a_1,\ldots,a_m;X_0,\ldots,X_n)$ have a common zero if and only if $R_j(a_1,\ldots,a_m)=0$ for all j. See, for example, Cox et al. 2007, Chap. 8, §5.

NOTES The history of plane algebraic curves goes back more than 2000 years — see the first chapter of Brieskorn and Knörrer 1986. Newton made the first systematic study of real cubic plane curves, classifying them into 72 cases (but missing 6).

2 Rational points on plane curves

For a projective plane curve C_F over $\mathbb{Q}$, the two fundamental questions are the following:

(a) Does C_F have a point with coordinates in $\mathbb{Q}$, i.e., does $F(X,Y,Z)$ have a nontrivial zero in $\mathbb{Q}$?

(b) If the answer to (a) is yes, can we describe the set of points?

One can also ask whether there are algorithms to answer these questions. For example, we may know that a curve has only finitely many points without having an algorithm to find the points or even determine their number.

In this section, we give an overview of the topic. For simplicity, we assume throughout that $C=C_F$ is geometrically irreducible, i.e., that $F(X,Y,Z)$ is irreducible in $\mathbb{Q}[X,Y,Z]$ and remains so over $\mathbb{Q}^{\mathrm{al}}$.

We first make an observation that will be useful throughout the book. Let

$$f(X,Y)=\sum a_{ij}X^iY^j$$

be a polynomial with coefficients a_{ij} in a field k, and let K be a Galois extension of k (possibly infinite). If $(a,b)\in K\times K$ is a zero of $f(X,Y)$, then so also is $(\sigma a,\sigma b)$ for all $\sigma\in\mathrm{Gal}(K/k)$ because

$$f(\sigma a,\sigma b)=\sum a_{ij}(\sigma a)^i(\sigma b)^j=\sigma(\sum a_{ij}a^ib^j)=\sigma f(a,b)=0.$$

Thus, $\mathrm{Gal}(K/k)$ acts on $C_f(K)$. More generally, if $C_1,C_2,\ldots$ are affine plane curves over k, then $\mathrm{Gal}(K/k)$ stabilizes the subset $C_1(K)\cap C_2(K)\cap\ldots$ of $K\times K$. On applying this remark to the curves

$$f=0,\quad \frac{\partial f}{\partial X}=0,\quad \frac{\partial f}{\partial Y}=0,$$

we see that $\mathrm{Gal}(K/\mathbb{Q})$ stabilizes the set of singular points of C_f. Similar remarks apply to projective plane curves.

Curves of degree one

In this case, the curve is a line

$$C : aX + bY + cZ = 0, \quad a,b,c \in \mathbb{Q}, \text{ not all zero.}$$

It always has points with coordinates in $\mathbb{Q}$, and it is possible to parameterize the points — for example, if $c \neq 0$, then the map

$$(s:t) \mapsto (s:t:-\tfrac{a}{c}s - \tfrac{b}{c}t)$$

is a bijection from $\mathbb{P}^1(K)$ onto $C(K)$ for all fields $K \supset \mathbb{Q}$.

Curves of degree two

In this case $F(X,Y,Z)$ is a quadratic form in 3 variables, and C is a conic. Note that C cannot be singular, because a nontangent line L through a point P of multiplicity $m \geq 2$ will meet C in a second point Q and the total intersection number

$$I(P, L \cap C) + I(Q, L \cap C) \overset{1.13}{\geq} m + 1$$

violates Bezout's theorem.

Sometimes it is easy to see that $C(\mathbb{Q})$ is empty. For example,

$$X^2 + Y^2 + Z^2$$

has no nontrivial zero in $\mathbb{Q}$ because it has no nontrivial zero in $\mathbb{R}$. Similarly,

$$X^2 + Y^2 - 3Z^2$$

has no nontrivial zero because, if it did, then it would have a zero (x,y,z) with $x,y,z \in \mathbb{Z}$ and $\gcd(x,y,z) = 1$; the only squares in $\mathbb{Z}/3\mathbb{Z}$ are 0 and 1, and so

$$x^2 + y^2 \equiv 0 \mod 3 \implies x \equiv 0 \equiv y \mod 3;$$

but then 3 divides z, which contradicts our assumption that $\gcd(x,y,z) = 1$.

In 1785, Legendre showed that such arguments can be made into an effective procedure for deciding whether $C_F(\mathbb{Q})$ is nonempty. He first shows that, by an elementary change of variables, F can be put in diagonal form

$$F = aX^2 + bY^2 + cZ^2, \quad a,b,c \in \mathbb{Z}, \quad a,b,c \text{ square free.}$$

He then shows that if $abc \neq 0$ and a,b,c are not all of the same sign, then $F(X,Y,Z)$ has a nontrivial zero in $\mathbb{Q}$ if and only if $-bc$, $-ca$, $-ab$ are squares modulo a,b,c respectively. It follows from the quadratic reciprocity law (IV, 8.5), that this implies that there exists an integer m, depending in a simple way on the coefficients of F, such that $C_F(\mathbb{Q}) \neq \emptyset$ if and only if $F(X,Y,Z) \equiv 0 \mod m$ has a solution in integers relatively prime to m. See 2.18.

Now suppose that C has a point P_0 with coordinates in $\mathbb{Q}$. Can we describe all the points? A line over $\mathbb{Q}$ always meets C in exactly two points counting multiplicities (Bezout's theorem). If the points coincide, then the line is tangent to C. Otherwise, the pair of points is stable under the action of $\mathrm{Gal}(\mathbb{Q}^{\mathrm{al}}/\mathbb{Q})$, and so, if one has coefficients in $\mathbb{Q}$, then both do. The lines through P_0 in $\mathbb{P}^2$ form a "$\mathbb{P}^1$", and the map sending a line through P_0 to its second point of intersection is a bijection from $\mathbb{P}^1(\mathbb{Q})$ to $C(\mathbb{Q})$. For example, take P_0 to be the point $(-1:0:1)$ on the curve $C : X^2+Y^2 = Z^2$. The line $bX - aY + bZ$, $a,b \in \mathbb{Q}$, of slope b/a through P_0 meets C at the point $(a^2-b^2 : 2ab : a^2+b^2)$. In this way, we obtain a parameterization $(a:b) \mapsto (a^2-b^2:2ab:a^2+b^2)$ of the points of C with coordinates in $\mathbb{Q}$.

Curves of degree 3

Let $C : F(X,Y,Z) = 0$ be a projective plane curve over $\mathbb{Q}$ of degree 3. If C is singular, then Bezout's theorem shows that it has only one singular point and that its multiplicity is 2. The singular point P_0 will have coordinates in some finite extension K of $\mathbb{Q}$, which we may take to be Galois over $\mathbb{Q}$, but $\mathrm{Gal}(K/\mathbb{Q})$ stabilizes the set of singular points in $C(K)$, hence fixes P_0, and so $P_0 \in C(\mathbb{Q})$. Now a line through P_0 will meet the curve in exactly one other point (possibly coincident with P_0), and so we again get a parameterization of the points of C with coordinates in $\mathbb{Q}$.

Nonsingular cubics will be the subject of the rest of the book. We shall see that Legendre's arguments fail for nonsingular cubic curves. For example,

$$3X^3 + 4Y^3 + 5Z^3$$

has nontrivial real zeros and nontrivial zeros modulo m for all integers m, but it has no nontrivial zero in $\mathbb{Q}$.

Let C be a nonsingular cubic curve over $\mathbb{Q}$. According to Bezout's theorem, a line meets the curve in three points counting multiplicities. If two of the points have coefficients in $\mathbb{Q}$, so also does the third.[4] Thus, the chord through two points in $C(\mathbb{Q})$ meets the curve in a third point in $C(\mathbb{Q})$, and the tangent line at a point in $C(\mathbb{Q})$ meets the curve at another point in $C(\mathbb{Q})$.

In a famous paper, published in 1922, Mordell proved the following theorem.

THEOREM 2.1 (FINITE BASIS THEOREM) *Let C be a nonsingular cubic curve over $\mathbb{Q}$. There exists a finite set of points in $C(\mathbb{Q})$ from which every other point in $C(\mathbb{Q})$ can be obtained by successive chord and tangent constructions.*

In fact, $C(\mathbb{Q})$, if nonempty, has a natural structure of a commutative group (see the next section), and the finite basis theorem says that $C(\mathbb{Q})$ is finitely generated. There is as yet no *proven* algorithm for finding the rank of the group.

REMARK 2.2 For a singular cubic curve C over $\mathbb{Q}$, the nonsingular points still form a group, which may be isomorphic to $(\mathbb{Q},+)$ or $(\mathbb{Q}^{\times},\times)$ (see II, §3), neither of which is finitely generated — the elements of any finitely generated subgroup of $(\mathbb{Q},+)$ have a common denominator, and a finitely generated subgroup of $(\mathbb{Q}^{\times},\times)$ can contain only finitely many prime numbers. Thus, the finite basis theorem fails for singular cubics.

Curves of genus > 1

The genus of a nonsingular projective curve C over $\mathbb{Q}$ is the genus of the Riemann surface $C(\mathbb{C})$. More generally, the genus of an arbitrary curve over $\mathbb{Q}$ is the genus of the nonsingular projective curve attached to its function field (Fulton 1969, §7.5, Cor. to Theorem 3). Mordell conjectured in his 1922 paper, and Faltings (1983) proved, that every curve of genus > 1 has only finitely many points with coordinates in $\mathbb{Q}$. This applies to every nonsingular projective plane curve of degree at least 4 and to every singular projective plane curve whose degree is sufficiently large compared to the multiplicities of its singularities (see 6.5 later in this chapter).

[4] On substituting $Y = aX + b$ into the polynomial defining the affine cubic, one obtains a cubic polynomial in X with coefficients in $\mathbb{Q}$, two of whose roots lie in $\mathbb{Q}$; it follows that the third root also lies $\mathbb{Q}$.

REMARK 2.3 Let $P \in \mathbb{P}^2(\mathbb{Q})$. Choose a representative $(a{:}b{:}c)$ for P with a,b,c integers having no common factor, and define the height $H(P)$ of P to be $\max(|a|,|b|,|c|)$. For any $B > 0$, there are only finitely many P with $H(P) < B$.

For a curve C of genus > 1, there is an effective bound for the number of points $P \in C(\mathbb{Q})$ but no known effective bound, in terms of the polynomial defining C, for the heights of the points $P \in C(\mathbb{Q})$. With such a bound, the points in $C(\mathbb{Q})$ could be found by a finite search. See Hindry and Silverman 2000, F.4.2, for a discussion of this problem.

ASIDE 2.4 There is a heuristic explanation for Mordell's conjecture. Let C be a curve of genus $g \geq 1$ over $\mathbb{Q}$. If $C(\mathbb{Q})$ is nonempty, then C embeds into its jacobian variety J, which is a projective variety of dimension g with a group structure, i.e., an abelian variety. A generalization of the finite basis theorem, due to Weil, says that $J(\mathbb{Q})$ is finitely generated. Hence, inside the g-dimensional set $J(\mathbb{C})$ we have the countable set $J(\mathbb{Q})$ and the (apparently unrelated) one-dimensional set $C(\mathbb{C})$. If $g > 1$, it would be an extraordinary coincidence if the second set contained more than a finite number of elements from the first set.

To continue our discussion, we shall need the field of p-adic numbers.

A brief introduction to the p-adic numbers

Let p be a prime number. A nonzero rational number a can be uniquely expressed as $a = p^r \frac{m}{n}$, where m,n,r are integers and m and n are not divisible by p. We set $|a|_p = \frac{1}{p^r}$, and we let $|0|_p = 0$. We call $|a|_p$ the ***p*-adic valuation** of a. It has the following properties:

(a) $|a|_p = 0$ if and only if $a = 0$,

(b) $|ab|_p = |a|_p |b|_p$,

(c) $|a+b|_p \leq \max\{|a|_p, |b|_p\} \quad (\leq |a|_p + |b_p|)$.

These statements imply that

$$d_p(a,b) \stackrel{\text{def}}{=} |a-b|_p$$

is a translation-invariant metric on $\mathbb{Q}$, called the ***p*-adic metric.** Note that for rational numbers a,b to be close for this metric means that their difference is divisible by a high power of p. The ***field* $\mathbb{Q}_p$ *of p-adic numbers*** is the completion of $\mathbb{Q}$ for this metric. We now explain what this means.

A sequence $a_1, a_2, a_3, \ldots$ of rational numbers is a ***Cauchy sequence*** for the p-adic metric if, for each real number $\varepsilon > 0$, there exists a natural number $N(\varepsilon)$ such that

$$|a_m - a_n|_p < \varepsilon \quad \text{whenever} \quad m, n > N(\varepsilon).$$

A sequence $a_1, a_2, a_3, \ldots$ is said to ***converge*** to $a \in \mathbb{Q}$ if, for each real number $\varepsilon > 0$, there exists a natural number $N(\varepsilon)$ such that

$$|a_n - a|_p < \varepsilon \quad \text{whenever } n > N(\varepsilon).$$

Let R be the set of all Cauchy sequences in $\mathbb{Q}$ for the p-adic metric. It becomes a ring with the obvious operations. An element of R is said to be ***null*** if it converges to zero. The set of null sequences is an ideal I in R, and $\mathbb{Q}_p$ is defined to be the quotient ring R/I. It is a field.

If $\alpha = (a_n)_{n \geq 1}$ is a Cauchy sequence in $\mathbb{Q}$, then property (c) implies that the sequence $|a_n|_p$ becomes constant for large n, and we set this constant value equal to $|\alpha|_p$. The map $\alpha \mapsto |\alpha|_p : R \to \mathbb{Q}$ factors through $\mathbb{Q}_p$ and has the properties (a), (b), (c) listed above. Therefore the p-adic metric on $\mathbb{Q}$ extends to $\mathbb{Q}_p$.

THEOREM 2.5 *(a) The field $\mathbb{Q}_p$ is complete, i.e., every Cauchy sequence in $\mathbb{Q}_p$ converges to a limit in $\mathbb{Q}_p$.*

(b) The map sending $a \in \mathbb{Q}$ to the equivalence class of the constant Cauchy sequence $a, a, a, \ldots$ is an injective homomorphism $\mathbb{Q} \hookrightarrow \mathbb{Q}_p$ with dense image.

PROOF. Exercise. □

We define the ring of p-adic integers $\mathbb{Z}_p$ to be the closure of $\mathbb{Z}$ in $\mathbb{Q}_p$. This is exactly the set of $\alpha \in \mathbb{Q}_p$ with $|\alpha|_p \leq 1$. Every p-adic integer a can be represented by a unique series

$$a_0 + a_1 p + \cdots + a_i p^i + \cdots, \quad 0 \leq a_i \leq p-1,$$

i.e., there is a unique sequence of integers $a_0, a_1, \ldots, 0 \leq a_i \leq p-1$, such that the partial sums of the above series form a Cauchy sequence representing a. The a_i, if they exist, satisfy congruences

$$a_0 \equiv a \bmod p\mathbb{Z}, \ldots, a_i \equiv \left(a - (a_0 + \cdots + a_{i-1} p^{i-1})\right) / p^i \bmod p\mathbb{Z}, \ldots.$$

This proves uniqueness, and the congruences can be used to inductively define the a_i. Using that $\mathbb{Z}$ is dense in $\mathbb{Z}_p$, one finds that

$$\mathbb{Z}/p^m\mathbb{Z} \simeq \mathbb{Z}_p/p^m\mathbb{Z}_p, \quad \text{all } m \geq 1.$$

On passing to the limit, we obtain isomorphisms

$$\varprojlim \mathbb{Z}/p^m\mathbb{Z} \simeq \varprojlim \mathbb{Z}_p/p^m\mathbb{Z}_p \simeq \mathbb{Z}_p.$$

REMARK 2.6 (a) The same construction as above, but with $|\cdot|_p$ replaced by the usual absolute value, yields $\mathbb{R}$ instead of $\mathbb{Q}_p$.

(b) Just as real numbers can be represented by decimals, p-adic numbers can be represented by infinite series of the form

$$a_{-n}p^{-n} + \cdots + a_0 + a_1 p + \cdots + a_m p^m + \cdots \quad 0 \leq a_i \leq p-1.$$

This can be proved by noting that $\mathbb{Q}_p^\times = p^{\mathbb{Z}} \cdot \mathbb{Z}_p^\times$ (direct product).

The ring of p-adic integers $\mathbb{Z}_p$ is compact, and it is open in $\mathbb{Q}_p$. Therefore, $\mathbb{Q}_p$ is locally compact. The relevance of the p-adic numbers to our present discussion is explained by the next statement.

Let $f(X_0,\ldots,X_n) \in \mathbb{Z}[X_0,\ldots,X_n]$ be a homogeneous form. Suppose that, for each natural number N, we have integers $a_0^{(N)},\ldots,a_n^{(N)}$, not all divisible by p, such that

$$f(a_0^{(N)},\ldots,a_n^{(N)}) \equiv 0 \mod p^N. \tag{4}$$

Because $\mathbb{Z}_p^{n+1}$ is compact, the collection of $n+1$-tuples $(a_0^{(N)},\ldots,a_n^{(N)})$, $N \in \mathbb{N}$, has a limit point $(\alpha_0,\ldots,\alpha_n)$ in $\mathbb{Z}_p^{n+1}$, and one sees easily that

$$f(\alpha_0,\ldots,\alpha_n) = 0. \tag{5}$$

Conversely, suppose that $(\alpha_0,\ldots,\alpha_n) \in \mathbb{Q}_p^{n+1}$ satisfies (5) and not all α_i are zero. After scaling, we may suppose that $\max|\alpha_i|_p = 1$. For each $N \in \mathbb{N}$, choose integers $a_0^{(N)},\ldots,a_n^{(N)}$ such that $|a_j^{(N)} - \alpha_j| \leq p^{-N}$ for all j. Then $(a_0^{(N)},\ldots,a_n^{(N)})$ satisfies (4).

NOTATION 2.7 For $a \in \mathbb{Q}^\times$, we let $\operatorname{ord}_p(a) = r$ if $a = p^r \frac{m}{n}$ with m and n not divisible by p, and we let $\operatorname{ord}_p(0) = \infty$. With the obvious conventions concerning ∞, we have

(a) $\operatorname{ord}_p(a) = \infty$ if and only if $a = 0$,

(b) $\mathrm{ord}_p(ab) = \mathrm{ord}_p(a) + \mathrm{ord}_p(b)$,

(c) $\mathrm{ord}_p(a+b) \geq \min\{\mathrm{ord}_p(a), \mathrm{ord}_p(b)\}$, with inequality unless $\mathrm{ord}_p(a) = \mathrm{ord}_p(b)$.

In particular, ord_p is a homomorphism $\mathbb{Q}^\times \to \mathbb{Z}$. This homomorphism extends by continuity to $\mathbb{Q}_p^\times$.

NOTES For a more leisurely exposition of p-adic numbers, see, for example, Chapter I of Koblitz 1977. According to a theorem of Ostrowski, $\mathbb{R}$ and the p-adic fields $\mathbb{Q}_p$ are the *only* completions of $\mathbb{Q}$ with respect to nontrivial valuations.

Curves of degree 2 and 3 continued

Clearly, a necessary condition for a curve to have a point with coordinates in $\mathbb{Q}$ is that it have points with coordinates in $\mathbb{R}$ and all the fields $\mathbb{Q}_p$. Indeed, we observed above that $X^2 + Y^2 + Z^2$ has no nontrivial zero in $\mathbb{Q}$ because it has no nontrivial real zero, and our argument for $X^2 + Y^2 - 3Z^2$ shows that it has no nontrivial zero in the field $\mathbb{Q}_3$ of 3-adic numbers. A modern interpretation of Legendre's theorem is that, for curves of degree 2, the condition is also sufficient.

THEOREM 2.8 (LEGENDRE) *A quadratic form $F(X,Y,Z)$ with coefficients in $\mathbb{Q}$ has a nontrivial zero in $\mathbb{Q}$ if and only if it has nontrivial zeros in $\mathbb{R}$ and in $\mathbb{Q}_p$ for all p.*

PROOF. See Cassels 1991, Chapter 3, or Serre 1973, Chapter IV. □

See 2.18 below for the relation of this statement to Legendre's original statement.

ASIDE 2.9 Theorem 2.8 is true for quadratic forms in any number of variables (Hasse–Minkowski theorem). See Serre 1973, Chap. IV, for a good exposition of the proof. The key cases are 3 and 4 variables (2 is easy, and for ≥ 5 variables, one uses induction on n), and the key result needed for its proof is the quadratic reciprocity law (IV, 8.5).

If for some class of polynomials (better algebraic varieties) it is known that a polynomial (or variety) has a zero in $\mathbb{Q}$ if and only if it has zeros in $\mathbb{R}$ and all $\mathbb{Q}_p$, then one says that the ***Hasse***, or ***local-global, principle*** holds for the class.

Hensel's lemma

LEMMA 2.10 *Let $f(X_1,\dots,X_n) \in \mathbb{Z}[X_1,\dots,X_n]$, and let $\underline{a} \in \mathbb{Z}^n$ have the property that, for some $m \geq 0$,*

$$f(\underline{a}) \equiv 0 \mod p^{2m+1}$$

but, for some i,

$$\left(\frac{\partial f}{\partial X_i}\right)(\underline{a}) \not\equiv 0 \mod p^{m+1}.$$

Then there exists a $\underline{b} \in \mathbb{Z}^n$ such that

$$\underline{b} \equiv \underline{a} \mod p^{m+1}$$

and

$$f(\underline{b}) \equiv 0 \mod p^{2m+2}.$$

PROOF. Consider the Taylor expansion

$$f(X_1,\dots,X_n) = f(a_1,\dots,a_n) + \sum_{i=1}^{n} \left(\frac{\partial f}{\partial X_i}\right)_{\underline{a}} (X_i - a_i)$$
$$+ \text{terms of degree} \geq 2.$$

Set $b_i = a_i + h_i\, p^{m+1}$, $h_i \in \mathbb{Z}$. Then

$$f(b_1,\dots,b_n) = f(a_1,\dots,a_n) + \sum_{i=1}^{n} \left(\frac{\partial f}{\partial X_i}\right)_{\underline{a}} h_i\, p^{m+1}$$
$$+ \text{ terms divisible by } p^{2m+2}.$$

We have to choose the h_i so that

$$f(a_1,\dots,a_n) + \sum_{i=1}^{n} \left(\frac{\partial f}{\partial X_i}\right)_{\underline{a}} h_i\, p^{m+1}$$

is divisible by p^{2m+2}. From the assumption, we know that there is a $k \leq m$ such that p^k divides $\left(\frac{\partial f}{\partial X_i}\right)_{\underline{a}}$ for all i but p^{k+1} does not divide all of them. Any h_i satisfying the following equation will suffice,

$$\frac{f(a_1,\dots,a_n)}{p^{k+m+1}} + \sum_{i=1}^{n} \frac{\left(\frac{\partial f}{\partial X_i}\right)_{\underline{a}}}{p^k} h_i \equiv 0 \mod p.$$

□

REMARK 2.11 If, in the lemma, $\underline{a}$ satisfies the condition

$$f(\underline{a}) \equiv 0 \mod p^{2m+r}$$

for some $r \geq 1$, then the construction in the proof gives a $\underline{b}$ such that

$$\underline{b} \equiv \underline{a} \mod p^{m+r}$$

and

$$f(\underline{b}) \equiv 0 \mod p^{2m+r+1}.$$

THEOREM 2.12 (HENSEL'S LEMMA) *Under the hypotheses of the lemma, there exists a* $\underline{b} \in \mathbb{Z}_p^n$ *such that* $f(\underline{b}) = 0$ *and* $\underline{b} \equiv \underline{a} \mod p^{m+1}$.

PROOF. On applying the lemma, we obtain an $\underline{a}_{2m+2} \in \mathbb{Z}^n$ such that $\underline{a}_{2m+2} \equiv \underline{a} \mod p^{m+1}$ and $f(\underline{a}_{2m+2}) \equiv 0 \mod p^{2m+2}$. The first congruence implies that

$$\left(\frac{\partial f}{\partial X_i}\right)(\underline{a}_{2m+2}) \equiv \left(\frac{\partial f}{\partial X_i}\right)(\underline{a}) \mod p^{m+1},$$

and so $\left(\frac{\partial f}{\partial X_i}\right)(\underline{a}_{2m+2}) \not\equiv 0 \mod p^{m+1}$ for some i. On applying the remark following the lemma, we obtain an $\underline{a}_{2m+3} \in \mathbb{Z}^n$ such that $\underline{a}_{2m+3} \equiv \underline{a}_{2m+2} \mod p^{m+2}$ and $f(\underline{a}_{2m+3}) \equiv 0 \mod p^{2m+3}$. Continuing in this fashion, we obtain a sequence $\underline{a}, \underline{a}_{2m+2}, \underline{a}_{2m+3}, \ldots$ of n-tuples of Cauchy sequences. Let $\underline{b}$ be the limit in $\mathbb{Z}_p^n$. The map $f: \mathbb{Z}^n \to \mathbb{Z}$ is continuous for the p-adic topologies, and so

$$f(\underline{b}) = f(\lim_r \underline{a}_{2m+r}) = \lim_r f(\underline{a}_{2m+r}) = 0.$$

□

EXAMPLE 2.13 Let $f(X) \in \mathbb{Z}[X]$, and let $\bar{f}(X) \in \mathbb{F}_p[X]$ be its reduction mod p. Let $a \in \mathbb{Z}$ be such that $\bar{a} \in \mathbb{F}_p$ is a simple root of $\bar{f}(X)$. Then $\frac{d\bar{f}}{dX}(\bar{a}) \neq 0$, and so the theorem shows that $\bar{a}$ lifts to a root of $f(X)$ in $\mathbb{Z}_p$.

EXAMPLE 2.14 Let $F(X,Y,Z) \in \mathbb{Z}[X,Y,Z]$ be a homogeneous polynomial, and let $(a:b:c) \in \mathbb{P}^2(\mathbb{F}_p)$ be a nonsingular point of the curve C_F over $\mathbb{F}_p$. Then, as in the last example, $(a:b:c)$ lifts to a point on the curve C_F with coordinates in $\mathbb{Z}_p$.

EXAMPLE 2.15 Let $f(X,Y,Z)$ be a quadratic form with coefficients in $\mathbb{Z}$ and discriminant $D \neq 0$. If p does not divide D, then $\bar{f}(X,Y,Z)$ is a quadratic form over $\mathbb{F}_p$ with nonzero discriminant, and it is known that such a form has a nontrivial zero in $\mathbb{F}_p$ (Serre 1973, I, §2). Therefore $f(X,Y,Z)$ has a nontrivial zero in $\mathbb{Q}_p$. If p divides D, then Hensel's lemma shows that $f(X,Y,Z)$ has a nontrivial zero in $\mathbb{Q}_p$ if it has an "approximate" zero.

EXERCISE 2.16 Let

$$F(X,Y,Z) = 5X^2 + 3Y^2 + 8Z^2 + 6(YZ + ZX + XY).$$

Find $(a,b,c) \in \mathbb{Z}^3$, not all divisible by 13, such that

$$F(a,b,c) \equiv 0 \bmod 13^2.$$

EXERCISE 2.17 Consider the affine plane curve $C : Y^2 = X^3 + p$. Prove that the point $(0,0)$ on the reduced curve over $\mathbb{F}_p$ does not lift to a point on C with coordinates in $\mathbb{Z}_p^2$. Why does this not violate Hensel's lemma?

ASIDE 2.18 Let $f(X,Y,Z)$ be a quadratic form with coefficients in $\mathbb{Q}$ and nonzero discriminant. After an elementary change of variables, we may suppose that $f = aX^2 + bY^2 + cZ^2$ with a,b,c square free integers. If a, b, and c have a prime factor p in common, we replace f with $p^{-1}f$. If two of the coefficients, say, a and b, have a prime factor p in common, we replace Z with pZ and divide by p. Thus, in studying the solutions of $f = 0$, we may suppose that

$$f(X,Y,Z) = aX^2 + bY^2 + cZ^2$$

with $a,b,c \in \mathbb{Z}$ and abc square free.

Recall that Legendre showed that $f = 0$ has a nonzero solution if and only if

(a) a,b,c are not all of the same sign, and

(b) $-bc$, $-ca$, $-ab$ are squares modulo a, b, c respectively.

We prove the necessity of the conditions (a) and (b). Suppose that $f = 0$ has a nonzero solution $(x,y,z) \in \mathbb{Q}^3$. Then certainly (a) holds. We may suppose that $x,y,z \in \mathbb{Z}$ and are relatively prime in pairs. If x and c had a prime factor p in common, then p would divide y, contradicting our assumption. Thus, x is invertible modulo c, and from $ax^2 + by^2 + cz^2 = 0$ we deduce that $-ab = (byx^{-1})^2$ modulo c. Similarly, $-bc$ and $-ca$ are squares modulo a and b.

We now prove that Legendre's original theorem implies Theorem 2.8 by showing that if f has nontrivial zeros in $\mathbb{R}$ and $\mathbb{Q}_p$ for all p dividing abc, then Legendre's conditions (a) and (b) hold. Certainly, (a) holds. Let $c = \prod p_i^{r_i}$.

Because f has a nontrivial zero in $\mathbb{Q}_{p_i}$, the argument in the last paragraph shows that $-ab$ is a square in $\mathbb{Z}_{p_i}/c\mathbb{Z}_{p_i} = \mathbb{Z}_{p_i}/p_i^{r_i}\mathbb{Z}_{p_i}$. As

$$\begin{aligned}\mathbb{Z}/c\mathbb{Z} &\simeq \prod\nolimits_i \mathbb{Z}/p_i^{r_i}\mathbb{Z} \quad \text{(Chinese remainder theorem)}\\ &\simeq \prod\nolimits_i \mathbb{Z}_{p_i}/p_i^{r_i}\mathbb{Z}_{p_i},\end{aligned}$$

we deduce that $-ab$ is a square in $\mathbb{Z}/c\mathbb{Z}$. Similarly, $-bc$ and $-ca$ are squares modulo a and b.

The proof that Theorem 2.8 implies Legendre's original theorem is just as elementary and only a little longer (Liu 2019).

NOTES The tangent process for constructing new rational points goes back to Diophantus (c250 A.D.) and was "much loved by Fermat"; the chord process was known to Newton. Hilbert and Hurwitz showed (in 1890) that, if a curve of genus zero has one rational point, then it has infinitely many, all given by rational values of a parameter. In 1901, Poincaré published a long article on rational points on curves in which he attempted to rescue the subject from being merely a collection of ad hoc results about individual equations. Although he is usually credited with conjecturing the finite basis theorem, he rather simply assumed it. Beppo Levi was the first to ask explicitly whether the finite basis theorem was true. In his remarkable 1922 paper, Mordell proved the finite basis theorem, and, in a rather off-handed way, conjectured that all curves of genus > 1 over $\mathbb{Q}$ have only finitely many rational points (the Mordell conjecture). Both were tremendously important, the former as the first general theorem in diophantine geometry, and the latter as one of the most important open conjectures in the subject until it was proved by Faltings in 1983.

For more on the history of these topics, see Cassels 1986, Schappacher 1990, and Schappacher and Schoof 1996, from which the above notes have largely been drawn.

3 The group structure on a plane cubic curve

Let C be a nonsingular projective plane curve of degree 3 over a field k, which, for simplicity, we take to be perfect. Assume that $C(k)$ is nonempty, and choose a point $O \in C(k)$. In this section, we give a geometric construction of a group structure on $C(k)$ having O as its zero element.

According to Bezout's theorem, a line meets the curve in three points counting multiplicities. Given $P, Q \in C(k)$, neither lying on the tangent line at the other, we let PQ denote the third point of intersection of the line through P and Q with C. Given $P \in C(k)$, we let PP denote the second point of intersection of the tangent line at P with C. The points PQ and

PP have coordinates in k.[5] Note that $PP = P$ if and only if P is a point of inflection.

For a pair $P, Q \in C(k)$, we define

$$P + Q = O(PQ),$$

i.e., we draw the line through P and Q to get PQ, and then the line through O and PQ to get $P + Q$, as in the following diagram:

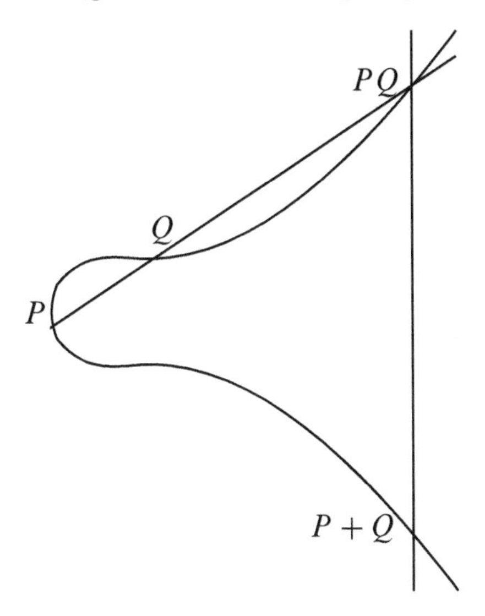

THEOREM 3.1 *The above construction makes $C(k)$ into a commutative group.*

First note that the definition does not depend on the order of P and Q; thus

$$P + Q = Q + P.$$

Next note that

$$O + P \stackrel{\text{def}}{=} O(OP) = P,$$

and so O is a neutral element.

Given $P \in C(k)$, let $P' = P(OO)$. Then $PP' = OO$, and $O(PP') = O(OO) = O$, i.e., $P + P' = O$.

[5]Certainly, there is a finite Galois extension K of k such that $PQ \in C(K)$, but then $\mathrm{Gal}(K/k)$ acts on the set $\{P, Q, PQ\}$, and so fixes PQ. Similarly, $PP \in C(k)$.

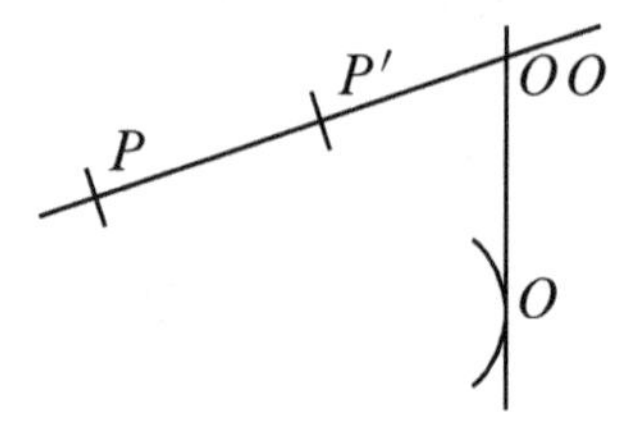

Thus the addition is commutative, has a zero element, and every element has a negative. It remains to check that it is associative, i.e., that

$$(P+Q)+R = P+(Q+R), \quad \text{all } P,Q,R \in C(k).$$

In Section 4 below, we shall explain how to deduce this from the Riemann–Roch theorem, but here we sketch an elegant geometric proof.

Geometric proof of associativity

In proving associativity, we may replace k with a larger field, and so we may assume k to be algebraically closed.

PROPOSITION 3.2 *If two cubic curves in $\mathbb{P}^2$ intersect in exactly nine points, then every cubic curve passing through eight of the points also passes through the ninth.*

PROOF. A cubic form

$$F(X,Y,Z) = a_1X^3 + a_2X^2Y + \cdots + a_{10}Z^3$$

has 10 coefficients $a_1,\ldots,a_{10}$. The condition that C_F pass through a point $P = (x\colon y\colon z)$ is a linear condition on $a_1,\ldots,a_{10}$, namely,

$$a_1x^3 + a_2x^2y + \cdots + a_{10}z^3 = 0.$$

If the eight points $P_1 = (x_1\colon y_1\colon z_1),\ldots,P_8$ are in "general position", specifically, if the vectors $(x_i^3, x_i^2y_i,\ldots,z_i^3)$, $i = 1,\ldots,8$, are linearly independent, then the cubic forms having $P_1,\ldots,P_8$ as zeros form a 2-dimensional space, and so there exist two such forms F and G such that the remainder can be written

$$\lambda F + \mu G, \quad \lambda,\mu \in k.$$

Now F and G have a ninth zero in common (by Bezout), and every curve $\lambda F + \mu G = 0$ passes through it.

When the P_i are not in general position, the proof can be completed by a case-by-case study (Walker 1950, III, 6.2). □

We now write $\ell(P,Q)$ for the line in $\mathbb{P}^2$ through the points P,Q. Let $P,Q,R \in C(k)$, and let

$$S = (P+Q)R, \quad T = P(Q+R).$$

Then $(P+Q)+R = OS$ and $P+(Q+R) = OT$. To prove the associativity, we have to show that $S = T$.

Consider the cubic curves,

$$\ell(P,Q)\cdot\ell(R,P+Q)\cdot\ell(QR,O) = 0,$$
$$\ell(P,Q+R)\cdot\ell(Q,R)\cdot\ell(PQ,O) = 0,$$

and C (see the picture below). All three pass through the eight points

$$O,P,Q,R,PQ,QR,P+Q,Q+R,$$

and the first two also pass through

$$U \stackrel{\text{def}}{=} \ell(P,Q+R)\cap\ell(R,P+Q).$$

Therefore, if the six lines $\ell(P,Q),\ldots,\ell(PQ,O)$ are distinct, then the proposition shows that C passes through U, which implies that $S = U = T$.

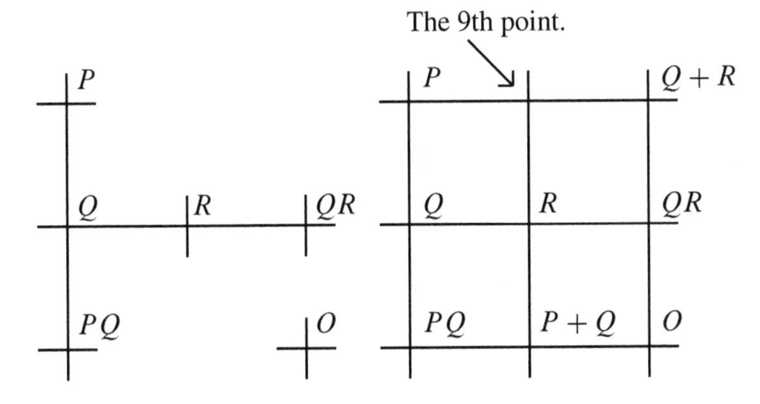

To handle the special case when two of the lines coincide, one uses the following stronger form of 3.2:

> Let C,C',C'' be cubic curves in $\mathbb{P}^2$ with C irreducible, and suppose that $C\cdot C' = \sum_{i=1}^{9}[P_i]$, where the P_i are nonsingular (not necessarily distinct) points on C; if $C\cdot C'' = \sum_{i=1}^{8}[P_i] + [Q]$, then $Q = P_9$ (Fulton 1969, §5.6, Proposition 3).

Here $C \cdot C'$ denotes the divisor $\sum_{P \in C \cap C'} I(P, C \cap C') \cdot [P]$ (see the next section).

EXERCISE 3.3 Find a necessary and sufficient condition for the line $L : Y = cX + d$ to be an inflectional tangent to the affine curve $C : Y^2 = X^3 + aX + b$, i.e., to meet C at a point P with $I(P, L \cap C) = 3$. Hence find a general formula for the elliptic curves C in canonical form having a rational point of order 3.

EXAMPLE 3.4 Consider the cubic curve

$$E : Y^2Z = X^3 + aXZ^2 + bZ^3, \quad \Delta \stackrel{\text{def}}{=} 4a^3 + 27b^2 \neq 0, \quad \text{char}(k) \neq 2.$$

We saw in 1.5 that E is nonsingular. As $O \stackrel{\text{def}}{=} (0{:}1{:}0)$ is a point of inflection (1.20), it is natural to take it to be the zero for the group law. The lines through it are the vertical lines in the affine plane, and so $(x{:}y{:}1) + (x'{:}y'{:}1) = O$ if and only if $x = x'$ and $y + y' = 0$. In particular, the points of order 2 on E are the points $(x{:}0{:}1)$ with x a root of $x^3 + ax + b$.

NOTES In his papers in the 1920s in which he generalized Mordell's finite basis theorem, Weil made systematic use of the commutative group structure on $E(\mathbb{Q})$, and (according to Schappacher 1990) was perhaps the first to do so — earlier mathematicians worked instead with the binary operation $P, Q \mapsto PQ = -(P + Q)$, which is not associative.

4 Regular functions; the Riemann–Roch theorem

Because it simplifies the exposition, we first explain the main results in the case of an algebraically closed base field.

Algebraically closed base fields

Throughout this section, k is algebraically closed.

REGULAR FUNCTIONS ON AFFINE CURVES

Let C be the affine plane curve over k defined by an irreducible polynomial $f(X,Y)$. Every polynomial $g(X,Y) \in k[X,Y]$ defines a function

$$(a,b) \mapsto g(a,b) \colon C(k) \to k$$

and the functions arising in this way are called the ***regular functions*** on C.

Obviously, every multiple of $f(X,Y)$ defines the zero function on $C(k)$, and the Nullstellensatz (CA, 13.8) implies these are the only such polynomials. Therefore the map sending g to the function $(a,b) \mapsto g(a,b)$ on $C(k)$ defines an isomorphism

$$k[X,Y]/(f) \to \{\text{ring of regular functions on } C\}.$$

Let

$$k[C] = k[X,Y]/(f) = k[x,y],$$

where $x = X \bmod(f)$ and $y = Y \bmod(f)$. Then x and y are the coordinate functions $P \mapsto x(P)$ and $P \mapsto y(P)$ on $C(k)$, and the elements of $k[C]$ are polynomials in x and y. A nonzero regular function g on C has only finitely many zeros in $C(k)$ because $C_f(k) \cap C_g(k)$ is finite (see 1.4c).

Because (f) is prime, $k[x,y]$ is an integral domain, and we let $k(C) = k(x,y)$ denote its field of fractions. An element $\varphi = g/h$ of $k(x,y)$ defines a function

$$(a,b) \mapsto \frac{g(a,b)}{h(a,b)} : C(k) \smallsetminus \{\text{ zeros of } h\} \to k.$$

We call such a φ a ***rational function*** on C, regular outside the finite set of zeros of h.

EXAMPLE 4.1 (a) Let C be the X-axis, i.e., the affine plane curve defined by the equation $Y = 0$. Then $k[C] = k[X,Y]/(Y) \simeq k[X]$ and $k(C) \simeq k(X)$. The rational functions on C are just the quotients $g(X)/h(X)$, $h \neq 0$, and such a function is regular outside the finite set of zeros of $h(X)$.

(b) Let C be the affine curve $Y^2 = X^3 + aX + b$. Then

$$k[C] = k[X,Y]/(Y^2 - X^3 - aX - b) = k[x,y].$$

Thus the regular functions on C are polynomials in the coordinate functions x and y, where x and y satisfy the relation

$$y^2 = x^3 + ax + b.$$

REGULAR FUNCTIONS ON PROJECTIVE CURVES

Let C be the projective plane curve over k (algebraically closed) defined by an irreducible homogeneous polynomial $F(X,Y,Z)$. If $G(X,Y,Z)$ and $H(X,Y,Z)$ are homogeneous polynomials of the same degree and H is not a multiple of F, then

$$(a:b:c) \mapsto \frac{G(a,b,c)}{H(a,b,c)}$$

is a well-defined function on the complement in $C(k)$ of the (finite) set of zeros of H. This is a rational function on C. More precisely, let

$$k[x,y,z] = k[X,Y,Z]/(F)$$

and let $k(x,y,z)$ be the field of fractions of $k[x,y,z]$. Because F is homogeneous, there is a well-defined decomposition

$$k[x,y,z] = \bigoplus_d k[x,y,z]_d,$$

where $k[x,y,z]_d$ consists of the elements of $k[x,y,z]$ having a representative in $k[X,Y,Z]$ that is homogeneous of degree d. Define

$$\begin{aligned} k(C) &= k(x,y,z)_0 \\ &= \{g/h \in k(x,y,z) \mid g,h \in k[x,y,z]_d, \text{ for some } d\}. \end{aligned}$$

It is a subfield of $k(x,y,z)$, and its elements are called the ***rational functions*** on C. A rational function g/h defines a ***regular*** function on the complement of the set of zeros of h in $C(k)$.

REMARK 4.2 Recall that there is a bijection

$$\begin{array}{ccccc} \mathbb{A}^2(k) & \leftrightarrow & U_2(k) & \subset & \mathbb{P}^2(k) \\ \left(\frac{a}{c},\frac{b}{c}\right) & \leftrightarrow & (a{:}b{:}c) & & \end{array}$$

To avoid confusion, write $k[X',Y']$ for the polynomial ring attached to $\mathbb{A}^2$ and $k[X,Y,Z]$ for the polynomial ring attached to $\mathbb{P}^2$. A polynomial $g(X',Y')$ defines a function $\mathbb{A}^2(k) \to k$, and the composite

$$U_2(k) \to \mathbb{A}^2(k) \xrightarrow{g} k$$

is

$$(a{:}b{:}c) \mapsto g\left(\tfrac{a}{c},\tfrac{b}{c}\right) = \frac{g^*(a,b,c)}{c^{\deg g}},$$

where $g^*(X,Y,Z) \stackrel{\text{def}}{=} g\left(\frac{X}{Z},\frac{Y}{Z}\right)\cdot Z^{\deg g}$ is $g(X,Y)$ made homogeneous by adding the fewest possible Z. Thus $g(X',Y')$ as a function on $\mathbb{A}^2 \simeq U_2$ agrees with $\frac{g^*(X,Y,Z)}{Z^{\deg g}}$. One see easily that the map

$$\frac{g(X',Y')}{h(X',Y')} \mapsto \frac{g^*(X,Y,Z)}{Z^{\deg g}} \frac{Z^{\deg h}}{h^*(X,Y,Z)} : k(X',Y') \to k(X,Y,Z)$$

is an injection, with image the subfield $k(X,Y,Z)_0$ of $k(X,Y,Z)$ of elements that can be expressed as a quotient of homogeneous polynomials of the same degree.

Now let C be an irreducible curve in $\mathbb{P}^2$, and assume that $C \cap U_2 \neq \emptyset$, i.e., that C is not the "line at infinity" $Z = 0$. Then the map

$$\frac{g(x',y')}{h(x',y')} \mapsto \frac{g^*(x,y,z)}{z^{\deg g}} \frac{z^{\deg h}}{h^*(x,y,z)} : k(x',y') \to k(x,y,z)_0$$

is a bijection from the field of rational functions on the affine curve $C \cap U_2$ to the field of rational functions on C. Moreover, if $\varphi' \mapsto \varphi$, then $\varphi(a{:}b{:}c) = \varphi'(\frac{a}{c},\frac{b}{c})$ for every point $(a{:}b{:}c) \in C(k) \cap U_2$ at which φ is defined.

EXAMPLE 4.3 The rational functions on $\mathbb{P}^1$ are the functions

$$(a{:}b) \mapsto \frac{G(a,b)}{H(a,b)},$$

where $G(X,Z)$ and $H(X,Z)$ are homogeneous polynomials of the same degree and $H(X,Z)$ is not the zero polynomial.

EXAMPLE 4.4 Let C be a nonsingular projective plane curve over $\mathbb{C}$. Then $C(\mathbb{C})$ has the structure of a compact Riemann surface, and the meromorphic functions on $C(\mathbb{C})$ coincide with the rational functions on C. In contrast, meromorphic functions on affine curves need not be rational, for example, e^z on $\mathbb{C} = \mathbb{A}^1(\mathbb{C})$ is not rational.

RIEMANN'S INEQUALITY

Let $C = C_F$ be the nonsingular projective curve over k (algebraically closed) defined by a homogeneous polynomial $F(X,Y,Z)$. As for meromorphic functions on Riemann surfaces, we try to understand the rational functions on C in terms of their zeros and poles.

The ***group of divisors*** $\mathrm{Div}(C)$ on C is defined to be the free abelian group on the set $C(k)$. Thus an element of $\mathrm{Div}(C)$ is a finite sum

$$D = \sum n_P[P], \quad n_P \in \mathbb{Z}, \quad P \in C(k).$$

The ***degree*** of D is $\sum n_P$. There is a partial ordering on $\mathrm{Div}(C)$,

$$\sum n_P[P] \geq \sum m_P[P] \iff n_P \geq m_P \text{ for all } P.$$

In particular, $\sum n_P[P] \geq 0$ if and only if all the n_p are nonnegative.

Let φ be a nonzero rational function on C_F. Then φ is defined by a quotient $\frac{G(X,Y,Z)}{H(X,Y,Z)}$ of two polynomials of the same degree, say, m, such that F does not divide H. Because φ is not identically zero, F does not divide G either. By Bezout's theorem

$$(\deg F)\cdot m = \sum_{P\in C_F\cap C_G} I(P, C_F\cap C_G)$$
$$(\deg F)\cdot m = \sum_{P\in C_F\cap C_H} I(P, C_F\cap C_H).$$

Define the divisor of φ to be

$$\operatorname{div}(\varphi) = \sum_{P\in C_F\cap C_G} I(P, C_F\cap C_G)[P] - \sum_{P\in C_F\cap C_H} I(P, C_F\cap C_H)[P].$$

If $[P]$ occurs in $\operatorname{div}(\varphi)$ with positive (resp. negative) coefficient, then P is called a ***zero*** (resp. ***pole***) of φ with ***multiplicity*** the coefficient. Note that $\operatorname{div}(\varphi)$ has degree zero, and so φ has as many zeros as poles (counted with multiplicity). Also, note that only the constant functions are without zeros and poles.

Given a divisor D, we define

$$L(D) = \{\varphi \mid \operatorname{div}(\varphi) + D \geq 0\} \cup \{0\}.$$

For example, if $D = [P] + 2[Q]$, then $L(D)$ consists of those rational functions having no poles outside $\{P, Q\}$ and having at worst a single pole at P and a double pole at Q. Each $L(D)$ is a vector space over k, and in fact a finite-dimensional vector space. We denote its dimension by $\ell(D)$.

THEOREM 4.5 (RIEMANN'S INEQUALITY) *There exists an integer g such that for all divisors D,*

$$\ell(D) \geq \deg D + 1 - g,$$

with equality for $\deg D$ sufficiently positive; in fact, equality holds if $\deg D \geq 2g - 2$.

PROOF. See Fulton 1969, Chap. 8. □

The integer g determined by the theorem is called the ***genus*** of C.

EXAMPLE 4.6 Let $a_1,\ldots,a_m \in k = \mathbb{A}^1(k) \subset \mathbb{P}^1(k)$, and let $D = \sum r_i[a_i]$ be a divisor with all $r_i > 0$. The rational functions φ on $\mathbb{A}^1$ with their poles in $\{a_1,\ldots,a_m\}$ and at worst a pole of order r_i at a_i are those of the form

$$\varphi = \frac{f(X)}{(X-a_1)^{r_1}\cdots(X-a_m)^{r_m}}, \quad f(X) \in k[X].$$

The function φ will not have a pole at ∞ if and only if $\deg f \leq \sum r_i = \deg D$. The dimension of $L(D)$ is therefore the dimension of the space of polynomials f of degree $\leq \deg D$, which is $\deg D + 1$. Thus the genus of $\mathbb{P}^1$ is 0.

REMARK 4.7 Riemann's inequality was originally proved for meromorphic functions on compact Riemann surfaces. Since the rational functions on a nonsingular projective curve over $\mathbb{C}$ coincide the meromorphic functions on its associated Riemann surface, the two have the same genus.

For Riemann's inequality to be useful, we need to be able to compute the genus of a curve. For a nonsingular projective plane curve, it is given by the formula,

$$g(C) = \frac{(\deg C - 1)(\deg C - 2)}{2} \tag{6}$$

(Fulton 1969, §8.3, Proposition 3). For example, a nonsingular projective plane curve has genus 0 if it is of degree 1 or 2 and it has genus 1 if it is of degree 3.

REMARK 4.8 Note that not all integers $g \geq 0$ can occur in the formula (6). Appropriately defined (see §6), there do exist nonsingular projective curves of arbitrary genus g — just not *nonsingular* projective *plane* curves.

PICARD GROUPS

Let C be a nonsingular projective plane curve over k (algebraically closed). The divisor of a rational function on C is said to be ***principal.*** Two divisors D and D' are said to be ***linearly equivalent***, denoted $D \sim D'$, if $D - D'$ is principal. We have groups

$$\operatorname{Div}(C) \supset \operatorname{Div}^0(C) \supset P(C),$$

where $\operatorname{Div}^0(C)$ is the group of divisors of degree 0 on C and $P(C)$ is the group of principal divisors.

Define ***Picard groups***,

$$\operatorname{Pic}(C) = \operatorname{Div}(C)/P(C), \quad \operatorname{Pic}^0(C) = \operatorname{Div}^0(C)/P(C).$$

ASIDE 4.9 The group $\mathrm{Pic}(C)$ is defined similarly for affine curves. When C is a nonsingular affine curve, the ring $k[C]$ is a Dedekind domain, and $\mathrm{Pic}(C)$ is its ideal class group.

PROPOSITION 4.10 *Let E be a nonsingular projective curve of genus 1, and let $O \in E(k)$. The map*

$$P \mapsto [P]-[O] : E(k) \to \mathrm{Pic}^0(E) \tag{7}$$

is bijective. It becomes a homomorphism when we endow $E(k)$ with the addition defined in the last section (hence the addition law on $E(k)$ is associative).

PROOF. For a divisor D on E, Riemann's inequality says that

$$\ell(D) = \deg D \text{ if } \deg D \geq 1.$$

If P and Q have the same image under the map (7), then $[P]-[Q]$ is principal, say, equal to $\mathrm{div}(\varphi)$. Then $\varphi \in L([Q])$, which consists only of the constant functions, and so $P = Q$. For the surjectivity, let D be a divisor of degree 0. Then $D+[O]$ has degree 1, and so there exists a rational function φ, unique up to multiplication by a nonzero constant, such that $\mathrm{div}(\varphi)+D+[O] \geq 0$. The only divisors ≥ 0 of degree 1 are those of the form $[P]$, and so there exists a unique point P such that $D+[O] \sim [P]$, i.e., such that $D \sim [P]-[O]$.

For the second statement, suppose that $P+Q = R$ in $E(k)$. Let L_1 be the line through P and Q, and L_2 the line through O and R. From the definition of $P+Q$, we know that both L_1 and L_2 have the point PQ as their third point of intersection with E. Regard L_1 and L_2 as linear forms in X, Y, Z, and let φ be the rational function $\frac{L_1}{L_2}$. Then φ has simple zeros at P, Q, PQ and simple poles at O, R, PQ, and so

$$\begin{aligned}\mathrm{div}(\varphi) &= [P]+[Q]+[PQ]-[O]-[R]-[PQ]\\ &= [P]+[Q]-[R]-[O].\end{aligned}$$

Hence

$$([P]-[O])+([Q]-[O]) \sim ([R]-[O]),$$

as required. □

REMARK 4.11 (a) It is clear from the proposition, that choosing a different zero $O \in E(k)$ only translates the addition law on E.

(b) The advantage of the geometric construction of the addition law is that it shows that the map $(P,Q) \mapsto P+Q\colon E \times E \to E$ is regular. The advantage of the construction in this section is that it immediately shows that addition is associative.

(c) The group law on (E,O) is intrinsically determined by the rule

$$\begin{cases} \sum_{i=1}^{n} P_i = \sum_{i=1}^{n} Q_i \iff \sum_{i=1}^{n}[P_i] \sim \sum_{i=1}^{n}[Q_i], \\ O+P = P \text{ for all } P. \end{cases}$$

When O is taken to be a point of inflection, then

$$P+Q+R=0 \iff [P]+[Q]+[R] \sim 3[O],$$

or, equivalently, $[P]+[Q]+[R]$ is the intersection divisor of a line with E. When P,Q,R are distinct, this last statement says that the points lie on a line; when $P=Q\neq R$, it says that the tangent to E at P meets E at R; when $P=Q=R$, it says that P is a point of inflction.

Perfect base fields

Throughout this section, k is a perfect field, for example, a field of characteristic zero or a finite field. All curves are required to be geometrically irreducible.

AFFINE CURVES

Let $C=C_f$ be a nonsingular affine plane curve over k. We define

$$k[C] = k[X,Y]/(f) = k[x,y],$$

and call it the ***ring of regular functions*** on C. We can no longer identify $k[C]$ with a ring of functions on $C(k)$ because, for example, $C(k)$ may be empty. However, every g in $k[C]$ defines a function $C(K) \to K$ for each field $K \supset k$, and $k[C]$ can be identified with the ring of families of such functions, compatible with inclusions $K \subset L$, defined by polynomials in the coordinate functions x and y. In other words, a curve C defines a functor from the category of fields containing k to sets, and the regular functions are the maps of functors $C \to \mathbb{A}^1$ expressible as polynomials in the coordinate functions. A ***rational function*** on C is an element of the field of fractions $k(C) = k(x,y)$ of $k[C]$.

A ***prime divisor*** on C is a nonzero prime ideal $\mathfrak{p}$ in $k[C]$, and the ***group of divisors*** $\mathrm{Div}(C)$ is the free abelian group on the set of prime divisors. When k is algebraically closed, the Nullstellensatz shows that the prime ideals in $k[C]$ are the ideals $(x-a, y-b)$ with $(a,b) \in C(k)$, and so, in this case, the definition agrees with that in the preceding section. The ***degree*** of a prime divisor $\mathfrak{p}$ is defined to be the dimension of $k[C]/\mathfrak{p}$ as a k-vector space, and

$$\deg(\textstyle\sum_{\mathfrak{p}} n_{\mathfrak{p}}\mathfrak{p}) = \sum_{\mathfrak{p}} n_{\mathfrak{p}} \deg(\mathfrak{p}).$$

For a prime divisor $\mathfrak{p}$, the localization of $k[C]$ at $\mathfrak{p}$,

$$k[C]_{\mathfrak{p}} = \{g/h \in k(C) \mid g,h \in k[C],\, h \notin \mathfrak{p}\},$$

is a ***discrete valuation ring***, i.e., a principal ideal domain with exactly one prime element $t_{\mathfrak{p}}$ up to associates (cf. Fulton 1969, §3.2). For $h \in k(C)^{\times}$, define $\mathrm{ord}_{\mathfrak{p}}(h)$ by the equation

$$h = h_0 t_{\mathfrak{p}}^{\mathrm{ord}_{\mathfrak{p}}(h)}, \quad h_0 \in k[C]_{\mathfrak{p}}^{\times}.$$

Then each $h \in k(C)^{\times}$ defines a divisor

$$\mathrm{div}(h) = \textstyle\sum_{\mathfrak{p}} \mathrm{ord}_{\mathfrak{p}}(h) \cdot \mathfrak{p}.$$

PROJECTIVE CURVES

Let $C = C_F$ be a nonsingular projective plane curve over k. We let

$$k[x,y,z] = k[X,Y,Z]/(F(X,Y,Z))$$

It is an integral domain, and remains so when tensored with any field $K \supset k$ (because C is geometrically irreducible). It is graded, $k[x,y,z] = \bigoplus_d k[x,y,z]_d$, and, as before, we define

$$\begin{aligned} k(C) &= k(x,y,z)_0 \\ &= \{g/h \in k(x,y,z) \mid g,h \in k[x,y,z]_d, \text{ for some } d\}. \end{aligned}$$

It is a subfield of $k(x,y,z)$, and its elements are called the ***rational functions*** on C.

Write C as a union of affine curves

$$C = C_0 \cup C_1 \cup C_2$$

in the usual way. As before, there are natural identifications of $k(C)$ with $k(C_i)$ for each i. Every prime divisor $\mathfrak{p}$ on one of the C_i defines a discrete valuation ring in $k(C)$, and the discrete valuation rings that arise in this way are exactly those containing k and having $k(C)$ as their field of fractions. We define a ***prime divisor*** on C to be such a discrete valuation ring. We shall use $\mathfrak{p}$ to denote a prime divisor on C, with $\mathcal{O}_\mathfrak{p}$ the corresponding discrete valuation ring and $\mathrm{ord}_\mathfrak{p}$ the corresponding valuation on $k(C)^\times$. The ***group of divisors*** on C is the free abelian group generated by the prime divisors on C. The ***degree*** of a prime divisor $\mathfrak{p}$ is the dimension of the residue field of $\mathcal{O}_\mathfrak{p}$ as a k-vector space, and $\deg(\sum n_\mathfrak{p}\mathfrak{p}) = \sum n_\mathfrak{p} \deg(\mathfrak{p})$. Every $h \in k(C)^\times$ defines a divisor

$$\mathrm{div}(h) = \sum_\mathfrak{p} \mathrm{ord}_\mathfrak{p}(h)\mathfrak{p},$$

which has degree zero, and is said to be ***principal***.

EXAMPLE 4.12 Consider the elliptic curve

$$E : Y^2Z = X^3 + aXZ^2 + bZ^3, \quad a, b \in k, \quad \Delta \neq 0,$$

over k. Write E_2 for the affine curve

$$Y^2 = X^3 + aX + b$$

and $k[x, y]$ for the ring of regular functions on E_2. A divisor on E is a finite sum

$$D = \sum n_\mathfrak{p}\mathfrak{p}$$

in which $n_\mathfrak{p} \in \mathbb{Z}$ and $\mathfrak{p}$ either corresponds to a nonzero prime ideal in $k[x, y]$ or is another symbol $\mathfrak{p}_\infty$ (the "prime divisor corresponding to the point at infinity"). The degree of $\mathfrak{p}$ is the degree of the field extension $[k[x, y]/\mathfrak{p} : k]$ if $\mathfrak{p} \neq \mathfrak{p}_\infty$, and is 1 if $\mathfrak{p} = \mathfrak{p}_\infty$.

THE RIEMANN–ROCH THEOREM

Let C be a nonsingular projective plane curve over k. For a divisor D on C, we define $L(D) = \{\varphi \mid \mathrm{div}(\varphi) + D \geq 0\} \cup \{0\}$, as before, and we let $\ell(D)$ denote its dimension as a k-vector space. There is a "canonical" divisor K_C on C, well defined up to linear equivalence, which equals the divisor of any differential one-form on C. The Riemann–Roch theorem states the following.

THEOREM 4.13 (RIEMANN–ROCH) *For any divisor D on C,*

$$\ell(D) = \deg(D) + 1 - g + \ell(K_C - D)$$

where K_C is a canonical divisor; moreover, K_C has degree $2g-2$, and so $\ell(D) = \deg(D) + 1 - g$ if $\deg(D) > 2g-2$.

The usual proofs of the Riemann–Roch theorem apply over arbitrary fields. As an alternative, we sketch a deduction of Theorem 4.13 from the same theorem over k^{al}. Fix an algebraic closure k^{al} and let $\bar{C}$ denote C regarded as a curve over k^{al}. The Galois group $\Gamma = \mathrm{Gal}(k^{\mathrm{al}}/k)$ acts on $C(k^{\mathrm{al}}) = \bar{C}(k^{\mathrm{al}})$ with finite orbits because each $P \in C(k^{\mathrm{al}})$ has coordinates in some finite extension of k.

Let C_i be one of the standard affine pieces of C, and let $\mathfrak{p}$ be a prime divisor on C_i. To give a k-homomorphism $k[C_i] = k[x,y] \to k^{\mathrm{al}}$ amounts to giving an element of $C_i(k^{\mathrm{al}})$, and the homomorphisms whose kernel contains $\mathfrak{p}$ correspond to the points in a single Γ-orbit in $C(k^{\mathrm{al}})$. In this way, we obtain a bijection from the set of prime divisors on C to the set of Γ-orbits in $C(k^{\mathrm{al}})$. Thus, we have a injective homomorphism

$$D \mapsto \bar{D} : \mathrm{Div}(C) \to \mathrm{Div}(\bar{C})$$

whose image consists of the divisors $\sum n_P P$ such that n_P is constant on each Γ-orbit. The map preserves principal divisors, the degrees of divisors, and the dimensions ℓ (because the condition for a function to lie in $L(D)$ is linear). It follows that C and $\bar{C}$ have the same genus, and that the Riemann–Roch theorem for $\bar{C}$ implies it for C.

PICARD GROUPS

Let C be a nonsingular projective plane curve over k, and let $\bar{C}$ denote C regarded as a curve over k^{al}. Then Γ acts $\mathrm{Pic}(\bar{C})$, and we define

$$\mathrm{Pic}(C) = \mathrm{Pic}(\bar{C})^{\Gamma}, \quad \mathrm{Pic}^0(C) = \mathrm{Pic}^0(\bar{C})^{\Gamma}.$$

Let C be a curve of genus 1 over k, and let $O \in C(k)$. We know (4.10) that the map

$$P \mapsto [P] - [O] : C(k^{\mathrm{al}}) \to \mathrm{Pic}^0(\bar{C})$$

is a bijection. Because $O \in C(k)$, this bijection commutes with the action of Γ, and so it defines a bijection $C(k^{\mathrm{al}})^{\Gamma} \to \mathrm{Pic}^0(\bar{C})^{\Gamma}$, i.e.,

$$C(k) \xrightarrow{\simeq} \mathrm{Pic}^0(C). \tag{8}$$

This is an isomorphism of abelian groups (because it is for k^{al}).

REMARK 4.14 The map from divisor classes on C to $\mathrm{Pic}(C)$ is injective, but it need not be surjective, i.e., not every element of $\mathrm{Pic}(C)$ need be represented by an element of $\mathrm{Div}(C)$, unless k has trivial Brauer group, for example, if k is finite (see IV, 1.10).

Regular maps of curves

Throughout this section, k is a perfect field. All curves are required to be geometrically irreducible. By an open subset of a curve, we mean the complement of a finite set of points.

AFFINE PLANE CURVES

A ***regular map*** $\varphi: C_{g_1} \to C_{g_2}$ of affine plane curves is a pair (f_1, f_2) of regular functions on C_{g_1} sending $C_{g_1}(K)$ into $C_{g_2}(K)$ for all fields K containing k, i.e., such that, for all $K \supset k$,

$$P \in C_{g_1}(K) \implies (f_1(P), f_2(P)) \in C_{g_2}(K).$$

Thus a regular map defines a map $C_{g_1}(K) \to C_{g_2}(K)$, functorial in K, and this functorial map determines the pair (f_1, f_2).

LEMMA 4.15 *Let (f_1, f_2) be a pair of regular functions on C_{g_1}. The map*

$$P \mapsto (f_1(P), f_2(P)): C_{g_1}(K) \to \mathbb{A}^2(K)$$

takes values in $C_{g_2}(K)$ for all fields $K \supset k$ if and only if $g_2(f_1, f_2) = 0$ (in $k[C_{g_1}]$).

PROOF. If $g_2(f_1, f_2) = 0$, then $g_2(f_1(P), f_2(P)) = 0$ for all $P \in C_{g_1}(K)$ and so $(f_1(P), f_2(P)) \in C_{g_2}(K)$. Conversely, if, for all $P \in C_{g_1}(k^{\mathrm{al}})$, $(f_1(P), f_2(P)) \in C_{g_2}(k^{\mathrm{al}})$, then $g_2(f_1, f_2)$ is the zero function on C_{g_1}. □

PROPOSITION 4.16 *Let C_{g_1} and C_{g_2} be geometrically irreducible affine plane curves over k. There are natural one-to-one correspondences between the following objects:*

(a) *regular maps $\varphi: C_{g_1} \to C_{g_2}$;*

(b) *pairs (f_1, f_2) of regular functions on C_{g_1} such that $g_2(f_1, f_2) = 0$ (in $k[C_{g_1}]$);*

(c) *functorial maps* $\varphi(K){:}C_{g_1}(K)\to C_{g_2}(K)$ *such that* $x\circ\varphi(K)$ *and* $y\circ\varphi(K)$ *are regular functions on* C_{g_1};

(d) *homomorphisms of* k*-algebras* $k[C_{g_2}]\to k[C_{g_1}]$.

PROOF. The lemma shows that the pair (f_1, f_2) defining φ in (a) satisfies the condition in (b). Moreover, it shows that the pair defines functorial map $P\mapsto(f_1(P), f_2(P))$ as in (c). Conversely, the regular functions $f_1\stackrel{\text{def}}{=}x\circ\varphi$ and $f_2=y\circ\varphi$ in (c) satisfy the condition in (b). Finally, write $k[C_{g_2}]=k[x,y]$. A homomorphism $k[C_{g_2}]\to k[C_{g_1}]$ is determined by the images f_1, f_2 of x, y, which can be any regular functions on C_{g_1} such that $g_2(f_1, f_2)=0$. □

PROJECTIVE PLANE CURVES

Consider polynomials $F_0(X,Y,Z), F_1(X,Y,Z), F_2(X,Y,Z)$ of the same degree. The map

$$(a_0{:}a_1{:}a_2)\mapsto(F_0(a_0,a_1,a_2){:}\,F_1(a_0,a_1,a_2){:}\,F_2(a_0,a_1,a_2))$$

defines a regular map to $\mathbb{P}^2$ on the subset of $\mathbb{P}^2$ where not all F_i vanish. Its restriction to any curve in $\mathbb{P}^2$ will also be regular where it is defined. It may be possible to extend the map to a larger set by representing it by different polynomials. Conversely, every regular map to $\mathbb{P}^2$ from an open subset of $\mathbb{P}^2$ arises in this way, at least "locally". Rather than give a precise definition, we give an example, and then state the criterion we shall use.

EXAMPLE 4.17 We prove that the circle $X^2+Y^2=Z^2$ over $\mathbb{C}$ is isomorphic to $\mathbb{P}^1$. This equation can be rewritten $(X+iY)(X-iY)=Z^2$, and so, after a change of variables, it becomes $C: XZ=Y^2$. Define

$$\varphi{:}\mathbb{P}^1\to C,\ (a{:}b)\mapsto(a^2{:}ab{:}b^2).$$

For the inverse, define

$$\psi{:}C\to\mathbb{P}^1\quad\text{by}\ \begin{cases}(a{:}b{:}c)\mapsto(a{:}b) & \text{if } a\neq 0\\ (a{:}b{:}c)\mapsto(b{:}c) & \text{if } b\neq 0\end{cases}.$$

Note that,

$$a\neq 0\neq b,\quad ac=b^2\implies\frac{c}{b}=\frac{b}{a}$$

and so the two maps agree on the set where they are both defined. Both φ and ψ are regular, and they define inverse maps on the sets of points.

Let $L = aX + bY + cZ$ be a nonzero linear form. The map

$$(x:y:z) \to (x/L(x,y,z), y/L(x,y,z), z/L(x,y,z))$$

is a bijection from the subset of $\mathbb{P}^2$ where $L \neq 0$ onto the plane $L = 1$ in $\mathbb{A}^3$. This last can be identified with $\mathbb{A}^2$ — for example, if $c \neq 0$, the projection $(x,y,z) \mapsto (x,y): \mathbb{A}^3 \to \mathbb{A}^2$ maps the plane $L = 1$ bijectively onto $\mathbb{A}^2$. Therefore, for every curve $C \subset \mathbb{P}^2$ not contained in the plane $L = 0$, $C_L \stackrel{\text{def}}{=} C \cap \{P \mid L(P) \neq 0\}$ is an affine plane curve. Note that, if $L = Z$, then $C_L = C_2$.

PROPOSITION 4.18 *Let C and C' be nonsingular projective plane curves over k. A regular map $C \to C'$ defines homomorphism $k(C') \to k(C)$ of fields which acts as the identity on k, and every such homomorphism arises from a unique regular map $C \to C'$*, i.e.,

$$\operatorname{Hom}(C,C') \simeq \operatorname{Hom}_k(k(C'),k(C)).$$

PROOF. See Fulton 1969, §7.5, Cor. to Theorem 3, □

REMARK 4.19 Let C and C' be as in the proposition, and let $U \subset C$ and $U' \subset C'$ be affine curves. Every regular map $U \to U'$ extends uniquely to a regular map $C \to C'$ (because it defines a k-homomorphism $k(C') \simeq k(U') \to k(U) \simeq k(C)$). The graph of the map $C \to C'$ is the closure of the graph of the map $U \to U'$.

This statement also holds for singular curves provided all the singularities are contained in U and U'.

EXAMPLE 4.20 Write

$$E(a,b): Y^2Z = X^3 + aXZ^2 + bZ^3$$
$$E(a,b)^{\text{aff}}: Y^2 = X^3 + aX + b.$$

Every regular map $\varphi^{\text{aff}}: E(a,b)^{\text{aff}} \to E(a',b')^{\text{aff}}$ extends uniquely to a regular map $\varphi: E(a,b) \to E(a',b')$. The curve $E(a,b)$ has exactly one point at infinity, namely, $(0:1:0)$, which is the third point of intersection of $E(a,b)$ with any "vertical line" $X = cZ$. If φ^{aff} sends vertical lines (lines $X = c$) to vertical lines, then φ must send the point at infinity on $E(a,b)$ to the point at infinity on $E(a',b')$.

DEFINITION 4.21 Let $\varphi: C \to C'$ be a regular map of curves over k.

(a) The map φ is ***constant*** if, for all fields $K \supset k$, the image of $\varphi(K)$ is a single point

(b) The map φ is ***dominant*** if the image of $\varphi(k^{\mathrm{al}})$ omits only finitely many points of $C'(k^{\mathrm{al}})$.

(c) The map φ is ***surjective*** if $\varphi(k^{\mathrm{al}})$ is surjective.

EXAMPLE 4.22 Let $C = \mathbb{A}^1 \smallsetminus \{0\}$.

(a) The regular map $x \mapsto x: C \to \mathbb{A}^1$ is dominant, but not surjective.

(b) For $n \in \mathbb{Z}$, the regular map $x \mapsto x^n: C \to C$ is constant if $n = 0$ and surjective otherwise (even though $x \mapsto x^n: k^\times \to k^\times$ need not be surjective).

PROPOSITION 4.23 *(a) A regular map of plane curves is either dominant or constant.*

(b) A dominant regular map of projective plane curves is surjective.

PROOF. (a) Omitted.

(b) See Fulton 1969, §8.3, Problem 8-18. □

Let C and C' be curves over k. A dominant regular map $\varphi: C \to C'$ defines (by composition) a homomorphism $k(C') \hookrightarrow k(C)$. The ***degree*** of φ is defined to the degree of $k(C)$ over the image of $k(C')$.

PROPOSITION 4.24 *Let $\varphi: C \to C'$ be a nonconstant map of plane curves.*

(a) *If the extension $k(C)/\varphi(k(C'))$ is separable of degree n, then*

$$\varphi(k^{\mathrm{al}}): C(k^{\mathrm{al}}) \to C'(k^{\mathrm{al}})$$

is $n{:}1$ outside a finite set.

(b) *The field extension $k(C)/\varphi(k(C'))$ is separable if, at some point, the map on tangent lines is an isomorphism.*

(c) *If the degree of φ is 1, and C and C' are nonsingular projective curves, then φ is an isomorphism.*

PROOF. (a) See Fulton 1969, §8.6, Problem 8.36.

(b) Omitted.

(c) If $\varphi: C \to C'$ induces an isomorphism $\varphi^*: k(C') \to k(\mathrm{C})$, then the inverse of φ^* arises from an regular map $C' \to C$ inverse to φ (apply 4.18). □

EXAMPLE 4.25 Consider the map $(x,y) \mapsto x\colon E^{\mathrm{aff}}(k) \to \mathbb{A}^1(k)$, where E^{aff} is the curve

$$E^{\mathrm{aff}} : Y^2 = X^3 + aX + b.$$

The map on the rings of regular functions is

$$X \mapsto x\colon k[X] \to k[x,y] \stackrel{\mathrm{def}}{=} k[X,Y]/(Y^2 - X^3 - aX - b).$$

Clearly $k(x,y) = k(x)[\sqrt{x^3+ax+b}]$, and so the map has degree 2. If $\mathrm{char}(k) \neq 2$, then the field extension is separable, and the map on points is 2:1 except over the roots of $X^3 + aX + b$,

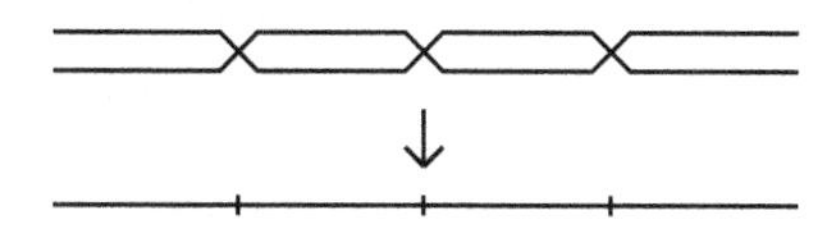

If $\mathrm{char}(k) = 2$, then the field extension is purely inseparable, and the map is 1:1 on points.

REMARK 4.26 A surjective regular map $\varphi\colon C \to C'$ of nonsingular projective curves, defines, in a natural way, a homomorphism $\varphi^*\colon \mathrm{Div}(C') \to \mathrm{Div}(C)$. This multiplies the degree of a divisor by $\deg(\varphi)$. In other words, when one counts multiplicities, $\varphi^{-1}(P)$ has $\deg(\varphi)$ points for all $P \in C'(k^{\mathrm{al}})$.

5 Defining algebraic curves over subfields

Let $\Omega \supset k$ be fields. A curve C over k is defined by polynomial equations with coefficients in k, and these same equations define a curve C_Ω over Ω. In this section, we examine the functor $C \rightsquigarrow C_\Omega$. This functor is faithful (the map on Hom sets is injective) but, in general, it is neither full (the map on Homs is not onto) nor essentially surjective (a curve over Ω need not be isomorphic to a curve of the form C_Ω).

EXAMPLE 5.1 (a) The curves

$$X^2 + Y^2 = 1, \quad X^2 + Y^2 = -1,$$

are not isomorphic over $\mathbb{R}$ (the second has no real points) but become isomorphic over $\mathbb{C}$ by the map $(x,y) \mapsto (ix, iy)$.

(b) An elliptic curve E over Ω with j-invariant $j(E)$ (see II, 2.1 below) arises from an elliptic curve over k if and only if $j(E) \in k$.

Let $\Gamma = \mathrm{Aut}(\Omega/k)$. We assume that $\Omega^{\Gamma} = k$. This is true, for example, if Ω is Galois over k or if Ω is algebraically closed and k is perfect (FT, 9.29).

Let C be a curve over Ω. When we apply an element $\sigma \in \Gamma$ to the coefficients of the equations defining C, we obtain a curve σC over Ω. A ***descent system on*** C is a family $(\varphi_\sigma)_{\sigma \in \Gamma}$ of isomorphisms $\varphi_\sigma\colon \sigma C \to C$ satisfying the cocycle condition,

$$\varphi_\sigma \circ (\sigma \varphi_\tau) = \varphi_{\sigma\tau} \text{ for all } \sigma, \tau \in \Gamma.$$

The pairs consisting of a curve over Ω and a descent system can be made into a category by defining a morphism $(C, (\varphi_\sigma)_\sigma) \to (C', (\varphi'_\sigma)_\sigma)$ to be a regular map $\alpha\colon C \to C'$ such that $\alpha \circ \varphi_\sigma = \varphi'_\sigma \circ \sigma\alpha$ for all $\sigma \in \Gamma$. For a curve C over k, C_Ω has a canonical descent system, because $\sigma(C_\Omega) = C_\Omega$, and so we can take $\varphi_\sigma = \mathrm{id}$.

Let C be a curve over Ω. A ***model*** of C over k is a curve C_0 over k together with an isomorphism $\varphi\colon C \to C_{0\Omega}$. Such a model (C_0, φ) is said to ***split*** a descent system $(\varphi_\sigma)_{\sigma \in \Gamma}$ on C if $\varphi_\sigma = \varphi^{-1} \circ \sigma\varphi$ for all $\sigma \in \Gamma$. A descent system is ***effective*** if it is split by some model over k.

PROPOSITION 5.2 *The functor sending a curve C over k to C_Ω endowed with its canonical descent system is fully faithful and its essential image consists of the pairs $(C, (\varphi_\sigma))$ with $(\varphi_\sigma)_{\sigma \in \Gamma}$ effective.*

PROOF. Left as an exercise. □

Let (φ_σ) be a descent system on C. For $P \in C(\Omega)$, define

$$^{\sigma}P = \varphi_\sigma(\sigma P).$$

Then

$$^{\sigma\tau}P = \varphi_{\sigma\tau}(\sigma\tau P) = (\varphi_\sigma \circ \sigma\varphi_\tau)(\sigma\tau P) = \varphi_\sigma(\sigma(^{\tau}P)) = {}^{\sigma}(^{\tau}P),$$

and so $(\sigma, P) \mapsto {}^{\sigma}P$ is an action of Γ on $C(\Omega)$. Conversely, an action $(\sigma, P) \mapsto {}^{\sigma}P$ of Γ on $C(\Omega)$ arises from a descent system if and only if, for every $\sigma \in \Gamma$, the map $\sigma P \mapsto {}^{\sigma}P\colon (\sigma C)(\Omega) \to C(\Omega)$ is regular; the action is then said to be ***regular***.

A finite set S of points in $C(\Omega)$ is said to ***rigidify*** C if no automorphism of C fixes every point in S except the identity map.

PROPOSITION 5.3 *A descent system (φ_σ) on C is effective if there exists a finite set S of points rigidifying C and a subfield K of Ω, finitely generated over k, such that ${}^\sigma P = P$ for every $P \in S$ and every σ fixing K.*

PROOF. See Milne 1999, 1.2. □

REMARK 5.4 Let C be a nonsingular projective curve over an algebraically closed field.

If C has genus zero, then it is isomorphic to $\mathbb{P}^1$. The automorphisms of $\mathbb{P}^1$ are the linear fractional transformations, and so any three distinct points rigidify $\mathbb{P}^1$.

If C has genus one, then it has only finitely many automorphisms fixing a given point O on C (see II, §2; IV, 7.13, 7.14). For each automorphism $\alpha \neq \mathrm{id}$, choose a P such that $\alpha(P) \neq P$. Then O and the points P rigidify C.

If C has genus greater than one, then it has only finitely many automorphisms (see, for example, Hartshorne 1977, IV, Exercise 5.2, when the field has characteristic zero). Thus, C is rigidified by some finite set.

Recall (FT, §9) that Γ has a natural topology under which the open subgroups of Γ correspond to the subfields of Ω that are finitely generated over k. An action of Γ on $C(\Omega)$ is continuous if and only if the stabilizer of each $P \in C(\Omega)$ is open.

PROPOSITION 5.5 *Assume Ω is algebraically closed. The functor sending a nonsingular projective curve C over k to C_Ω endowed with the natural action of Γ on $C(\Omega)$ is fully faithful, with essential image the nonsingular projective curves endowed with a continuous regular action of Γ.*

PROOF. Because Ω is algebraically closed, to give a descent system on a curve C' over Ω is the same as giving a regular action of Γ on $C'(\Omega)$. Moreover, a regular map of curves with descent systems will preserve the descent systems if and only if it commutes with the actions. According to 5.4, C' is rigidified by a finite set S, and because the action is continuous the condition in 5.3 holds. Thus the statement follows from the preceding propositions. □

6 General algebraic curves

We saw in Remark 4.8 that not all algebraic curves can be realized as nonsingular plane curves. In this section, we define more general algebraic curves. These will be needed mainly in Chapter V. Throughout, the field k is perfect.

Affine algebraic curves

An ideal $\mathfrak{a}$ in a ring R is radical if $R/\mathfrak{a}$ has no nilpotent elements, and it is prime if $R/\mathfrak{a}$ is an integral domain. For example, the ideal (f) in $k[X,Y]$ is radical if and only if f has no multiple factors, and it is prime if and only if f is irreducible. An ideal $\mathfrak{a} = (f_1, \ldots, f_r)$ in $k[X_1, \ldots, X_n]$ defines an ideal in $k^{\mathrm{al}}[X_1, \ldots, X_n]$ with the same generators.

The ***affine n-space*** over k is $\mathbb{A}^n(k) = k^n$. An ideal $\mathfrak{a}$ in $k[X_1, \ldots, X_n]$, assumed to be radical and remain radical in $k^{\mathrm{al}}[X_1, \ldots, X_n]$, defines an ***affine algebraic variety*** $V_{\mathfrak{a}}$ over k whose points in any field $K \supset k$ are its zeros, i.e., $V_{\mathfrak{a}}(K)$ is the set of $(x_1, \ldots, x_n) \in K^n$ such that $f(x_1, \ldots, x_n) = 0$ for all $f \in \mathfrak{a}$ (equivalently, for all f in a set of generators for $\mathfrak{a}$).

The variety $V_{\mathfrak{a}}$ is ***irreducible*** if $\mathfrak{a}$ is prime, and ***geometrically irreducible*** if $\mathfrak{a}$ remains prime in $k^{\mathrm{al}}[X_1, \ldots, X_n]$. A radical ideal $\mathfrak{a}$ can be written as a finite intersection $\mathfrak{a} = \mathfrak{p}_1 \cap \cdots \cap \mathfrak{p}_r$ of prime ideals, and then

$$V_{\mathfrak{a}}(K) = V_{\mathfrak{p}_1}(K) \cup \cdots \cup V_{\mathfrak{p}_r}(K), \quad K \supset k,$$

with each $V_{\mathfrak{p}_i}$ irreducible. The varieties $V_{\mathfrak{p}_i}$ are called the (irreducible) ***components of*** $V_{\mathfrak{a}}$.

The dimension of an irreducible variety $V_{\mathfrak{p}}$ is zero if $V_{\mathfrak{p}}(k^{\mathrm{al}})$ consists of a single point. It is one, and the variety is called a curve, if its maximal proper closed subvarieties have dimension zero. Inductively, the dimension of $V_{\mathfrak{p}}$ is m if its maximal proper closed subvarieties have dimension $m-1$.

Let $P = (a_1, \ldots, a_n) \in V_{\mathfrak{p}}(K)$. The tangent space $T_P(V_{\mathfrak{a}})$ to $V_{\mathfrak{a}}$ at P is the linear subspace (centred at P) defined by the equations

$$\sum \left(\frac{\partial f}{\partial X_i} \right)_P (X_i - a_i) = 0$$

where f runs over the elements of $\mathfrak{a}$ (equivalently, over a set of generators for $\mathfrak{a}$). Then $\dim_K(T_P(V_{\mathfrak{a}})) \geq \dim V$, with equality if P ***nonsingular.***

Projective algebraic curves

An ideal $\mathfrak{a}$ in $k[X_0, \ldots, X_n]$ is homogeneous if, with each f, it contains the homogeneous components of f. A ***projective algebraic variety*** in $\mathbb{P}^n$

is defined by a homogeneous ideal $\mathfrak{a}$ in $k[X_0,\ldots,X_n]$. Each projective algebraic variety in $\mathbb{P}^n$ is, in a natural way, a union of $n+1$ affine algebraic varieties. The definitions and statements in Section §4 carry over in an obvious way to projective algebraic curves. For example, the Riemann–Roch theorem holds for all nonsingular projective algebraic curves, and Propositions 4.18, 4.23, 4.24 hold with "plane" omitted. It is not necessary to assume that the curves in §5 are plane.

Nonsingular projective curves, plane curves, and function fields

PROPOSITION 6.1 *Let C_f be an irreducible affine plane curve over k. The field $k(C_f)$ is finitely generated of transcendence degree 1 over k, and every such field arises from a curve. The curve C_f is geometrically irreducible if and only if k is algebraically closed in $k(C_f)$.*

PROOF. By definition, $k(C_f)$ is the field of fractions of $k[X.Y]/(f)$, and so it is obviously finitely generated of transcendence degree 1. Conversely, let F be finitely generated of transcendence degree 1 over k. By definition, this means that F is a finite extension of $k(X)$ for some transcendental X in F. Using that k is perfect, one shows that X can be chosen so that F is a separable extension of $k(X)$, and so $F = k(X)(y)$ for some y. Let $f(X,Y) \in k(X)[Y]$ be the minimum polynomial of y over $k(X)$. After multiplying f by a power of X, we may suppose that it lies in $k[X.Y]$. Now $F = k(C_f)$.

Let k' be the algebraic closure of k in $k(C_f)$. Then

$$k^{\mathrm{al}} \otimes_k k' \subset k^{\mathrm{al}} \otimes_k k[X,Y]/(f) \simeq k^{\mathrm{al}}[X,Y[/(f),$$

and so $k^{\mathrm{al}}[X,Y]/(f)$ is not an integral domain if $k' \neq k$. We omit the proof of the converse. □

DEFINITION 6.2 A ***function field (in one variable)*** over k, is a field F containing k and such that

(a) k is algebraically closed in F;

(b) F is finitely generated (of transendence degree 1) over k.

Thus, the functor $C \rightsquigarrow k(C)$ from irreducible affine algebraic curves over k to function fields in one variable over k is essentially surjective.

THEOREM 6.3 *The functor $C \rightsquigarrow k(C)$ is a contravariant equivalence from the category of geometrically irreducible nonsingular projective curves over k to the category of function fields in one variable over k.*

PROOF. In particular, when we allow nonsingular projective curves other than plane curves, then the functor $C \rightsquigarrow k(C)$ becomes essentially surjective (cf. 4.18). Let F be a function field in one variable over k. From Proposition 6.1, we know that $F = k(C)$ for C an affine curve over k. The next theorem allows us to replace C with a nonsingular projective curve. □

THEOREM 6.4 *Let C be a projective plane curve over a perfect field k. There exists a nonsingular projective curve C' over k and a surjective morphism $\varphi: C' \to C$ such that*

(a) *the homomorphism $k(C) \to k(C')$ defined by φ is an isomorphism;*

(b) *$\varphi(k^{\mathrm{al}}): C'(k^{\mathrm{al}}) \to C(k^{\mathrm{al}})$ restricts to a bijection $U' \to U$, where U is the set of nonsingular points in $C(k^{\mathrm{al}})$ and U' is its inverse image in $C'(k^{\mathrm{al}})$.*

If $\varphi': C'' \to C$ is a second such curve and morphism, then there is a unique isomorphism $\psi: C' \to C''$ such that $\varphi' \circ \psi = \varphi$. Every nonsingular projective curve over k arises in this way.

PROOF. See Lorenzini 1996 or Fulton 1969, Chap. 7. □

ASIDE 6.5 For any nonsingular projective curve C, there exists a regular map $\varphi: C \to C'$ from C onto a plane projective curve C' such that φ is an isomorphism outside a finite set and C' has only ordinary multiple points as singularities (Fulton 1969, Chap. 7). The genus of C is given by

$$g(C) = \frac{(\deg C' - 1)(\deg C' - 2)}{2} - \sum_{P \in C(k)} \frac{m_P(C')(m_P(C') - 1)}{2},$$

where $m_P(C')$ is the multiplicity of P on C' (ibid., Chap. 8, §3).

ASIDE 6.6 Theorem 6.3 shows that the study of nonsingular projective curves is essentially equivalent to that of function fields in one variable. This observation, which goes back to Dedekind and Weber 1882 in the complex case,[6] allows one to study algebraic curves by the methods of algebraic number theory, and makes the subject quite elementary. It was the approach used by the German school from Dedekind and Weber in the 1880s to Deuring and Hasse in the 1950s. Although elementary, this approach misses the geometry, and works well only in dimension one.

[6]They used algebraic methods to give rigorous proofs of statements about Riemann surfaces that had only previously been proved, in a nonrigorous way, using analysis and topology.

Chapter II

Basic Theory of Elliptic Curves

1 Definition of an elliptic curve

DEFINITION 1.1 An ***elliptic curve*** over k can be defined, according to taste, as

(a) a nonsingular projective plane curve E over k of degree 3 together with a point $O \in E(k)$;

(b) same as (a) except that O is required to be a point of inflection;

(c) a nonsingular projective plane curve over k of the form

$$Y^2Z + a_1XYZ + a_3YZ^2 = X^3 + a_2X^2Z + a_4XZ^2 + a_6Z^3; \quad (9)$$

(d) a nonsingular complete curve E over k of genus 1 together with a point $O \in E(k)$.

(a)→(d). Let E be as in (a). Then E has genus 1 by formula (6), p. 41.

(d)→(c). Let (E,O) be as in (d). We shall see (p. 59) that the Riemann–Roch theorem implies that the k-vector space $L(3O)$ has a basis $1,x,y$ such that the map

$$P \mapsto (x(P):y(P):1): E \setminus \{O\} \to \mathbb{P}^2$$

defines an isomorphism from E onto the curve (9) sending O to $(0:1:0)$.

(b)↔(c). Let (E,O) be as in (b). We shall show (Proposition 1.2) that a linear change of variables will carry E into the form (c) and O into the point $(0\!:\!1\!:\!0)$. Conversely, let E be as in (c); then $O = (0\!:\!1\!:\!0) \in E(k)$ and is a point of inflection (I, 1.20).

(b)→(a). Obvious.

On combining these statements, we see that a curve E as in (a) can be embedded in $\mathbb{P}^2$ in such a way that O becomes a point of inflection. We also give a direct proof of this (p. 61).

For definiteness, we adopt (d) as our definition.

Transforming a cubic equation into standard form

PROPOSITION 1.2 *Let C be a nonsingular cubic projective plane curve over k, and let O be a point of inflection in $C(k)$.*

(a) *After an invertible linear change of variables with coefficients in k, the point O will have coordinates $(0\!:\!1\!:\!0)$ and the tangent to C at O will be the line $Z = 0$.*

(b) *If $O = (0\!:\!1\!:\!0)$ and the tangent to C at O is the line $Z = 0$, then the equation of C has the form (9).*

PROOF. We first prove (a). Let $(a\!:\!b\!:\!c) \in \mathbb{P}^2(k)$, and assume that $b \neq 0$. The regular map

$$(x\!:\!y\!:\!z) \mapsto (bx - ay\!:\!by\!:\!bz - cy)\colon \mathbb{P}^2 \to \mathbb{P}^2$$

sends $(a\!:\!b\!:\!c)$ to $(0\!:\!b^2\!:\!0) = (0\!:\!1\!:\!0)$ and is an isomorphism (it has an inverse of a similar form). If $b = 0$, but $c \neq 0$, say, we first interchange the y and z coordinates. Thus, we may suppose that $O = (0 : 1 : 0)$.

Let

$$L : aX + bY + cZ = 0, \quad a,b,c \in k,$$

be the tangent line at $(0\!:\!1\!:\!0)$. Let $A = (a_{ij})$ be any invertible 3×3 matrix whose first two columns are orthogonal to (a,b,c), and define a change of variables by

$$A \begin{pmatrix} X' \\ Y' \\ Z' \end{pmatrix} = \begin{pmatrix} X \\ Y \\ Z \end{pmatrix}.$$

With respect to the variables X', Y', Z', the equation of the line L becomes

$$0 = (a,b,c) \begin{pmatrix} X \\ Y \\ Z \end{pmatrix} = (a,b,c) A \begin{pmatrix} X' \\ Y' \\ Z' \end{pmatrix} = (0,0,d) \begin{pmatrix} X' \\ Y' \\ Z' \end{pmatrix} = dZ'.$$

Moreover, $d \neq 0$, and so we may take the equation of the line to be $Z' = 0$. This completes the proof of (a).

Now assume that C and O satisfy the conditions in (b). The general cubic form is

$$c_1X^3 + c_2X^2Y + c_3X^2Z + c_4XY^2 + c_5XYZ + c_6XZ^2 + \\ c_7Y^3 + c_8Y^2Z + c_9YZ^2 + c_{10}Z^3.$$

Let $F(X,Y,Z)$ be the polynomial defining C. Because C is nonsingular, F is geometrically irreducible (I, 1.21).

Because $O = (0{:}1{:}0) \in C(k)$, $\boxed{c_7 = 0}$.

Recall that $U_1 = \{(x{:}y{:}z) \mid y \neq 0\}$ and that we identify U_1 with $\mathbb{A}^2$ by $(x{:}1{:}z) \leftrightarrow (x,z)$. Moreover $C \cap U_1$ is the affine curve defined by $F(X,1,Z)$,

$$c_1X^3 + c_2X^2 + c_3X^2Z + c_4X + c_5XZ + c_6XZ^2 + c_8Z + c_9Z^2 + c_{10}Z^3.$$

The tangent at $(0{:}1{:}0) \leftrightarrow (0,0)$ is the line

$$c_4X + c_8Z = 0.$$

Because this equals the line $Z = 0$, $\boxed{c_4 = 0}$. As C is nonsingular, $\boxed{c_8 \neq 0}$.

Let L denote the line $Z = 0$. The intersection number

$$I(O, L \cap C) = I(Z, F(X,1,Z)) \overset{1.8\text{d}}{=} I(Z, c_1X^3 + c_2X^2).$$

Because O is a point of inflection, $I(O, L \cap C) \geq 3$, and so $\boxed{c_2 = 0}$.

On combining the boxed statements, we find that our cubic form has become

$$c_1X^3 + c_3X^2Z + c_5XYZ + c_6XZ^2 + c_8Y^2Z + c_9YZ^2 + c_{10}Z^3, \quad c_8 \neq 0.$$

Moreover, $c_1 \neq 0$ because otherwise the polynomial is divisible by Z. After dividing through by c_1 and replacing Z with $-c_1Z/c_8$, we obtain an equation of the form (9). □

Projective embedding of a curve of genus 1 with a rational point.

Let E be a complete nonsingular curve of genus 1 over a field k and let $O \in E(k)$. According to the Riemann–Roch theorem (I, 4.13), the rational functions on E having no poles except at O and at worst a pole of order

$m \geq 1$ at O, form a vector space of dimension m over k, i.e., $L(m[O])$ has dimension m for $m \geq 1$. The constant functions lie in $L([O])$, and so $\{1\}$ is a basis for $L([O])$. Choose x so that $\{1, x\}$ is a basis for $L(2[O])$, and then y so that $\{1, x, y\}$ is a basis for $L(3[O])$. The set $\{1, x, y, x^2\}$ is linearly independent, hence a basis for $L(4[O])$, because otherwise x^2 would be a linear combination of $1, x, y$ and could not have a quadruple pole at O. Similarly, $\{1, x, y, x^2, xy\}$ is a basis for $L(5[O])$.

The subset $\{1, x, y, x^2, xy, x^3, y^2\}$ of $L(6[O])$ contains seven elements, and so it must be linearly dependent, say,

$$a_0 y^2 + a_1 xy + a_3 y = a_0' x^3 + a_2 x^2 + a_4 x + a_6, \quad a_i \in k,$$

as regular functions on $E \smallsetminus \{O\}$. Both a_0 and a_0' are nonzero because the set with either x^3 or y^2 omitted is linearly independent, and so, after replacing y with $a_0 y / a_0'$ and x with $a_0 x / a_0'$ and multiplying through by $a_0'^2 / a_0^3$, we can suppose that both equal 1. The map $P \mapsto (x(P), y(P))$ sends $E \smallsetminus \{O\}$ onto the affine plane curve

$$C : Y^2 + a_1 XY + a_3 Y = X^3 + a_2 X^2 + a_4 X + a_6.$$

The function x has a double pole at O and no other pole, and so it has only two zeros. Similarly, $x + c$ has two zeros for every $c \in k$ (counting multiplicities), and so the composite of the maps

$$E \smallsetminus \{O\} \to C \to \mathbb{A}^1, \quad P \mapsto (x(P), y(P)) \mapsto x(P),$$

has degree 2 (see I, 4.24, 4.26). Similarly, the composite of the maps

$$E \smallsetminus \{O\} \to C \to \mathbb{A}^1, \quad P \mapsto (x(P), y(P)) \mapsto y(P)$$

has degree 3. Therefore the degree of

$$P \mapsto (x(P), y(P)) \colon E \smallsetminus \{O\} \to C$$

divides both 2 and 3, and so equals 1, i.e., $k(C) \simeq k(E)$. If C were singular, then its associated nonsingular curve E would have genus 0, contradicting our assumption. Therefore C is nonsingular, and so the map is an isomorphism, and it extends to an isomorphism of E onto

$$\bar{C} : Y^2 Z + a_1 XYZ + a_3 YZ^2 = X^3 + a_2 X^2 Z + a_4 XZ^2 + a_6 Z^3$$

(see I, 4.24c).

Transforming a point to a point of inflection

Let C be a nonsingular projective plane cubic curve over k, and let $O \in C(k)$. Write

$$C : F_1(X,Y)Z^2 + F_2(X,Y)Z + F_3(X,Y) = 0,$$

where $F_i(X,Y)$ is homogeneous of degree i, and let $C^{\text{aff}} = C \cap \{Z \neq 0\}$, so that

$$C^{\text{aff}}: F_1(X,Y) + F_2(X,Y) + F_3(X,Y) = 0.$$

As we wish to transform O into a point of inflection, we may suppose that it is not already one. Therefore, the tangent to C at O meets C in a point $P \neq O$. After a linear change of variables, we may suppose that $P, O \in C^{\text{aff}}(k)$, that $P = (0{:}0{:}1)$, and that $O = (0{:}y{:}1)$, $y \neq 0$. Then

$$F_1(0,1)Y + F_2(0,1)Y^2 + F_3(0,1)Y^3 = 0,$$

has $Y = 0$ as a root and $Y = y$ as a double root, and so

$$F_2(0,1)^2 - 4F_1(0,1)F_3(0,1) = 0. \tag{10}$$

The line $Y = tX$ intersects C^{aff} at the points whose x-coordinates satisfy

$$xF_1(1,t) + x^2F_2(1,t) + x^3F_3(1,t) = 0.$$

The solution $x = 0$ gives the origin, and so

$$F_1(1,t) + xF_2(1,t) + x^2F_3(1,t) = 0$$

gives a relation between the functions $P \mapsto x(P)$ and $P \mapsto t(P) = \frac{y(P)}{x(P)}$ on $C^{\text{aff}} \smallsetminus \{(0,0)\}$. Let $s = 2F_3(1,t)x + F_2(1,t)$. On exanding s^2 and using that

$$F_1(1,t)F_3(1,t) + xF_2(1,t)F_3(1,t) + x^2F_3(1,t)^2 = 0,$$

we find that

$$s^2 = G(t), \quad G(t) = F_2(1,t)^2 - 4F_1(1,t)F_3(1,t).$$

Now

$$\begin{cases} s \mapsto 2F_3(1,y/x)x + F_2(1,y/x) \\ t \mapsto y/x \end{cases}$$

defines a homomorphism $k[s,t] \to k[x,y][x^{-1}]$, and hence a regular map $C^{\text{aff}} \smallsetminus \{O\} \to E$, where E is the affine curve $s^2 = G(t)$ (see I, 4.16). The polynomial $G(t)$ has degree 3 because of (10), and the map $C^{\text{aff}} \smallsetminus \{O\} \to E$ defines an isomorphism of C onto the projective closure of E sending O to $(0:1:0)$ (see I, 4.19).

NOTES The above argument was found by Nagell (1928–29). Cassels (1991, p. 34) notes that while "older geometrical techniques (adjoint curves etc.) had shown that every elliptic curve is birationally equivalent to a cubic, Nagell was the first to show that it can be reduced to the canonical form."

ASIDE 1.3 The ***Hessian*** of a projective plane curve C_F is

$$H(X,Y,Z) = \begin{vmatrix} \frac{\partial^2 F}{\partial X^2} & \frac{\partial^2 F}{\partial X \partial Y} & \frac{\partial^2 F}{\partial X \partial Z} \\ \frac{\partial^2 F}{\partial X \partial Y} & \frac{\partial^2 F}{\partial Y^2} & \frac{\partial^2 F}{\partial Y \partial Z} \\ \frac{\partial^2 F}{\partial X \partial Z} & \frac{\partial^2 F}{\partial Y \partial Z} & \frac{\partial^2 F}{\partial Z^2} \end{vmatrix}.$$

Assume that $\text{char}(k) \neq 2$. A nonsingular point $P = (a:b:c)$ on the curve C_F is a point of inflection if and only if $H(a,b,c) = 0$, in which case $I(P, C_F \cap H) = 1$ if and only if the point of inflection is ordinary (Fulton 1969, §5.3, Theorem). If F has degree d, then H has degree $3(d-2)$, and so a nonsingular cubic has 9 points of inflection in k^{al} by Bezout's theorem. Unfortunately, it might have no point of inflection with coordinates in k. An invertible linear change of variables will not change this (it will only multiply the Hessian by a nonzero constant). This explains why a nonlinear change of variables was needed in the above argument.

ASIDE 1.4 An equation $Y^2 = f(X)$, $\deg f = 4$, defines a singular curve whose associated nonsingular curve has genus 1, and so is an elliptic curve if it has a rational point. Similarly, an intersection of two quadric surfaces defines an elliptic curve if it has a rational point. See Cassels 1991, p. 35.

Isogenies

Let (E,O) and (E',O') be elliptic curves over k. An ***isogeny*** from E to E' is a nonconstant regular map $\varphi\colon E \to E'$ such that $\varphi(O) = \varphi(O')$. Then $\varphi(k^{\text{al}})\colon E(k^{\text{al}}) \to E'(k^{\text{al}})$ is surjective (4.23).

PROPOSITION 1.5 *If $\varphi\colon (E,O) \to (E',O')$ is an isogeny, then the map $\varphi(k^{\text{al}})\colon E(k^{\text{al}}) \to E'(k^{\text{al}})$ is a homomorphism of groups.*

PROOF. We may suppose that k is algebraically closed. Then φ defines a homomorphism

$$\varphi_*\colon \text{Div}(E) \to \text{Div}(E'), \quad \sum n_P[P] \mapsto \sum n_P[\varphi(P)],$$

which obviously preserves degrees. It also preserves principal divisors (specifically, $\varphi_*(\mathrm{div}(f)) = \mathrm{div}(f_*)$ where f_* is the norm of f from $k(E)$ to $\varphi^* k(E')$; see ANT 4.1(c)), and so it passes to the Pic groups. We have a commutative diagram

$$\begin{array}{ccc} E(k) & \xrightarrow{P \mapsto [P]-[O]} & \mathrm{Pic}^0(E) \\ \downarrow{\scriptstyle \varphi(k)} & & \downarrow{\scriptstyle \varphi_*} \\ E'(k) & \xrightarrow{P \mapsto [P]-[O']} & \mathrm{Pic}^0(E'). \end{array}$$

As the horizontal maps are isomorphisms and all the maps except possibly $\varphi(k)$ are homomorphisms, $\varphi(k)$ must also be a homomorphism. □

1.6 The proposition implies that the diagram of regular maps

$$\begin{array}{ccc} E \times E & \xrightarrow{+} & E \\ \downarrow{\scriptstyle \varphi \times \varphi} & & \downarrow{\scriptstyle \varphi} \\ E' \times E' & \xrightarrow{+} & E' \end{array}$$

commutes. Therefore, $\varphi(K): E(K) \to E'(K)$ is a homomorphism of groups for all fields K containing k.

1.7 Let $\varphi: E \to E'$ and $\varphi': E \to E''$ be isogenies. If $\mathrm{Ker}(\varphi) \subset \mathrm{Ker}(\varphi')$, then there exists an isogeny $\psi: E' \to E''$ such that $\psi \circ \varphi = \varphi'$. When the degrees of φ and φ' are prime to $\mathrm{char}(k)$, the hypothesis means that $\mathrm{Ker}(\varphi(k^{\mathrm{al}})) \subset \mathrm{Ker}(\varphi'(k^{\mathrm{al}}))$; otherwise it is a condition on the group-scheme kernels.

1.8 Elliptic curves E and E' are said to be ***isogenous*** if there exists an isogeny $E \to E'$. Isogeny is an equivalence relation. It is reflexive because the identity map is an isogeny, and it is transitive because a composite of isogenies is an isogeny. Let $\varphi: E \to E'$ be an isogeny. The kernel of φ is finite, and so it is contained in $E_n \overset{\mathrm{def}}{=} \mathrm{Ker}(E \xrightarrow{n} E)$ for some n. Now the isogeny $n: E \to E$ factors into a composite of isogenies

$$E \xrightarrow{\varphi} E' \xrightarrow{\varphi'} E.$$

2 The Weierstrass equation for an elliptic curve

In this section, we list many of the equations describing an elliptic curve and its structure.

Characteristic $\neq 2,3$

Let E be an elliptic curve over k, described by an equation (9). When k has characteristic $\neq 2,3$, a change of variables

$$X' = X, \quad Y' = Y + \frac{a_1}{2}X, \quad Z' = Z$$

will eliminate the XYZ term in (9), and a change of variables

$$X' = X + \frac{a_2}{3}, \quad Y' = Y + \frac{a_3}{2}, \quad Z' = Z$$

will then eliminate the X^2Z and YZ^2 terms. Thus we arrive at the equation

$$Y^2Z = X^3 + aXZ^2 + bZ^3.$$

THEOREM 2.1 *Let k be a field of characteristic $\neq 2,3$.*

(a) Every elliptic curve (E,O) over k is isomorphic to a curve of the form

$$E(a,b) : Y^2Z = X^3 + aXZ^2 + bZ^3, \qquad a,b \in k,$$

pointed by $(0{:}1{:}0)$. Conversely, the curve $E(a,b)$ is nonsingular (and so, when equipped with $(0{:}1{:}0)$, is an elliptic curve) if and only if $\Delta \stackrel{\text{def}}{=} 4a^3 + 27b^2 \neq 0$.

(b) Let $\varphi\colon E(a',b') \to E(a,b)$ be an isomorphism sending $(0{:}1{:}0)$ to $(0{:}1{:}0)$; then there exists a $c \in k^\times$ such that $a' = c^4a$, $b' = c^6b$ and φ is the map $(x{:}y{:}z) \mapsto (c^2x{:}c^3y{:}z)$. Conversely, if $a' = c^4a$, $b' = c^6b$ for some $c \in k^\times$, then $(x{:}y{:}z) \mapsto (c^2x{:}c^3y{:}z)$ is an isomorphism $E(a',b') \to E(a,b)$ sending $(0{:}1{:}0)$ to $(0{:}1{:}0)$.

(c) When (E,O) is isomorphic to $(E(a,b),O)$, we let

$$j(E) = \frac{1728(4a^3)}{4a^3 + 27b^2}.$$

Then $j(E)$ depends only on (E,O), and two elliptic curves (E,O) and (E',O') become isomorphic over k^{al} if and only if $j(E) = j(E')$.

PROOF. (a) The first statement was proved above. The point $(0\colon 1\colon 0)$ is always nonsingular on $E(a,b)$, and we saw in (I, 1.5) that the affine curve

$$Y^2 = X^3 + aX + b$$

is nonsingular if and only if $4a^3 + 27b^2 \neq 0$.

(b) The regular function $x \circ \varphi$ on $E(a',b')$ has a double pole at $O' = (0\colon 1\colon 0)$, and so $x \circ \varphi = u_1 x' + r$ for some $u_1 \in k^\times$ and $r \in k$ (see the proof of 1.1d →1.1c, p. 59). Similarly, $y \circ \varphi = u_2 y' + sx' + t$ for some $u_2 \in k^\times$ and $s,t \in k$. But $f \mapsto f \circ \varphi$ is a homomorphism $k[x,y] \to k[x',y']$, where x,y and x',y' are the coordinate functions on $E(a,b)$ and $E(a',b')$. As x and y satisfy

$$Y^2 = X^3 + aX + b,$$

so also do $x \circ \varphi$ and $y \circ \varphi$, i.e.,

$$(u_2 y' + sx' + t)^2 = \left(u_1 x' + r\right)^3 + a\left(u_1 x' + r\right) + b.$$

But any polynomial satisfied by x', y' is a multiple of

$$Y^2 - X^3 - a'X - b'$$

from which it follows that $u_2^2 = u_1^3$ and $r,s,t = 0$. Let $c = u_1/u_2$; then $a' = c^4 a$ and $b' = c^6 b$, and φ is as described. The converse is obvious.

(c) If (E,O) is isomorphic to both $(E(a,b),O)$ and $(E(a',b'),O')$, then there exists a $c \in k^\times$ such that $a' = c^4 a$ and $b' = c^6 b$, and so obviously the two curves give the same j. Conversely, suppose that $j(E) = j(E')$. Note first that

$$a = 0 \iff j(E) = 0 \iff j(E') = 0 \iff a' = 0.$$

As any two curves of the form $Y^2 Z = X^3 + bZ^3$ are isomorphic over k^{al}, it remains to consider the case that a and a' are both nonzero. After replacing (a,b) with $(c^4 a, c^6 b)$, where $c = (a'/a)^{1/4}$, we will have that $a = a'$. Now $j(E) = j(E') \Longrightarrow b = \pm b'$. A minus sign can be removed by a change of variables with $c = \sqrt{-1}$. □

REMARK 2.2 Two elliptic curves can have the same j-invariant and yet not be isomorphic over k. For example, if c is not a square in k, then the curve

$$Y^2 Z = X^3 + ac^2 XZ^2 + bc^3 Z^3$$

has the same j invariant as $E(a,b)$, but is not isomorphic to it.

REMARK 2.3 For every $j \in k$, there exists an elliptic curve E over k with $j(E) = j$, for example,

$$\begin{aligned} Y^2Z &= X^3 + Z^3 & j &= 0, \\ Y^2Z &= X^3 + XZ^2 & j &= 1728, \\ Y^2Z &= X^3 - \tfrac{27}{4}\tfrac{j}{j-1728}XZ^2 - \tfrac{27}{4}\tfrac{j}{j-1728}Z^3 & j &\neq 0, 1728. \end{aligned}$$

We next give the formulas for the addition and doubling of points on the elliptic curve

$$E : Y^2Z = X^3 + aXZ^2 + bZ^3, \quad a, b \in k.$$

These can be proved by a straightforward calculation using the description of the group law in I, §3 (see Cassels 1991, Formulary).

ADDITION FORMULA

Let $P = (x : y : 1)$ be the sum of $P_1 = (x_1 : y_1 : 1)$ and $P_2 = (x_2 : y_2 : 1)$. If $P_2 = -P_1$, then $P = O$, and if $P_1 = P_2$, we can apply the duplication formula below. Otherwise, $x_1 \neq x_2$, and (x, y) is determined by the formulas

$$\begin{aligned} x(x_1 - x_2)^2 &= x_1x_2^2 + x_1^2x_2 - 2y_1y_2 + a(x_1 + x_2) + 2b \\ y(x_1 - x_2)^3 &= W_2y_2 - W_1y_1, \end{aligned}$$

where

$$\begin{aligned} W_1 &= 3x_1x_2^2 + x_2^3 + a(x_1 + 3x_2) + 4b \\ W_2 &= 3x_1^2x_2 + x_1^3 + a(3x_1 + x_2) + 4b. \end{aligned}$$

DUPLICATION FORMULA

Let $P = (x : y : 1)$ and $2P = (x_2 : y_2 : 1)$. If $y = 0$, then $2P = 0$. Otherwise $y \neq 0$, and (x_2, y_2) is determined by the formulas

$$\begin{aligned} x_2 &= \frac{(3x^2 + a)^2 - 8xy^2}{4y^2} = \frac{x^4 - 2ax^2 - 8bx + a^2}{4(x^3 + ax + b)} \\ y_2 &= \frac{x^6 + 5ax^4 + 20bx^3 - 5a^2x^2 - 4abx - a^3 - 8b^2}{(2y)^3}. \end{aligned}$$

General base field

Let E be an elliptic curve over an arbitrary field k. An equation of the form

$$Y^2Z + a_1XYZ + a_3YZ^2 = X^3 + a_2X^2Z + a_4XZ^2 + a_6Z^3 \quad (11)$$

for E is called a ***Weierstrass equation***. Attached to an equation (11), there are the following quantities:

$$\begin{aligned} b_2 &= a_1^2 + 4a_2 & c_4 &= b_2^2 - 24b_4 \\ b_4 &= a_1a_3 + 2a_4 & c_6 &= -b_2^3 + 36b_2b_4 - 216b_6 \\ b_6 &= a_3^2 + 4a_6 \\ b_8 &= b_2a_6 - a_1a_3a_4 + a_2a_3^2 - a_4^2 \\ \Delta &= -b_2^2b_8 - 8b_4^3 - 27b_6^2 + 9b_2b_4b_6 \\ j &= c_4^3/\Delta. \end{aligned}$$

The subscripts indicate weights.

THEOREM 2.4 *Let k be an arbitrary field.*

(a) Every elliptic curve (E, O) over k is isomorphic to a curve of the form

$$E(a_1,\ldots): Y^2Z + a_1XYZ + a_3YZ^2 = X^3 + a_2X^2Z + a_4XZ^2 + a_6Z^3$$

pointed by $(0{:}1{:}0)$. Conversely, the curve $E(a_1,\ldots)$ is nonsingular (and so, when equipped with $(0{:}1{:}0)$, is an elliptic curve) if and only if $\Delta \neq 0$.

(b) Let $\varphi: E(a_1',\ldots) \to E(a_1,\ldots)$ be an isomorphism sending O to O'; then there exist $u \in k^\times$ and $r,s,t \in k$ such that

$$\begin{aligned} ua_1' &= a_1 + 2s \\ u^2a_2' &= a_2 - sa_1 + 3r - s^2 \\ u^3a_3' &= a_3 + ra_1 + 2t \\ u^4a_4' &= a_4 - sa_3 + 2ra_2 - (t+rs)a_1 + 3r^2 - 2st \\ u^6a_6' &= a_6 + ra_4 + r^2a_2 + r^3 - ta_3 - rta_1 - t^2 \\ u^{12}\Delta' &= \Delta \end{aligned}$$

and φ is the map sending $(x{:}y{:}z)$ to $(u^2x + rz{:}u^3y + su^2x + tz{:}z)$. Conversely, if there exist $u \in k^\times$ and $r,s,t \in k$ satisfying these equations, then

$$(x{:}y{:}z) \mapsto (u^2x + rz{:}u^3y + su^2x + tz{:}z)$$

is an isomorphism $E(a_1',\ldots) \to E(a_1,\ldots)$ sending O to O'.

(c) When (E,O) is isomorphic to $(E(a_1,\ldots),O)$, we let

$$j(E) = c_4^3/\Delta.$$

Then $j(E)$ depends only on (E,O), and two elliptic curves (E,O) and (E',O') become isomorphic over k^{al} if and only if $j(E) = j(E')$.

PROOF. The proof is the same as that of Theorem 2.1, only more complicated. □

REMARK 2.5 In realizing a curve of genus 1 as a nonsingular plane cubic, it was crucial that the curve have a point in $E(k)$. Without this assumption, it may only be possible to realize the curve as a singular plane curve of possibly much higher degree.

NOTES About 1965, Tate worked out the formulas in Theorem 2.4 and included them in a letter to Cassels, which was published as Tate 1975. The formulas have been copied, and, on occasion, miscopied, by all later authors. The name "Weierstrass equation" for (11) is a little misleading since Weierstrass wrote his elliptic curves as

$$(\wp')^2 = 4\wp^3 - g_2\wp - g_3, \tag{12}$$

but, as Tate (1974, §2) writes: We call (11) a Weierstrass equation because in characteristics $\neq 2,3$, we can replace x and y by

$$\wp = x + \frac{a_1^2 + 4a_2}{12}, \quad \wp' = 2y + a_1x + a_3,$$

and (11) becomes (12) (with $c_4 = 12g_2$, $c_6 = 216g_3$, and $\Delta = g_2^2 - 27g_3^2$).

Expansion near zero; the formal group of an elliptic curve

An elliptic curve, including its group structure, can be described "near" O by a single power series in two variables. This is the formal group of the curve.

DEFINITION 2.6 A ***one-parameter commutative formal group*** over a commutative ring R is a power series

$$F(X,Y) = \sum_{i,j\geq 0} a_{i,j}X^iY^j \in R[[X,Y]]$$

satisfying the following conditions:

(a) $F(X,Y) = X + Y +$ terms of degree ≥ 2;

(b) $F(X,F(Y,Z)) = F(F(X,Y),Z)$;

(c) $F(X,Y) = F(Y,X)$.

These conditions imply that $F(X,0) = X$ and $F(0,Y) = Y$ and that there exists a unique power series $i(T) = -T + \sum_{n\geq 2} a_n T^n \in R[[T]]$ such that $F(T,i(T)) = 0$.

Let F be a one-parameter formal group over a complete discrete valuation ring R with maximal ideal $\mathfrak{m}$, for example, $R = \mathbb{Z}_p$ and $\mathfrak{m} = (p)$. For all $a,b \in \mathfrak{m}$, the power series $F(a,b)$ converges to an element of $\mathfrak{m}$, and F defines a commutative group structure $+_F$ on $\mathfrak{m}$.

2.7 Consider an elliptic curve E, O over a field k, given by a Weierstrass equation (11). On the affine curve $E^{\mathrm{aff}} \stackrel{\mathrm{def}}{=} E \cap \{y \neq 0\}$, O becomes the origin $(0,0)$. The equation for E^{aff}, when rewritten in terms of the symbols $Z = -X/Y$ and $W = -1/Y$, becomes

$$W = Z^3 + a_1 ZW + a_2 Z^2 W + a_3 W^2 + a_4 ZW^2 + a_6 W^3.$$

By repeatedly substituting for W on the right, we obtain an expression for W as a power series in Z,

$$W = Z^3 + a_1 Z^4 + (a_1^2 + a_2) Z^5 + (a_1^3 + 2a_1 a_2 + a_3) Z^6 + \cdots$$

The coefficients of the power series are polynomials in a_1, a_2, a_3, a_4, a_6 with coefficients in $\mathbb{N}$. More formally, when we localize the ring $k[z,w]$ at the ideal (z,w), and complete, then the complete local ring we obtain contains the power series ring in the symbol z as a subring, and equals it. From $X = Z/W$ and $Y = -1/W = -X/Z$, we find that

$$X = Z^{-2} - a_1 Z^{-1} - a_2 - a_3 Z - (a_4 + a_1 a_3) Z^2 + \cdots,$$
$$Y = -Z^{-3} + a_1 Z^{-2} + \cdots,$$

in the ring $k((Z))$ of Laurent series. Finally, if $P_1 + P_2 = P_3$ in $E(k)$, i.e., $(z_1, w_1) + (z_2, w_2) = (z_3, w_3)$, then z_3 can be expressed as a power series $F(z_1, z_2)$ in z_1, z_2, where

$$F(Z_1, Z_2) = Z_1 + Z_2 - a_1 Z_1 Z_2 - a_2 (Z_1^2 Z_2 + Z_1 Z_2^2) + \cdots.$$

More formally, the homomorphism $k[[Z]] \to k[[Z]] \hat{\otimes} k[[Z]] \simeq k[[Z_1, Z_2]]$ defined by addition on E sends Z to $F(Z_1, Z_2)$. The power series F is the one-parameter commutative formal group of E. It describes the structure of E near O. See Tate 1974, §3.

3 Reduction of an elliptic curve modulo p

Consider an elliptic curve over $\mathbb{Q}$. After a change of variables, the coefficients of its Weierstrass equation will lie in $\mathbb{Z}$, and so we may look at it modulo p to obtain a curve over the field $\mathbb{F}_p \overset{\text{def}}{=} \mathbb{Z}/p\mathbb{Z}$. The curve over $\mathbb{F}_p$ will still be a plane cubic curve, but it may have a singular point. However, once the singular point is removed, the curve acquires a canonical group structure. In this section, we first describe the algebraic curves having a group structure defined by regular maps. Then we discuss singular cubic curves and their canonical group structures. Finally, we describe the reduction of elliptic curves.

Throughout, k is a perfect field.

Algebraic groups of dimension 1

Let C be a curve over k admitting a group structure defined by regular maps. Because k is perfect, there exists a nonsingular point P in $C(k^{\mathrm{al}})$ (see 1.4d). It follows that every point Q in $C(k^{\mathrm{al}})$ is nonsingular because the map $R \mapsto R+(Q-P)$ is a regular isomorphism $C_{k^{\mathrm{al}}} \to C_{k^{\mathrm{al}}}$ sending P to Q. Hence the curve is nonsingular.

An ***algebraic group of dimension*** **1** over k is an algebraic curve over k equipped with a group structure defined by regular maps. The following is a complete list of connected algebraic groups of dimension 1 over k.

ELLIPTIC CURVES

These are the only projective curves admitting a group structure.

THE ADDITIVE GROUP

The affine line $\mathbb{A}^1$ becomes a group under addition,

$$\mathbb{A}^1(k) = k, \quad (x,y) \mapsto x+y : k \times k \to k.$$

We write $\mathbb{G}_a$ for $\mathbb{A}^1$ endowed with this group structure.

THE MULTIPLICATIVE GROUP

The affine line with the origin removed becomes a group under multiplication,

$$\mathbb{A}^1(k) \smallsetminus \{0\} = k^\times, \quad (x,y) \mapsto xy : k^\times \times k^\times \to k^\times.$$

We write $\mathbb{G}_m$ for $\mathbb{A}^1 \smallsetminus \{0\}$ endowed with this group structure. Note that the map $x \mapsto (x, x^{-1})$ realizes $\mathbb{G}_m$ as an affine plane curve $XY = 1$.

TWISTED MULTIPLICATIVE GROUPS

Let a be a nonsquare in $k^\times$, and let $L = k[\sqrt{a}]$. We define an algebraic group $\mathbb{G}_m[a]$ over k such that

$$\mathbb{G}_m[a](k) = \{\gamma \in L^\times \mid \mathrm{Nm}_{L/k}\, \gamma = 1\}.$$

Let $\alpha = \sqrt{a}$, so that $\{1, \alpha\}$ is a basis for L as a k-vector space. Then

$$\begin{aligned}(x+\alpha y)(x'+\alpha y') &= xx' + ayy' + \alpha(xy' + x'y)\\ \mathrm{Nm}(x+\alpha y) &= (x+\alpha y)(x-\alpha y) = x^2 - ay^2.\end{aligned}$$

We define $\mathbb{G}_m[a]$ to be the affine plane curve $X^2 - aY^2 = 1$ with the group structure

$$(x, y)\cdot(x', y') = (xx' + ayy', xy' + x'y).$$

For example, when $k = \mathbb{R}$ and $a = -1$, this is the circle group $X^2 + Y^2 = 1$.

An invertible change of variables transforms $\mathbb{G}_m[a]$ into $\mathbb{G}_m[ac^2]$, $c \in k^\times$, and so, up to such a change, $\mathbb{G}_m[a]$ depends only on the field $k[\sqrt{a}]$. The equations defining $\mathbb{G}_m[a]$ still define an algebraic group when a is a square in k, say, $a = \alpha^2$, but then $X^2 - aY^2 = (X + \alpha Y)(X - \alpha Y)$, and so the change of variables $X' = X + \alpha Y$, $Y' = X - \alpha Y$ transforms the group into $\mathbb{G}_m$. In particular, this shows that $\mathbb{G}_m[a]$ becomes isomorphic to $\mathbb{G}_m$ over $k[\sqrt{a}]$, and so it is a "twist" of $\mathbb{G}_m$.

REMARK 3.1 Let $k = \mathbb{F}_q$, the field with q-elements. Then

$$\begin{aligned}\#\mathbb{G}_a(k) &= q,\\ \#\mathbb{G}_m(k) &= q-1,\\ \#\mathbb{G}_m[a](k) &= q+1 \quad (a \text{ not a square in } k).\end{aligned}$$

Only the last is not obvious. As $\mathbb{F}_q[\sqrt{a}]$ is the field $\mathbb{F}_{q^2}$, there is an exact sequence

$$1 \longrightarrow \mathbb{G}_m[a](\mathbb{F}_q) \longrightarrow \mathbb{F}_{q^2}^\times \xrightarrow{\mathrm{Nm}} \mathbb{F}_q^\times.$$

The norm map is surjective because every quadratic form in at least three variables over a finite field has a nontrivial zero (Serre 1973, I, §2), and so

$$\#\mathbb{G}_m[a](\mathbb{F}_q) = (q^2-1)/(q-1) = q+1.$$

We make a few remarks concerning the proofs of the above statement. We have seen that if a nonsingular projective curve has genus 1, then it has a group structure, but why is the converse true? The simplest explanation when $k = \mathbb{C}$ comes from topology. If M is a compact oriented manifold and $\alpha: M \to M$ is a continuous map, then the Lefschetz fixed point theorem says that

$$(\Delta \cdot \Gamma_\alpha) = \sum\nolimits_i (-1)^i \operatorname{Trace}(\alpha | H^i(M, \mathbb{Q})).$$

Here Δ is the diagonal in $M \times M$ and Γ_α is the graph of α, so that $(\Delta \cdot \Gamma_\alpha)$ is the number of "fixed points of α counting multiplicities". Let $L(\alpha)$ denote the integer on the right, and assume that M has a group structure. For a nonzero a in M, the translation map $\tau_a = (x \mapsto x + a)$ has no fixed points, and so

$$L(\tau_a) = (\Delta \cdot \Gamma_\alpha) = 0.$$

The map $a \mapsto L(\tau_a): M \to \mathbb{Z}$ is continuous, and hence constant on each connected component of M. On letting a tend to zero, we find that $L(\tau_0) = 0$. But τ_0 is the identity map, and so

$$L(\tau_0) \stackrel{\text{def}}{=} \sum\nolimits_i (-1)^i \operatorname{Tr}(\mathrm{id} \,|\, H^i(M, \mathbb{Q})) = \sum\nolimits_i (-1)^i \dim_{\mathbb{Q}} H^i(M, \mathbb{Q}).$$

We have shown that if M has a group structure, then its Euler–Poincaré characteristic is zero. The Euler–Poincaré characteristic of a compact Riemann surface of genus g is $2 - 2g$, and so $g = 1$ if it has a group structure.

A similar argument works over any field. If C is a projective curve with a group structure, then one proves directly that for the diagonal Δ in $C \times C$,

$$(\Delta \cdot \Gamma_{\tau_a}) = 0 \text{ if } a \neq 0,$$
$$(\Delta \cdot \Delta) = 2 - 2g,$$

and then "by continuity" that $(\Delta \cdot \Delta) = (\Delta \cdot \Gamma_{\tau_a})$.

To prove that $\mathbb{G}_a$ and $\mathbb{G}_m$ are the only affine curves C over k^{al} admitting a group structure, embed C in a nonsingular projective curve C' as the complement of a finite set S of points. The translations by elements of $C(k^{\mathrm{al}})$ define an infinite set of automorphisms of C' stabilizing the set S, from which one can deduce that $C' = \mathbb{P}^1$ and that S consists of one or two points (see, for example, Milne 2017, 16.16). The extension of the statement to k is an exercise in Galois cohomology (IV, 7.17).

Singular cubic curves

Let E be a singular cubic projective plane curve over k. Then E has exactly one singular point S, and, because k is perfect, S has coordinates in k(see the argument p. 23). Assume that $E(k)$ contains a point $O \neq S$. Then the same definition as in the nonsingular case turns $E^{\text{ns}}(k) \stackrel{\text{def}}{=} E(k) \smallsetminus \{S\}$ into a group with O as its zero element. Namely, consider the line through two nonsingular points P and Q. According to Bezout's theorem and (I, 1.13), it will intersect the curve in exactly one additional point PQ, which cannot be singular. Define $P+Q$ to be the third point of intersection with the cubic of the line through PQ and O. This gives a group structure.

We first examine this in two basic examples.

A CUBIC CURVE WITH A CUSP

The projective plane curve

$$E: Y^2Z = X^3$$

has a cusp at $S = (0{:}0{:}1)$ because the affine curve $Y^2 = X^3$ has a cusp at $(0,0)$ (see I, 1.12). Note that S is the only point on E with Y-coordinate zero, and so E^{ns} is the affine plane curve

$$E_1: Z = X^3.$$

The line $Z = \alpha X + \beta$ intersects E_1 at points (x_1,z_1), (x_2,z_2), (x_3,z_3) with x_1,x_2,x_3 roots of

$$X^3 - \alpha X - \beta.$$

Because the coefficient of X^2 in this polynomial is zero, the sum $x_1+x_2+x_3$ is zero. Therefore the map

$$P \mapsto x(P): E_1(k) \to k$$

has the property that

$$P_1+P_2+P_3 = 0 \implies x(P_1)+x(P_2)+x(P_3) = 0.$$

As $O = (0,0)$, the map $P \mapsto -P$ is $(x,z) \mapsto (-x,-z)$, and so the map $P \mapsto x(P)$ has the property that

$$x(-P) = -P.$$

These two properties imply that $P \mapsto x(P): E_1(k) \to k$ is a homomorphism. In fact, it defines an isomorphism $E_1 \to \mathbb{G}_a$. In other words, the map

$$(x:y:z) \mapsto \frac{x}{y}: E^{\text{ns}} \to \mathbb{G}_a$$

is an isomorphism of algebraic groups.

A CUBIC CURVE WITH A NODE

The projective plane curve

$$E: Y^2Z = X^3 + cX^2Z, \quad c \neq 0,$$

has a node at $S = (0:0:1)$ because the affine curve

$$Y^2 = X^3 + cX^2, \quad c \neq 0,$$

has a node at $(0,0)$ (see I, 1.6). The tangent lines at $(0,0)$ are given by the equation

$$Y^2 - cX^2 = 0.$$

If c is a square in k, say, $c = \gamma^2$, this factors as

$$(Y + \gamma X)(Y - \gamma X) = 0$$

and the tangent lines are defined over k (we say that the tangents are ***rational***). The equation for E can be written $(Y^2 - cX^2)Z = X^3$. When we let $U = Y + \gamma X$ and $V = Y - \gamma X$, this becomes

$$8\gamma^3 UVZ = (U - V)^3.$$

A line $Z = \alpha X + \beta Y$ intersects E at points $(u_i : v_i : z_i)$, $i = 1,2,3$, satisfying

$$(U - V)^3 - 8\gamma^3 UV(\alpha X + \beta Y) = 0.$$

As the coefficient of V^3 is -1, we see that

$$\frac{u_1}{v_1}\frac{u_2}{v_2}\frac{u_3}{v_3} = 1.$$

As $(x:y:z) \mapsto (x:-y:z)$ inverts u/v, we see that the map $(x:y:z) \mapsto u/v: E^{\text{ns}}(k) \to k^{\times}$ is a homomorphism. In fact, the map

$$(x:y:z) \mapsto \frac{y + \gamma x}{y - \gamma x}: E^{\text{ns}} \to \mathbb{G}_m \tag{13}$$

is an isomorphism of algebraic groups.

If c is not a square, then the tangent lines are not defined over k, and the map (13) is defined only over $k[\gamma]$, $\gamma = \sqrt{c}$. This is a Galois extension of k, and (13) commutes with the actions of $\mathrm{Gal}(k[\gamma]/k)$ when we regard it as a map $(E^{\mathrm{ns}})_{k[\gamma]} \to \mathbb{G}_m[c]_{k[\gamma]}$. Therefore, (13) defines an isomorphism $E^{\mathrm{ns}} \to \mathbb{G}_m[c]$ of algebraic groups over k (I, 5.5).

THE GENERAL CASE

Assume first that $\mathrm{char}(k) \neq 2, 3$, so that we can write the curve E as

$$E : Y^2Z = X^3 + aXZ^2 + bZ^3, \quad a, b \in k, \quad \Delta = 4a^3 + 27b^2 = 0.$$

As the point $(0{:}1{:}0)$ is always nonsingular, we only need to study the affine curve

$$Y^2 = X^3 + aX + b.$$

We try to find a t such that this equation becomes

$$\begin{aligned} Y^2 &= (X-t)^2(X+2t) \\ &= X^3 - 3t^2X + 2t^3. \end{aligned}$$

For this, we need to choose t so that

$$t^2 = -\frac{a}{3}, \quad t^3 = \frac{b}{2}.$$

Hence $t = \frac{b/2}{-a/3} = -\frac{3}{2}\frac{b}{a}$. Using that $\Delta = 0$, one checks that this works.

With this value for t, the equation for E becomes

$$Y^2 = (X-t)^3 + 3t(X-t)^2.$$

This has a singularity at $(t, 0)$, which is a cusp if $3t = 0$, a node with rational tangents if $3t$ is a square in $k^\times$, and a node with nonrational tangents if $3t$ is a nonsquare in $k^\times$. Note that

$$-2ab = -2(-3t^2)(2t^3) = (2t^2)^2(3t),$$

and so $3t$ is zero or nonzero, a square or a nonsquare, according as $-2ab$ is. In each case, the curve becomes one of the above examples after a linear change of variables. We can summarize the situation as follows.

If $ab = 0$, then the curve has a cusp S, and the map

$$P \mapsto \frac{1}{(\text{slope of } PS) - (\text{slope of the tangent at } S)}$$

is an isomorphism of algebraic groups $E^{\mathrm{ns}} \to \mathbb{G}_a$.

If $-2ab \neq 0$, then the curve has a node S. Let $Y = \alpha_1 X + \beta_1$ and $Y = \alpha_2 X + \beta_2$ be the tangents at S. If $-2ab$ is a square in k, then the tangents are rational, and the map

$$P = (x, y) \mapsto \frac{y - \alpha_1 x - \beta_1}{y - \alpha_2 x - \beta_2}$$

is an isomorphism of algebraic groups $E^{\mathrm{ns}} \to \mathbb{G}_m$. If $-2ab$ is not a square in k, then this map is defined only over $k[\sqrt{-2ab}]$, and determines an isomorphism of algebraic groups $E^{\mathrm{ns}} \to \mathbb{G}_m[-2ab]$.

When we allow the characteristic of k to be arbitary, and write the equation for E in general Weierstrass form (11), p. 67, with $\Delta = 0$, the singularity is cusp if $c_4 = 0$ and a node if $c_4 \neq 0$. The isomorphisms of E^{ns} with $\mathbb{G}_a$ or a (twist of) $\mathbb{G}_m$ are as described above (Tate 1974, p. 182).

Reduction of an elliptic curve

Let E be an elliptic curve over $\mathbb{Q}$, and let

$$Y^2Z + a_1XYZ + a_3YZ^2 = X^3 + a_2X^2Z + a_4XZ^2 + a_6Z^3 \tag{14}$$

be a Weierstrass equation for E. We make a change a variables

$$\begin{cases} X \mapsto u^2X + r \\ Y \mapsto u^3Y + su^2X + t \\ Z \mapsto Z \end{cases} \tag{15}$$

with $u, r, s, t \in \mathbb{Q}$, $u \neq 0$, chosen so that the new a_i are integers and $|\Delta|$ is minimal — the equation is then said to be ***minimal.*** Two minimal equations are related by a change of variables (15) with $u, r, s, t \in \mathbb{Z}$, $u \in \mathbb{Z}^\times$. Most of the theory is independent of the choice of the minimal equation. For example, the curve $\bar{E}$ obtained by reducing a minimal equation (14) for E modulo a prime p is well defined up to a change of variables (15) with $u, r, s, t \in \mathbb{F}_p$, $u \neq 0$; it is called the ***reduction of E modulo*** p

When we are interested only in reductions modulo primes $p \neq 2, 3$, we can work instead with an equation

$$Y^2Z = X^3 + aXZ^2 + bZ^3, \quad a, b \in \mathbb{Q}. \tag{16}$$

In this case, we make a change of variables $X \mapsto X/c^2$, $Y \mapsto Y/c^3$ with $c \in \mathbb{Q}^\times$ chosen so that the new a, b are integers and $|\Delta|$ is minimal.

An elliptic curve may have an equation of the form (14) that is "more minimal", i.e., has smaller discriminant, than any of the form (16). This does not matter for primes $\neq 2, 3$, but, for example,

$$Y^2 + Y = X^3 - X^2$$

defines a nonsingular curve over $\mathbb{F}_2$ whereas all equations of the form (16) define singular curves (I, 1.5).

The different types of reduction

Let E be an elliptic curve over $\mathbb{Q}$, given by a minimal equation (14), and let $\bar{E}$ be its reduction modulo p, given by the equation

$$Y^2Z + \bar{a}_1XYZ + \bar{a}_3YZ^2 = X^3 + \bar{a}_2X^2Z + \bar{a}_4XZ^2 + \bar{a}_6Z^3,$$

where $\bar{a}_i = a_i$ modulo p.

Good reduction

If p does not divide Δ, then $\bar{E}$ is an elliptic curve over $\mathbb{F}_p$. For a point $P = (x{:}y{:}z)$ on E, we can choose a representative (x, y, z) for P with $x, y, z \in \mathbb{Z}$ and having no common factor, and then $\bar{P} \stackrel{\text{def}}{=} (\bar{x} : \bar{y} : \bar{z})$ is a well-defined point on $\bar{E}$. As $(0{:}1{:}0)$ reduces to $(0{:}1{:}0)$ and lines reduce to lines, the map $E(\mathbb{Q}) \to \bar{E}(\mathbb{F}_p)$ is a homomorphism.

Cuspidal, or additive, reduction

This is the case where $\bar{E}$ has a cusp, and so $\bar{E}^{\text{ns}} \simeq \mathbb{G}_a$. It occurs exactly when p divides both Δ and c_4. For $p \neq 2, 3$, it occurs exactly when p divides both $4a^3 + 27b^2$ and $-2ab$.

Nodal, or multiplicative, reduction

This is the case where $\bar{E}$ has a node. It occurs exactly when p divides Δ but not c_4. For $p \neq 2, 3$, it occurs exactly when p divides $4a^3 + 27b^2$ but not $-2ab$; then the tangents at the node are rational over $\mathbb{F}_p$ if and only if $-2ab$ becomes a square in $\mathbb{F}_p$, in which case $\bar{E}^{\text{ns}} \simeq \mathbb{G}_m$ and E is said to have *split multiplicative reduction*. If $-2ab$ is not a square modulo p, then $\bar{E}^{\text{ns}} \simeq \mathbb{G}_m[\sqrt{-2ab}]$ and E is said to have *nonsplit multiplicative reduction*.

In the nodal and cuspidal cases, there is a well-defined reduction map $E(\mathbb{Q}) \to \bar{E}(\mathbb{F}_p)$, which becomes a homomorphism when we omit the singular point from $\bar{E}(\mathbb{F}_p)$ and any points in $E(\mathbb{Q})$ mapping to it.

SEMISTABLE REDUCTION

If E has good or nodal reduction, then the minimal equation remains minimal after replacing the ground field by a larger field. This is not so for cuspidal reduction. Consider, for example, the curve

$$E : Y^2 Z = X^3 + pXZ^2 + pZ^3.$$

After passing to an extension field in which p becomes a sixth power, say, $\pi^6 = p$, we can make a change of variables so that the equation becomes

$$E : Y^2 Z = X^3 + \pi^2 XZ^2 + Z^3.$$

Modulo π this becomes

$$Y^2 Z = X^3 + Z^3,$$

which is nonsingular in characteristic $\neq 2, 3$. In fact, for any curve E with cuspidal reduction at p, there exists a finite extension of the ground field such that E has either good or nodal reduction at the primes over p.

In summary: good and nodal reduction are not changed by a field extension (in fact, the minimal equation remains minimal) but cuspidal reduction always becomes good or nodal reduction in an appropriate finite extension (and the minimal equation changes). For this reason, a curve is said to have ***semistable reduction*** at p if it has good or nodal reduction.

Type	tangents	$\Delta \bmod p$	$-2ab \bmod p$	$\bar{E}^{\mathrm{ns}}$	$\#E(\mathbb{F}_p)$
good		$\neq 0$		$\bar{E}$	IV, §9
cusp		0	0	$\mathbb{G}_a$	p
node	rational	0	$\square$	$\mathbb{G}_m$	$p-1$
node	not rational	0	$\neq \square$	$\mathbb{G}_m[\overline{-2ab}]$	$p+1$

Other fields

Throughout this section, we can replace $\mathbb{Q}$ and $\mathbb{Z}$ with $\mathbb{Q}_p$ and $\mathbb{Z}_p$, or, in fact, with any finite extension of $\mathbb{Q}_p$ and its ring of integers. A minimal model for E over $\mathbb{Q}$ will remain minimal over $\mathbb{Q}_p$. Also, we can replace $\mathbb{Q}$ and $\mathbb{Z}$ with a number field K and its ring of integers, with the caution that, if the ring of integers in K is not a principal ideal domain, then it may not be possible to find an equation for the elliptic curve that is minimal for all primes simultaneously, i.e., that remains minimal over all the local fields.

EXERCISE 3.2 Show that the curve

$$E : Y^2 + Y = X^3 - X^2 - 10X - 20$$

has good reduction at all primes except 11.

EXERCISE 3.3 (a) Find examples of elliptic curves E over $\mathbb{Q}$ and primes p such that

i) E_p has a cusp S that lifts to a point in $E(\mathbb{Q}_p)$;

ii) E_p has a node S that lifts to a point in $E(\mathbb{Q}_p)$;

iii) E_p has a node S that does not lift to a point in $E(\mathbb{Q}_p)$.

Here E_p is the reduction of the curve modulo a prime $p \neq 2, 3$. The equation you give for E should be a minimal equation of the standard form $Y^2Z = X^3 + aXZ^2 + bZ^3$.

(b) For the example you gave in (a)(i), decide whether it acquires good or nodal reduction in a finite extension of $\mathbb{Q}$.

EXERCISE 3.4 Let α, β, γ be nonzero relatively prime integers such that

$$\alpha^\ell + \beta^\ell = \gamma^\ell,$$

where ℓ is a prime $\neq 2, 3$, and consider the elliptic curve

$$E : Y^2Z = X(X - \alpha^\ell Z)(X - \gamma^\ell Z).$$

(a) Show that E has discriminant $\Delta = 16\alpha^{2\ell}\beta^{2\ell}\gamma^{2\ell}$.

(b) Show that if p does not divide $\alpha\beta\gamma$, then E has good reduction at p.

(c) Show that if p is an odd prime dividing $\alpha\beta\gamma$, then E has at worst nodal reduction at p.

(d) Show that (the minimal equation for) E has at worst nodal reduction at 2.

[After possibly re-ordering α, β, γ, we may suppose, first that γ is even, and then that $\alpha^\ell \equiv 1 \mod 4$. Make the change of variables $x = 4X$, $y = 8Y + 4X$, and verify that the resulting equation has integer coefficients.]

ASIDE 3.5 The ***conductor*** of an elliptic curve E over $\mathbb{Q}$ is $\prod_p p^{f_p}$ where $f_p = 0$, 1, or ≥ 2 according as E has good, nodal, or cuspidal reduction at p. Note that (b),(c),(d) in Exercise 3.4 show that the conductor N of E divides $\prod_{p|\alpha\beta\gamma} p$, and hence is much smaller than Δ. This will be enough for us to show in Chapter V that E does not exist, but enthusiasts may wish to verify that $N = \prod_{p|\alpha\beta\gamma} p$. [Hint: First show that if p does not divide c_4, then the equation is minimal at p.]

4 Elliptic curves over $\mathbb{Q}_p$

Consider an elliptic curve

$$E: Y^2Z = X^3 + aXZ^2 + bZ^3, \quad a, b \in \mathbb{Q}_p, \quad 4a^3 + 27b^2 \neq 0.$$

After a change of variables $X \mapsto X/c^2$, $Y \mapsto Y/c^3$, $Z \mapsto Z$, we may suppose that $a, b \in \mathbb{Z}_p$. We may also suppose that $\mathrm{ord}_p(\Delta)$ is minimal, but that is not necessary. As in the last section, we obtain from E a curve $\bar{E}$ over $\mathbb{F}_p$ and a reduction map

$$P \mapsto \bar{P}: E(\mathbb{Q}_p) \to \bar{E}(\mathbb{F}_p).$$

We shall define a filtration

$$E(\mathbb{Q}_p) \supset E^0(\mathbb{Q}_p) \supset E^1(\mathbb{Q}_p) \supset \cdots \supset E^n(\mathbb{Q}_p) \supset \cdots$$

and identify the quotients. First, define

$$E^0(\mathbb{Q}_p) = \{P \mid \bar{P} \text{ is nonsingular}\}.$$

It is a subgroup because (0: 1: 0) is always nonsingular and a line through two nonsingular points on a cubic (or tangent to a nonsingular point) will meet the cubic again at a nonsingular point.

Write $\bar{E}^{\mathrm{ns}}$ for $\bar{E}$ with the singular point (if any) removed. The reduction map

$$P \mapsto \bar{P}: E^0(\mathbb{Q}_p) \to \bar{E}^{\mathrm{ns}}(\mathbb{F}_p)$$

is a homomorphism, and we define $E^1(\mathbb{Q}_p)$ be its kernel. Thus $E^1(\mathbb{Q}_p)$ consists of the points P that can be represented as $(x: y: z)$ with x and z divisible by p but y prime to p. In particular, $P \in E^1(\mathbb{Q}_p) \implies y(P) \neq 0$. Define

$$E^n(\mathbb{Q}_p) = \left\{ P \in E^1(\mathbb{Q}_p) \;\middle|\; \frac{x(P)}{y(P)} \in p^n\mathbb{Z}_p \right\}.$$

THEOREM 4.1 *The filtration*

$$E(\mathbb{Q}_p) \supset E^0(\mathbb{Q}_p) \supset E^1(\mathbb{Q}_p) \supset \cdots \supset E^n(\mathbb{Q}_p) \supset \cdots$$

has the following properties:

(a) *the quotient $E(\mathbb{Q}_p)/E^0(\mathbb{Q}_p)$ is finite;*

(b) *the map $P \mapsto \bar{P}$ defines an isomorphism*

$$E^0(\mathbb{Q}_p)/E^1(\mathbb{Q}_p) \to \bar{E}^{\text{ns}}(\mathbb{F}_p);$$

(c) *for $n \geq 1$, $E^n(\mathbb{Q}_p)$ is a subgroup of $E(\mathbb{Q}_p)$, and the map*

$$P \mapsto p^{-n}\frac{x(P)}{y(P)} \quad \text{mod } p\colon E^n(\mathbb{Q}_p)/E^{n+1}(\mathbb{Q}_p) \to \mathbb{F}_p$$

is an isomorphism of groups;

(d) *the filtration is exhaustive, i.e., $\bigcap_n E^n(\mathbb{Q}_p) = \{0\}$.*

PROOF. (a) We prove that $E(\mathbb{Q}_p)$ has a natural topology with respect to which it is compact and $E^0(\mathbb{Q}_p)$ is an open subgroup. Since $E(\mathbb{Q}_p)$ is a union of the cosets of $E^0(\mathbb{Q}_p)$, it follows that there can only be finitely many of them.

Endow $\mathbb{Q}_p \times \mathbb{Q}_p \times \mathbb{Q}_p$ with the product topology, $\mathbb{Q}_p^3 \smallsetminus \{(0,0,0)\}$ with the subspace topology, and $\mathbb{P}^2(\mathbb{Q}_p)$ with the quotient topology via

$$\mathbb{Q}_p^3 \smallsetminus \{(0,0,0)\} \to \mathbb{P}^2(\mathbb{Q}_p).$$

Then $\mathbb{P}^2(\mathbb{Q}_p)$ is the union of the images of the sets $\mathbb{Z}_p^\times \times \mathbb{Z}_p \times \mathbb{Z}_p$, $\mathbb{Z}_p \times \mathbb{Z}_p^\times \times \mathbb{Z}_p$, $\mathbb{Z}_p \times \mathbb{Z}_p \times \mathbb{Z}_p^\times$, each of which is compact and open. Therefore $\mathbb{P}^2(\mathbb{Q}_p)$ is compact. Its subset $E(\mathbb{Q}_p)$ is closed because it is the zero set of a polynomial, and so it also is compact. Relative to this topology on $\mathbb{P}^2(\mathbb{Q}_p)$, two points that are close have the same reduction modulo p. Therefore $E^0(\mathbb{Q}_p)$ is the intersection of $E(\mathbb{Q}_p)$ with an open subset of $\mathbb{P}^2(\mathbb{Q}_p)$.

(b) Hensel's lemma (I, 2.12) implies that the reduction map $E^0(\mathbb{Q}_p) \to \bar{E}^{\text{ns}}(\mathbb{F}_p)$ is surjective, and we defined $E^1(\mathbb{Q}_p)$ to be its kernel.

(c) We assume inductively that $E^n(\mathbb{Q}_p)$ is a subgroup of $E(\mathbb{Q}_p)$. If $P = (x\colon y\colon 1)$ lies in $E^1(\mathbb{Q}_p)$, then $y \notin \mathbb{Z}_p$. Set $x = p^{-m}x_0$ and $y = p^{-m'}y_0$ with x_0 and y_0 units in $\mathbb{Z}_p$ and $m' \geq 1$. Then

$$p^{-2m'}y_0^2 = p^{-3m}x_0^3 + ap^{-m}x_0 + b.$$

On taking ord_p of the two sides, we find that $2m' = 3m$. Since m' and m are integers, this implies that there is an integer $n \geq 1$ such $m = 2n$ and $m' = 3n$; in fact, $n = m' - m$.

The above discussion shows that if

$$P = (x{:}\, y{:}\, z) \in E^n(\mathbb{Q}_p) \smallsetminus E^{n+1}(\mathbb{Q}_p), \quad n \geq 1,$$

then

$$\begin{cases} \mathrm{ord}_p(x) &= \mathrm{ord}_p(z) - 2n \\ \mathrm{ord}_p(y) &= \mathrm{ord}_p(z) - 3n. \end{cases}$$

Hence P can be expressed as $P = (p^n x_0{:}\, y_0{:}\, p^{3n} z_0)$ with $\mathrm{ord}_p(y_0) = 0$ and $x_0, z_0 \in \mathbb{Z}_p$; in fact, this is true for all $P \in E^n(\mathbb{Q}_p)$. Since P lies on E,

$$p^{3n} y_0^2 z_0 = p^{3n} x_0^3 + a p^{7n} x_0 z_0^2 + b p^{9n} z_0^3,$$

and so $P_0 \stackrel{\text{def}}{=} (\bar{x}_0 : \bar{y}_0 : \bar{z}_0)$ lies on the curve

$$E_0 : Y^2 Z = X^3.$$

As $\bar{y}_0 \neq 0$, P_0 is not the singular point of E_0. From the description of the group laws in terms of chords and tangents, we see that the map

$$P \mapsto P_0 : E^n(\mathbb{Q}_p) \to E_0(\mathbb{F}_p)$$

is a homomorphism. Its kernel is $E^{n+1}(\mathbb{Q}_p)$, which is therefore a subgroup, and it follows from Hensel's lemma that its image is the set of nonsingular points of $E_0(\mathbb{F}_p)$. We know from the preceding section (p. 73) that $Q \mapsto \frac{x(Q)}{y(Q)}$ is an isomorphism $E_0^{\mathrm{ns}}(\mathbb{F}_p) \to \mathbb{F}_p$. The composite $P \mapsto P_0 \mapsto \frac{x(P_0)}{y(P_0)}$ is $P \mapsto \frac{p^{-n} x(P)}{y(P)} \mod p$.

(d) If $P \in \bigcap_n E^n(\mathbb{Q}_p)$, then $x(P) = 0$, $y(P) \neq 0$. This implies that either $z(P) = 0$ or $y(P)^2 = b z(P)^3$, but the second equality contradicts $P \in E^1(\mathbb{Q}_p)$. Hence $z(P) = 0$ and $P = (0{:}\,1{:}\,0)$. □

COROLLARY 4.2 *Let m be an integer prime to p.*

(a) *The map*

$$P \mapsto mP{:}\, E^1(\mathbb{Q}_p) \to E^1(\mathbb{Q}_p)$$

is a bijection.

(b) *If E has good reduction, then the reduction map $E(\mathbb{Q}_p) \to \bar{E}(\mathbb{Q}_p)$ is an isomorphism on points killed by m, i.e., $E(\mathbb{Q}_p)_m \simeq \bar{E}(\mathbb{Q}_p)_m$.*

PROOF. (a) Let $P \in E^1(\mathbb{Q}_p)$ be such that $mP = 0$. If $P \neq 0$, then $P \in E^n(\mathbb{Q}_p) \smallsetminus E^{n+1}(\mathbb{Q}_p)$ for some n (by 4.1d), but $E^n(\mathbb{Q}_p)/E^{n+1}(\mathbb{Q}_p) \simeq \mathbb{Z}/p\mathbb{Z}$ (by 4.1c). The image of P in $\mathbb{Z}/p\mathbb{Z}$ is nonzero, and so m times it is also nonzero, which contradicts the fact that $mP = 0$. Therefore the map is injective.

Let $P \in E^1(\mathbb{Q}_p)$. Because $E^1(\mathbb{Q}_p)/E^2(\mathbb{Q}_p) \simeq \mathbb{Z}/p\mathbb{Z}$ and p does not divide m, multiplication by m is an isomorphism on $E^1(\mathbb{Q}_p)/E^2(\mathbb{Q}_p)$. Therefore there exists a $Q_1 \in E^1(\mathbb{Q}_p)$ such that

$$P = mQ_1 \mod E^2(\mathbb{Q}_p).$$

Similarly, there exists a $Q_2 \in E^2(\mathbb{Q}_p)$ such that

$$(P - mQ_1) = mQ_2 \mod E^3(\mathbb{Q}_p).$$

Continuing in this fashion, we obtain a sequence $Q_1, Q_2, \ldots$ of points in $E(\mathbb{Q}_p)$ such that

$$Q_i \in E^i(\mathbb{Q}_p), \quad P - m\sum\nolimits_1^n Q_i \in E^{n+1}(\mathbb{Q}_p).$$

The first condition implies that the series $\sum Q_i$ converges to a point in $E(\mathbb{Q}_p)$ (recall that $E(\mathbb{Q}_p)$ is compact), and the second condition implies that its limit Q has the property that $P = mQ$.

(b) Because E has good reduction, $E^0(\mathbb{Q}_p) = E(\mathbb{Q}_p)$. Now apply the snake lemma to the diagram

$$\begin{array}{ccccccccc} 0 & \longrightarrow & E^1(\mathbb{Q}_p) & \longrightarrow & E(\mathbb{Q}_p) & \longrightarrow & \bar{E}(\mathbb{F}_p) & \longrightarrow & 0 \\ & & \downarrow \simeq & & \downarrow m & & \downarrow m & & \\ 0 & \longrightarrow & E^1(\mathbb{Q}_p) & \longrightarrow & E(\mathbb{Q}_p) & \longrightarrow & \bar{E}(\mathbb{F}_p) & \longrightarrow & 0. \end{array}$$

□

REMARK 4.3 Implicitly, we have assumed that $p \neq 2, 3$. By working with the full Weierstrass equation, it is possible to avoid that assumption.

Similar statements hold for elliptic curves over any field K complete with respect to a discrete valuation except that, when the residue field k is infinite, it is necessary to work with a minimal equation in order for the quotient $E(K)/E^0(K)$ to be finite, and the proof of its finiteness is more difficult. Otherwise, the same arguments apply. In particular, if E has good reduction and m is prime to the residue characteristic, then $E(K)_m \simeq \bar{E}(k)_m$.

ASIDE 4.4 Instead of proving the above results directly, it is possible to deduce them from the properties of the associated formal group.

Consider an elliptic curve E over field K complete with respect to a discrete valuation, and let R be the ring of integers in K, e.g., $K = \mathbb{Q}_p$ and $R = \mathbb{Z}_p$. Choose a minimal Weierstrass equation for E. In particular, the equation has coefficients $a_1, a_2, \ldots$ in R, which means that all the power series in 2.7 have coefficients in R. Thus, we obtain a formal group law over R. The map $P \mapsto z(P) = -x(P)/y(P)$ defines an isomorphism of groups $(\mathfrak{m}, +_F) \to E^1(K)$. See Tate 1974, §6.

ASIDE 4.5 Elliptic curves over $\mathbb{Q}_p$ with split multiplicative reduction can all be described as quotients $\mathbb{Q}_p^\times/q^\mathbb{Z}$, i.e., as "Tate curves". See Chapter III, p. 132.

5 Torsion points

For an elliptic curve E over $\mathbb{Q}$ (resp. $\mathbb{Q}_p$), we fix an equation

$$Y^2Z = X^3 + aXZ^2 + bZ^3, \quad \Delta = 4a^3 + 27b^2 \neq 0,$$

with coefficients a, b in $\mathbb{Z}$ (resp. $\mathbb{Z}_p$). In particular, this allows us to speak of the coordinates of a point on E. The torsion subgroup $E(\mathbb{Q})_{\text{tors}}$ of $E(\mathbb{Q})$ is the subgroup of points of finite order.

THEOREM 5.1 (LUTZ–NAGELL) *Let E be an elliptic curve over $\mathbb{Q}$. If $P = (x\colon y\colon 1) \in E(\mathbb{Q})_{\text{tors}}$, then $x, y \in \mathbb{Z}$ and either $y = 0$ or $y|\Delta$.*

The theorem follows from the Propositions 5.3 and 5.4 below, the first of which says that, if P and $2P$ have integer coordinates, then either $y = 0$ or $y|\Delta$, and the second of which implies that the torsion points all have integer coordinates.

REMARK 5.2 (a) The theorem provides an algorithm for finding all the torsion points on E: for each $y = 0$ or $y|\Delta$, find the integer roots x of

$$X^3 + aX + b - y^2$$

— they divide $b - y^2$ — and then check whether $(x\colon y\colon 1)$ is a torsion point. This is faster if the equation of E has been chosen to be minimal.

(b) The converse of the theorem is not true: a point $P \in E(\mathbb{Q})$ can satisfy the conditions in the theorem without being a torsion point.

(c) The theorem can often be used to prove that a point $P \in E(\mathbb{Q})$ is of infinite order: compute multiples nP of P until one has coordinates that are not integers, or better, just compute the x-coordinates of $2P, 4P, 8P$, using the duplication formula p. 66.

PROPOSITION 5.3 *Let E be an elliptic curve over $\mathbb{Q}$ and $P=(x_1\colon y_1\colon 1)$ a point of $E(\mathbb{Q})$. If P and $2P$ have integer coordinates (when we set $z=1$), then either $y_1=0$ or $y_1|\Delta$.*

PROOF. Assume that $y_1 \neq 0$, and set $2P=(x_2\colon y_2\colon 1)$. Then $-2P=(x_2\colon -y_2\colon 1)$ is the second point of intersection of the tangent line at P with the affine curve $Y^2=X^3+aX+b$. Let $f(X)=X^3+aX+b$. The tangent line at P is $Y=\alpha X+\beta$, where $\alpha=\left(\frac{dY}{dX}\right)_P=\frac{f'(x_1)}{2y_1}$, and so the X-coordinates of its points of intersection with the curve satisfy

$$0=(\alpha X+\beta)^2-(X^3+aX+b)=-X^3+\alpha^2X^2+\cdots.$$

But we know that these X-coordinates are x_1, x_1, x_2, and so

$$x_1+x_1+x_2=\alpha^2.$$

By assumption, x_1, x_2 are integers, and so α^2 and $\alpha=f'(x_1)/2y_1$ are integers. Thus $y_1|f'(x_1)$, and directly from the equation $y_1^2=f(x_1)$ we see that $y_1|f(x_1)$. Hence y_1 divides both $f(x_1)$ and $f'(x_1)$. The theory of resultants (I, §1) shows that [1]

$$\Delta=r(X)f(X)+s(X)f'(X), \quad \text{some } r(X),s(X)\in\mathbb{Z}[X],$$

and so this implies that $y_1|\Delta$. □

PROPOSITION 5.4 *Let E be an elliptic curve over $\mathbb{Q}_p$. The group $E^1(\mathbb{Q}_p)$ is torsion-free.*

Before proving the proposition, we derive some consequences.

COROLLARY 5.5 *If $P=(x\colon y\colon 1)\in E(\mathbb{Q}_p)_{\text{tors}}$, then $x,y\in\mathbb{Z}_p$.*

PROOF. Recall that $\bar{P}$ is obtained from P by choosing primitive coordinates $(x\colon y\colon z)$ for P (i.e., coordinates such that $x,y,z\in\mathbb{Z}_p$ but not all of $x,y,z\in p\mathbb{Z}_p$), and setting $\bar{P}=(\bar{x}\colon\bar{y}\colon\bar{z})$, and that $E^1(\mathbb{Q}_p)=\{P\in E(\mathbb{Q}_p)\mid \bar{P}=(0\colon 1\colon 0)\}$.

If $P=(x\colon y\colon 1)$ with $(x,y)\notin\mathbb{Z}_p^2$ and $(x'\colon y'\colon z')$ is a system of primitive coordinates for P, then $z'\in p\mathbb{Z}_p$. Hence $z(\bar{P})=0$, which implies that $\bar{P}=(0\colon 1\colon 0)$, and so $P\in E^1(\mathbb{Q}_p)$. Thus (contrapositively) if $P=(x\colon y\colon 1)\notin E^1(\mathbb{Q}_p)$, then $x,y\in\mathbb{Z}_p$.

According to the proposition, if $P=(x\colon y\colon 1)$ is nonzero and torsion, then $P\notin E^1(\mathbb{Q})$, and so $x,y\in\mathbb{Z}_p$. □

[1] In fact, $\Delta=\left(-27(X^3+aX-b)\right)\cdot f(X)+(3X^2+4a)(3X^2+a)\cdot f'(X)$.

COROLLARY 5.6 *Let E be an elliptic curve over $\mathbb{Q}$. If $P = (x:y:1) \in E(\mathbb{Q})_{\text{tors}}$, then $x, y \in \mathbb{Z}$.*

PROOF. This follows from Corollary 5.5, because a rational number is an integer if it lies in $\mathbb{Z}_p$ for all p. □

COROLLARY 5.7 *Let E be an elliptic curve over $\mathbb{Q}$. If E has good reduction at p, then the reduction map*

$$E(\mathbb{Q})_{\text{tors}} \to \bar{E}(\mathbb{F}_p)$$

is injective.

PROOF. Because E has good reduction, $E^0(\mathbb{Q}_p) = E(\mathbb{Q}_p)$. The reduction map $E(\mathbb{Q}_p) \to \bar{E}(\mathbb{Q}_p)$ has kernel $E^1(\mathbb{Q}_p)$, which intersects $E(\mathbb{Q})_{\text{tors}}$ in $\{O\}$. □

REMARK 5.8 This puts a serious restriction on the size of $E(\mathbb{Q})_{\text{tors}}$. For example, if E has good reduction at 5, then, according to the Riemann hypothesis (IV, §9), $\bar{E}$ has at most $5+1+2\sqrt{5}$ points with coordinates in $\mathbb{F}_5$, and so E has at most 10 torsion points with coordinates in $\mathbb{Q}$.

PROOF OF PROPOSITION 5.4.

After Corollary 4.2, it remains to show that $E^1(\mathbb{Q}_p)$ contains no point $P \neq 0$ such that $pP = 0$. For this, we analyse the filtration on $E^1(\mathbb{Q}_p)$ (see §4).

For $P \in E^1(\mathbb{Q}_p)$, we have $y(P) \neq 0$, which suggests that we look at the affine curve $E \cap \{(x:y:z) \mid y \neq 0\}$:

$$E_1 : Z = X^3 + aXZ^2 + bZ^3. \tag{17}$$

A point $P = (x:y:z)$ on E has coordinates $x'(P) \stackrel{\text{def}}{=} \frac{x(P)}{y(P)}$, $z'(P) \stackrel{\text{def}}{=} \frac{z(P)}{y(P)}$ on E_1. For example, $O = (0:1:0)$ becomes the origin on E_1, and so $P \mapsto -P$ becomes reflection in the origin $(x',z') \mapsto (-x',-z')$. As before, $P+Q+R=0$ if and only if P, Q, R lie on a line.

In terms of our new picture,

$$E^n(\mathbb{Q}_p) = \{P \in E^1(\mathbb{Q}_p) \mid x'(P) \in p^n\mathbb{Z}_p\}.$$

Thus the $E^n(\mathbb{Q}_p)$ form a neighbourhood base of zero in $E_1(\mathbb{Q}_p)$. The key lemma is the following:

LEMMA 5.9 *Let $P_1, P_2, P_3 \in E(\mathbb{Q}_p)$ be such that $P_1 + P_2 + P_3 = O$. If $P_1, P_2 \in E^n(\mathbb{Q}_p)$, then $P_3 \in E^n(\mathbb{Q}_p)$, and*

$$x'(P_1) + x'(P_2) + x'(P_3) \in p^{5n}\mathbb{Z}_p.$$

PROOF. We saw in the proof of Theorem 4.1 that if

$$P = (x\colon y\colon 1) \in E^n(\mathbb{Q}_p) \smallsetminus E^{n+1}(\mathbb{Q}_p),$$

then $\mathrm{ord}_p(x) = -2n$, $\mathrm{ord}_p(y) = -3n$. In terms of homogeneous coordinates $P = (x\colon y\colon z)$, this means that

$$P \in E^n(\mathbb{Q}_p) \smallsetminus E^{n+1}(\mathbb{Q}_p) \Rightarrow \begin{cases} \mathrm{ord}_p \frac{x(P)}{z(P)} = -2n \\ \mathrm{ord}_p \frac{y(P)}{z(P)} = -3n \end{cases}$$

$$\Rightarrow \begin{cases} \mathrm{ord}_p \frac{x(P)}{y(P)} = n \\ \mathrm{ord}_p \frac{z(P)}{y(P)} = 3n. \end{cases}$$

Thus

$$P \in E^n(\mathbb{Q}_p) \Longrightarrow x'(P) \in p^n\mathbb{Z}_p, \quad z'(P) \in p^{3n}\mathbb{Z}_p.$$

Let $x_i' = x'(P_i)$ and $z_i' = z'(P_i)$ for $i = 1, 2, 3$. The line through P_1, P_2 (assumed distinct) is $Z = \alpha X + \beta$, where

$$\begin{aligned} \alpha &= \frac{z_2' - z_1'}{x_2' - x_1'} \\ &= \frac{(x_2'^3 - x_1'^3) + a(x_2' - x_1')(z_2'^2 - z_1'^2) + b(z_2'^3 - z_1'^3)}{x_2' - x_1'} \qquad \text{(by (17))} \\ &\cdots \\ &= \frac{x_2'^2 + x_1'x_2' + x_1'^2 + az_2'^2}{1 - ax_1'(z_2' + z_1') - b(z_2'^2 + z_1'z_2 + z_1'^2)}. \end{aligned}$$

The bottom line is a unit in $\mathbb{Z}_p$, and so $\alpha \in p^{2n}\mathbb{Z}_p$. Moreover

$$\beta = z_1' - \alpha x_1' \in p^{3n}\mathbb{Z}_p.$$

On substituting $\alpha X + \beta$ for Z in the equation for E_1, we obtain the equation

$$\alpha X + \beta = X^3 + aX(\alpha X + \beta)^2 + b(\alpha X + \beta)^3.$$

We know that the solutions of this equation are x_1', x_2', x_3', and so

$$x_1' + x_2' + x_3' = \frac{2a\alpha\beta + 3b\alpha^2\beta}{1 + a\alpha^2 + b\alpha^3} \in p^{5n}\mathbb{Z}_p.$$

The proof when $P_1 = P_2$ is similar. □

We now complete the proof of Proposition 5.4. For $P \in E^n(\mathbb{Q}_p)$, let $\bar{x}(P) = x'(P) \mod p^{5n}\mathbb{Z}_p$. The lemma shows that the map

$$P \mapsto \bar{x}(P) : E^n(\mathbb{Q}_p) \to p^n\mathbb{Z}_p/p^{5n}\mathbb{Z}_p$$

has the property

$$P_1 + P_2 + P_3 = 0 \implies \bar{x}(P_1) + \bar{x}(P_2) + \bar{x}(P_3) = 0.$$

As $\bar{x}(-P) = -\bar{x}(P)$, $P \mapsto \bar{x}(P)$ is a homomorphism of abelian groups. Suppose that $P \in E^1(\mathbb{Q}_p)$ has order p. As P is nonzero, it lies in $E^n(\mathbb{Q}_p) \smallsetminus E^{n+1}(\mathbb{Q}_p)$ for some n. Then $\bar{x}(P) \in p^n\mathbb{Z}_p \smallsetminus p^{n+1}\mathbb{Z}_p \mod p^{5n}\mathbb{Z}_p$, and so

$$\bar{x}(pP) = p\bar{x}(P) \in p^{n+1}\mathbb{Z}_p \smallsetminus p^{n+2}\mathbb{Z}_p \mod p^{5n}\mathbb{Z}_p.$$

This contradicts the fact that $pP = 0$.

Complements

5.10 For an elliptic curve E over a number field K, the torsion points need not have coordinates that are algebraic integers (when z is taken to be 1). Let K_v be the completion of K at a prime lying over p, and let π be a prime element in K_v. The same argument as above shows that there is an isomorphism

$$E^n(K_v)/E^{5n}(K_v) \to \pi^n\mathcal{O}_v/\pi^{5n}\mathcal{O}_v.$$

However, if p is a high power of π (i.e., the extension $K/\mathbb{Q}$ is highly ramified at v) and n is small, this no longer excludes the possibility that $E^n(K_v)$ may contain an element of order p.

5.11 It was conjectured by Beppo Levi[2] at the International Congress of Mathematicians in 1908, and proved by Mazur (1977, III, 5.1), that the torsion subgroup of $E(\mathbb{Q})$ is isomorphic to one of the following groups:

$$\begin{array}{ll} \mathbb{Z}/m\mathbb{Z} & \text{for} \quad m = 1,2,\ldots,10,12; \\ \mathbb{Z}/2\mathbb{Z} \times \mathbb{Z}/m\mathbb{Z} & \text{for} \quad m = 2,4,6,8. \end{array}$$

This can be interpreted as a statement about the curves $X_1(N)$ considered in Chapter V (see V, 2.8). The 15 curves in the exercise below exhibit all possible torsion subgroups (in order). By contrast, $E(\mathbb{Q}^{\text{al}})_{\text{tors}} \approx \mathbb{Q}/\mathbb{Z} \times \mathbb{Q}/\mathbb{Z}$.

[2] The conjecture was forgotten, and then re-conjectured by Ogg in 1975 (Schappacher and Schoof 1996).

5.12 The fact that $E(\mathbb{Q})_{\text{tors}}$ is so much smaller than $E(\mathbb{Q}^{\text{al}})_{\text{tors}}$ indicates that the image of the Galois group in the automorphism group of $E(\mathbb{Q}^{\text{al}})_{\text{tors}}$ is large. Indeed, for every elliptic curve E over $\mathbb{Q}$ without complex multiplication over $\mathbb{Q}^{\text{al}}$, there exists an integer N (not depending on n) such that the image of the homomorphism

$$\rho_n\colon \text{Gal}(\mathbb{Q}^{\text{al}}/\mathbb{Q}) \to \text{Aut}(E(\mathbb{Q}^{\text{al}})_n) \approx \text{GL}_2(\mathbb{Z}/n\mathbb{Z})$$

has index at most N in $\text{GL}_2(\mathbb{Z}/n\mathbb{Z})$ (Serre 1972).

EXERCISE 5.13 For each of the following elliptic curves, compute the torsion subgroups of $E(\mathbb{Q})$,

$$\begin{array}{ll}
Y^2 = X^3+2, & Y^2 = X^3+X, \\
Y^2 = X^3+4, & Y^2 = X^3+4X, \\
Y^2+Y = X^3-X^2, & Y^2 = X^3+1, \\
Y^2-XY+2Y = X^3+2X^2, & Y^2+7XY-6Y = X^3-6X^2, \\
Y^2+3XY+6Y = X^3+6X^2, & Y^2-7XY-36Y = X^3-18X^2, \\
Y^2+43XY-210Y = X^3-210X^2, & Y^2 = X^3-X, \\
Y^2 = X^3+5X^2+4X, & Y^2+5XY-6Y = X^3-3X^2, \\
Y^2 = X^3+337X^2+20736X. &
\end{array}$$

6 Endomorphisms

Let E be an elliptic curve over k. By an ***endomorphism*** of E, we mean a regular map $\alpha\colon E \to E$ such that $\alpha(O) = O$. Then $\alpha(K)\colon E(K) \to E(K)$ is a homomorphism for all fields K containing k (see 1.6). The endomorphisms of E become a ring (not necessarily commutative) when we define

$$(\alpha+\beta)(P) = \alpha(P)+\beta(P)$$
$$(\alpha\beta)(P) = \alpha(\beta(P)).$$

The first is a regular map because it is the composite of the regular maps

$$E \xrightarrow{(\alpha,\beta)} E\times E \xrightarrow{+} E,$$

and the second is simply the composite of the regular maps α and β. Note that

$$\deg(\alpha\beta) = \deg(\alpha)\cdot\deg(\beta). \tag{18}$$

For $m \in \mathbb{Z}$, we let m denote the endomorphism $m\,\text{id}_E = (P \mapsto mP)$ of E.

THEOREM 6.1 *Let α and β be endomorphisms of E. There exist $r,s,t \in \mathbb{Z}$, depending only on α and β, such that*

$$\deg(m\beta + n\alpha) = rm^2 + smn + tn^2$$

for all $m,n \in \mathbb{Z}$. Moreover,

$$r = \deg(\beta) \geq 0, \quad t = \deg(\alpha) \geq 0, \quad 4rt - s^2 \geq 0.$$

PROOF. We defer the proof of the first statement to later in this section. For the second statement, obviously $r = \deg(\beta) \geq 0$ and $t = \deg(\alpha) \geq 0$. By definition,

$$rm^2 + smn + tn^2 = \deg(m\beta + n\alpha) \geq 0$$

for all $m,n \in \mathbb{Z}$, and so

$$rq^2 + sq + t \geq 0$$

for all $q \in \mathbb{Q}$. As

$$rq^2 + sq + t = r\left(q + \frac{s}{2r}\right)^2 + \left(t - \frac{s^2}{4r}\right),$$

on taking $q = -s/2r$, we see that $4rt - s^2 \geq 0$. □

COROLLARY 6.2 *Every endomorphism α of E satisfies a quadratic equation*

$$\alpha^2 - s\alpha + t = 0$$

with $s,t \in \mathbb{Z}$ and $t = \deg(\alpha)$. Moreover, $4t - s^2 \geq 0$.

PROOF. According to the theorem, there exist integers s and $t = \deg(\alpha)$ such that

$$\deg(\alpha + m) = m^2 + sm + t = \deg(\alpha - m - s),$$

and so (by (18))

$$\deg((\alpha + m)(\alpha - m - s)) = \left(m^2 + sm + t\right)^2.$$

But

$$(\alpha + m)(\alpha - m - s) = \alpha^2 - s\alpha - m(m + s),$$

and so

$$\deg(\alpha^2 - s\alpha + n) = (-n + t)^2$$

for all integers n of the form $-m(s+m)$ with $m \in \mathbb{Z}$. Theorem 6.1 applied to $\alpha^2 - s\alpha$ and 1 shows that $\deg(\alpha^2 - s\alpha + n)$ is a polynomial of degree 2 in n with integer coefficients, and so the last equality holds for all integers n. In particular,

$$\deg(\alpha^2 - s\alpha + t) = 0.$$

But the only endomorphism of degree 0 is the constant endomorphism 0. □

The polynomial $f(T) = T^2 - sT + t$ is called the ***characteristic polynomial*** of α. It is the unique polynomial in $\mathbb{Z}[T]$ such that $f(m) = \deg(m - \alpha)$ for all $m \in \mathbb{Z}$.

PROPOSITION 6.3 *Let E be an elliptic curve over a field k. For all integers $m > 0$ prime to the characteristic of k, $E(k^{\mathrm{al}})_m$ is a free $\mathbb{Z}/m\mathbb{Z}$-module of rank 2.*

PROOF. When $k = \mathbb{C}$, there is a lattice Λ in $\mathbb{C}$ such that $E(\mathbb{C}) \approx \mathbb{C}/\Lambda$ (see[3] III, 3.10), and so

$$E(\mathbb{C})_m \approx \frac{1}{m}\Lambda/\Lambda \simeq \Lambda/m\Lambda \approx (\mathbb{Z}/m\mathbb{Z})^2.$$

This implies the statement for all fields of characteristic zero (see III, 3.12).

Let k be a perfect field of characteristic $p \neq 0$, and let $W(k)$ be the ring of Witt vectors with coefficients in k (Serre 1979, Chapter II). This is a complete discrete valuation ring containing $\mathbb{Z}$ and such that $W(k)/pW(k) = k$. For example, $W(k) = \mathbb{Z}_p$ if $k = \mathbb{F}_p$. Let E be an elliptic curve over k. Choose a Weierstrass equation for E, and lift it to $W(k)$. In this way, we get an elliptic curve $\tilde{E}$ over the field of fractions K of $W(k)$. For every finite extension K' of K, the reduction map $\tilde{E}(K')_m \to E(k')_m$, where k' is the residue field of K', is an isomorphism (4.2, 4.3). As K has characteristic 0, $\tilde{E}(K')_m$ is a free $\mathbb{Z}/m\mathbb{Z}$-module of rank 2 for all sufficiently large K', and it follows that $E(k')_m$ is a free $\mathbb{Z}/m\mathbb{Z}$-module of rank 2 for all sufficiently large finite extensions k' of k. □

Let E be an elliptic curve over a perfect field k, and let ℓ be a prime not equal to the characteristic of k. The ***Tate module*** $T_\ell E$ of E is defined to be

$$T_\ell E = \varprojlim E(k^{\mathrm{al}})_{\ell^n}.$$

---

[3]Chapter III uses nothing from Chapter II after §2.

In other words, an element of $T_\ell E$ is a sequence $a_1, a_2, a_3, \ldots$ with $\ell a_n = a_{n-1}$ for all $n \geq 0$ and $\ell a_1 = 0$. There is a natural action of $\mathbb{Z}_\ell$ on $T_\ell E$ that makes it into a free $\mathbb{Z}_\ell$-module of rank 2 such that $T_\ell E/\ell^n T_\ell E \simeq E(k^{\mathrm{al}})_{\ell^n}$. Let $V_\ell E = T_\ell E \otimes_{\mathbb{Z}_\ell} \mathbb{Q}_\ell$.

PROPOSITION 6.4 *Let E and ℓ be as above, and let α be a nonzero endomorphism of E. The characteristic polynomial of α as an endomorphism of E is equal to the characteristic polynomial of α acting on $V_\ell E$. In particular*

$$\det(\alpha | T_\ell E) = \deg \alpha.$$

PROOF. When $k = \mathbb{C}$, this follows from the fact that $T_\ell E \simeq \mathbb{Z}_\ell \otimes \Lambda$ — see III, §3. In the general case, if $\alpha \in \mathbb{Z}$, then both polynomials equal $(T-\alpha)^2$, and so we may suppose that $\alpha \notin \mathbb{Z}$. Let $P(T)$ be the characteristic polynomial of α as an endomorphism of E, and regard it as an element of $\mathbb{Q}[T]$. If $P(T)$ is irreducible in $\mathbb{Q}_\ell[T]$, then both polynomials are monic of degree 2 and satisfied by α, and so are equal. Otherwise

$$\mathbb{Q}[\alpha] \otimes_{\mathbb{Q}} \mathbb{Q}_\ell \simeq \mathbb{Q}_\ell \times \mathbb{Q}_\ell, \quad \alpha \otimes 1 \leftrightarrow (\alpha_1, \alpha_2)$$

where α_1 and α_2 are the roots of $P(T)$ in $\mathbb{Q}_\ell$. If $\mathbb{Q}[\alpha]$ acts on $V_\ell(E)$ through one of the factors $\mathbb{Q}_\ell$, this would contradict the next proposition. Therefore α acts on $V_\ell E$ with characteristic polynomial $(T-\alpha_1)(T-\alpha_2) = P(T)$. □

COROLLARY 6.5 *Let $T^2 - sT + t$ be the characteristic polynomial of α acting on $T_\ell E$. Then*

(a) *s, t lie in $\mathbb{Z}$ and are independent of ℓ,*

(b) *$4t - s^2 \geq 0$, and*

(c) *$\alpha^2 - s\alpha + t = 0$ (as an endomorphism of E).*

PROOF. The characteristic polynomial of α as an endomorphism of E has these properties. □

LEMMA 6.6 *Let α be an isogeny $E \to E'$. If α is divisible by ℓ^n in* $\mathrm{Hom}(T_\ell E, T_\ell E')$, *then it is divisible by ℓ^n in* $\mathrm{Hom}(E, E')$.

PROOF. The hypothesis implies that α is zero on $E_{\ell^n}(k^{\mathrm{al}})$, and so (see 1.7) α factors as $\alpha = \alpha' \circ \ell^n$,

$$\begin{array}{ccccccccc} 0 & \longrightarrow & E_{\ell^n} & \longrightarrow & E & \xrightarrow{\ell^n} & E & \longrightarrow & 0 \\ & & & & & \searrow^{\alpha} & \downarrow{\scriptstyle \alpha'} & & \\ & & & & & & E'. & & \end{array}$$

□

PROPOSITION 6.7 *For all primes $\ell \neq \mathrm{char}(k)$, the natural map*

$$\mathbb{Z}_l \otimes \mathrm{End}(E) \to \mathrm{End}(T_l E)$$

is injective with torsion-free cokernel.

PROOF. Lemma 6.6 shows that the map

$$\mathbb{Z}_\ell \otimes \mathrm{End}(E) \to \mathrm{End}(T_\ell E)$$

has torsion-free cokernel. It remains to show that it is injective. The elements of $\mathbb{Z}_\ell \otimes \mathrm{End}(E)$ are finite sums

$$\sum c_i \otimes a_i, \quad c_i \in \mathbb{Z}_\ell, \quad a_i \in \mathrm{End}(E),$$

and so it suffices to show that the map $\mathbb{Z}_\ell \otimes M \to \mathrm{End}(T_\ell E)$ is injective for every finitely generated submodule M of $\mathrm{End}(E)$. Let $e_1,\ldots,e_m$ be a basis for such a M; we have to show that $T_\ell(e_1),\ldots,T_\ell(e_m)$ are linearly independent over $\mathbb{Z}_\ell$ in $\mathrm{End}(T_\ell E)$. We extend deg by linearity to a function $\mathbb{Q}\otimes\mathrm{End}(E) \to \mathbb{Q}$. It follows from Theorem 6.1 that the map $\deg\colon \mathbb{Q}M \to \mathbb{Q}$ is continuous for the real topology, and so $U = \{v \mid \deg(v) < 1\}$ is an open neighbourhood of 0. As

$$(\mathbb{Q}M \cap \mathrm{End}(E)) \cap U \subset \mathrm{End}(E) \cap U = 0,$$

we see that $\mathbb{Q}M \cap \mathrm{End}(E)$ is discrete in $\mathbb{Q}M$, and therefore is a finitely generated $\mathbb{Z}$-module (ANT, 4.15). Hence there is a common denominator for the elements of $\mathbb{Q}M \cap \mathrm{End}(E)$:

(*) there exists an integer N such that $N(\mathbb{Q}M \cap \mathrm{End}(E)) \subset M$.

Suppose that $T_\ell(e_1),\ldots,T_\ell(e_m)$ are linearly dependent, so that there exist $a_i \in \mathbb{Z}_\ell$, not all zero, such that $\sum a_i T_\ell(e_i) = 0$. For every $n \in \mathbb{N}$, there exist $n_i \in \mathbb{Z}$ such that $\ell^n | (a_i - n_i)$ in $\mathbb{Z}_\ell$ for all i. Then $\sum n_i T_\ell(e_i)$ is divisible by ℓ^n in $\mathrm{End}(T_\ell E)$, and so $\sum n_i e_i$ is divisible by ℓ^n in $\mathrm{End}(E)$ (by 6.6). Hence $N\left(\sum n_i e_i/\ell^n\right) \in N(\mathbb{Q}M \cap \mathrm{End}(E))$. When n is sufficiently large, $|n_i|_\ell = |a_i|_\ell$ and $|Na_i|_\ell > 1/\ell^n$ for some i with $a_i \neq 0$. Then $|Nn_i/\ell^n|_\ell = |Na_i|_\ell \cdot \ell^n > 1$, and so $Nn_i/\ell^n \notin \mathbb{Z}$. Therefore $N\left(\sum n_i e_i/\ell^n\right)$ does not lie in M, which contradicts (*). This completes the proof. □

The proposition shows that $\mathrm{End}^0(E) \overset{\text{def}}{=} \mathrm{End}(E) \otimes \mathbb{Q}$ is a finite-dimensional $\mathbb{Q}$-algebra.

For a nonzero endomorphism of E with characteristic polynomial $T^2 - sT + t$, we let

$$\alpha' = s - \alpha.$$

PROPOSITION 6.8 *Let α and β be endomorphisms of E. Then $(\alpha+\beta)' = \alpha'+\beta'$, $(\alpha\beta)' = \beta'\alpha'$, and*

$$\alpha\alpha' = \deg(\alpha) = \alpha'\alpha.$$

PROOF. Let ℓ be a prime number $\neq \mathrm{char}(k)$, and let α act on $T_\ell E$ as the matrix $A = \begin{pmatrix} a & b \\ c & d \end{pmatrix}$. Then s is the trace $a+d$ of A, and α' act on $T_\ell E$ as the matrix $A' = \begin{pmatrix} d & -b \\ -c & a \end{pmatrix}$, and so all statements are obvious. □

An ***involution*** of a finite-dimensional $\mathbb{Q}$-algebra R is a $\mathbb{Q}$-linear map $\alpha \mapsto \alpha^*: R \to R$ such that $(\alpha\beta)^* = \beta^*\alpha^*$ for all $\alpha, \beta \in R$. It is ***positive*** if $\mathrm{Tr}_{R/\mathbb{Q}}(\alpha^*\alpha) > 0$ for all $\alpha \neq 0$. The proposition shows that $\alpha \mapsto \alpha'$ defines a positive involution of $\mathrm{End}^0(E)$.

THEOREM 6.9 *Let E be an elliptic curve over k, and let $R = \mathrm{End}^0(E)$. Then $(R,')$ is isomorphic to one of the following:*

(a) *$\mathbb{Q}$ with $'$ the identity map;*

(b) *an imaginary quadratic field F with $'$ complex conjugation;*

(c) *a quaternion[4] division algebra Q over $\mathbb{Q}$.*

The third case occurs only in characteristic $p \neq 0$, and Q is nonsplit at p and ∞ and split at all other primes.

PROOF. We begin by listing some properties of $(R,')$, and then we show that any pair with these properties is isomorphic to one on the above list.

(i) R is a division algebra over $\mathbb{Q}$ (because, for every nonzero element α of $\mathrm{End}(E)$, there exists a $\beta \in \mathrm{End}(E)$ such that $\alpha\beta = m$ for some nonzero integer m; see 1.8).

(ii) Every element of R satisfies a quadratic equation (see 6.2).

(iii) The map $\alpha \mapsto \alpha'$ is a positive involution of R (see 6.8).

(iv) If $\mathrm{char}(k) = 0$, then R acts faithfully on a 2-dimensional vector space over $\mathbb{Q}$ (embed k in $\mathbb{C}$, and let $E(\mathbb{C}) = \mathbb{C}/\Lambda$; then $\mathrm{End}^0(E)$ acts faithfully on $\Lambda \otimes \mathbb{Q}$; see III, 3.3).

---

[4]A ***quaternion algebra*** R over a field k is a simple k-algebra of degree 4 with centre k. A field $K \supset k$ is said to ***split*** R if $R \otimes_k K \approx M_2(K)$. For a quaternion algebra R over $\mathbb{Q}$, let $S(R)$ denote the set of prime numbers or ∞ such that $\mathbb{Q}_l$ splits R. Then $S(R)$ is a finite set that determines R up to isomorphism. The sets of the form $S(R)$ are exactly those with a finite even number of elements. The set $S(R)$ is empty if and only if $R \approx M_2(\mathbb{Q})$. (See Wikipedia: quaternion algebra.)

(v) For all primes $\ell \neq \mathrm{char}(k)$, $R \otimes_{\mathbb{Q}} \mathbb{Q}_\ell$ acts faithfully on a 2-dimensional vector space over $\mathbb{Q}_\ell$ (see 6.7).

From (v) we see that $[R:\mathbb{Q}] \leq 4$. Let F be the centre of R. Then

$$[R:\mathbb{Q}] = [R:F][F:\mathbb{Q}],$$

and it follows from a theorem of Wedderburn that $[R:F]$ is a square. There are only four possibilities, namely,

(a) $[R:\mathbb{Q}] = 1$ and $R = \mathbb{Q}$;

(b) $[R:\mathbb{Q}] = 2$ and R a field (meaning a commutative field);

(c) $[R:\mathbb{Q}] = 4$ and R a quaternion division algebra;

(d) $[R:\mathbb{Q}] = 3$ or 4 and R a field.

In case (a), the only (positive) involution on $\mathbb{Q}$ is the identity map. In case (b), there is no positive involution on R unless it is imaginary, in which case the only positive involution is complex conjugation. Because of (iv), case (c) can only occur in nonzero characteristic p. As R is a division algebra, it must be nonsplit at a nonzero even number of primes. From (v) we see that it is split at all primes $\ell \neq p, \infty$, and so it must be nonsplit at p and ∞. Case (d) is ruled out by (ii) and the primitive element theorem (FT, 5.1). □

All possibilities in the theorem occur.

Proof of Theorem 6.1

It remains to prove the first statement of Theorem 6.1. According to the elementary lemma (IV, 4.8), for this it suffices to prove that deg satisfies the parallelogram law,

$$\deg(\alpha+\beta)+\deg(\alpha-\beta) = 2\deg(\alpha)+2\deg(\beta), \quad \text{all } \alpha,\beta \in \mathrm{End}(E).$$

For a geometric proof of the theorem, which applies to all abelian varieties, see Milne 1986a, §12.

Alternatively, it is possible to prove directly that, for any endomorphism α of E, there exists a unique endomorphism α' such that $\alpha \circ \alpha' = \deg(\alpha) = \alpha' \circ \alpha$; moreover, $(\alpha+\beta)' = \alpha' + \beta'$ for all α, β (Silverman 2009, III, 6.2c). But now

$$\begin{aligned}\deg(\alpha+\beta)+\deg(\alpha-\beta) &= (a+\beta)(\alpha+\beta)' + (\alpha-\beta)(\alpha-\beta)' \\ &= (\alpha+\beta)(\alpha'+\beta') + (\alpha-\beta)(\alpha'-\beta') \\ &= 2\alpha\alpha' + 2\beta\beta' \\ &= 2\deg(a) + 2\deg(\beta)\end{aligned}$$

as required.

Finally, in the rest of this section, following Cassels 1991, §24, we give an elementary proof of parallelogram law.

Let $k = k_0(T)$, and let $P = (w_0 : w_1 : \ldots) \in \mathbb{P}^n(k)$. We call $(w_0 : w_1 : \ldots)$ a ***primitive representative*** for P if $w_0, w_1, \ldots k_0[T]$ and have gcd $= 1$. The ***height*** $h(P)$ of P is then defined to be

$$h(P) = \max(\deg(w_0), \deg(w_1), \ldots).$$

For an elliptic curve E over k and a point $P = (x : y : z) \in E(K)$ such that $(x : z) \in \mathbb{P}^1(k)$, define

$$h(P) = h(x(P) : z(P)).$$

LEMMA 6.10 *Let $P_1, P_2 \in E(K)$ be such that $h(P_1)$ and $h(P_2)$ are defined. Then $h(P_1 + P_2)$ and $h(P_1 - P_2)$ are defined, and*

$$h(P_1 + P_2) + h(P_1 - P_2) \leq 2h(P_1) + 2h(P_2) + C$$

for some constant C (independent of P_1, P_2).

PROOF. Let $P_1 + P_2 = P_3$ and $P_1 - P_2 = P_4$, and let $P_i = (x_i : y_i : z_i)$. Then

$$(x_3 x_4 : x_3 z_4 + x_4 z_3 : z_3 z_4) = (w_0 : w_1 : w_2),$$

where

$$\begin{aligned}
w_0 &= x_1^2 x_2^2 - 2a x_1 x_2 z_1 z_2 - 4b(x_1 z_1 z_2^2 + x_2 z_1^2 z_2) + a^2 z_1^2 z_2^2. \\
w_1 &= 2(x_1 x_2 + a z_1 z_2)(x_1 z_2 + x_2 z_1) + 4b z_1^2 z_2^2 \\
w_2 &= (x_2 z_1 - x_1 z_2)^2
\end{aligned}$$

After choosing $(x_1 : y_1 : z_1)$ and $(x_2 : y_2 : z_2)$ to be primitive representatives for P_1 and P_2, it is clear that

$$h(w_0 : w_1 : w_2) \leq h(P_1)^2 + h(P_2)^2 + C$$

for some constant C. Here $h(w_0 : w_1 : w_2)$ is the height of $(w_0 : w_1 : w_2)$ as a point of $\mathbb{P}^2(k)$. On the other hand, one sees easily that

$$h(w_0 : w_1 : w_2) \geq h(P_3) + h(P_4).$$

On combining these two inequalities, we obtain the required inequality. □

Now consider an elliptic curve E over k_0, defined by a Weierstrass polynomial, so that $k(E) = k_0(x, y)$ with x transcendental over k_0. Let α be an endomorphism of E over k, and consider $\alpha^*: k(E) \to k(E)$.

LEMMA 6.11 *Let P be a generic point on E (so P lies in $E(k)$). We have $\alpha^*(x) \in k_0(x)$, and*

$$\deg(\alpha) = h(\alpha P).$$

PROOF. Exercise. □

LEMMA 6.12 *Let α, β be endomorphisms of E. Then*

$$\deg(\alpha+\beta) + \deg(\alpha-\beta) = 2\deg(\alpha) + 2\deg(\beta).$$

PROOF. Let P be a generic point on E (so lying in $E(k)$). Then

$$(\alpha+\beta)^*(P) = \alpha^*(P) + \beta^*(P)$$
$$(\alpha-\beta)^*(P) = \alpha^*(P) - \beta^*(P)$$

and so the last two lemmas show that there exists a constant C such that

$$\deg(\alpha+\beta) + \deg(\alpha-\beta) \le 2\deg(\alpha) + 2\deg(\beta) + C.$$

Now $\deg(2) = 4$, and so $\deg(2^n\alpha) = 4^n \deg(\alpha)$. On replacing α and β with $2^n\alpha$ and $2^n\beta$, dividing through by 4^n, and letting $n \to \infty$, we obtain the inequality

$$\deg(\alpha+\beta) + \deg(\alpha-\beta) \le 2\deg(\alpha) + 2\deg(\beta).$$

Putting $\alpha' = \alpha + \beta$ and $\beta' = \alpha - \beta$ in this gives the reverse inequality. □

NOTES Most of the results in this section go back to Hasse in the 1930s. They play an essential role in his proof of the Riemann hypothesis for elliptic curves (see IV, §9). Hasse (1936) proved that the endomorphism algebra $\mathrm{End}^0(E)$ of an elliptic curve E is $\mathbb{Q}$, an imaginary quadratic field, or a definite quaternion algebra. Deuring (1941) determined the structure of the endomorphism rings of elliptic curves in nonzero characteristic.

Over $\mathbb{C}$, $j(\tau)$ is said to be a singular value of j if the corresponding elliptic curve E has $\mathrm{End}(E) \neq \mathbb{Z}$. Thus, an elliptic curve E with $\mathrm{End}(E) \neq \mathbb{Z}$ is an "elliptic curve with singular j-invariant". Eventually this was shortened to "singular elliptic curve". By extension, Deuring called an elliptic curve supersingular if its endomorphism ring has rank 4 over $\mathbb{Z}$. In this case, he showed that the endomorphism ring is a maximal order.

7 Néron models

Recall that an elliptic curve E over $\mathbb{Q}_p$ has a Weierstrass equation

$$Y^2Z + a_1XYZ + a_3YZ^2 = X^3 + a_2X^2Z + a_4XZ^2 + a_6Z^3, \quad a_i \in \mathbb{Q}_p,$$

which is uniquely determined up to a change of variables of the form

$$\begin{aligned} X &= u^2X' + r \\ Y &= u^3Y' + su^2X' + t \end{aligned}$$

with $u, r, s, t \in \mathbb{Q}_p$ and $u \neq 0$. Under such a change, the discriminant Δ transforms according to $\Delta = u^{12}\Delta'$. We make a change of variables so that the $a_i \in \mathbb{Z}_p$ and $\mathrm{ord}_p(\Delta)$ is as small as possible. The new "minimal" equation is uniquely determined up to a change of variables of the above form with $u, r, s, t \in \mathbb{Z}_p$ and $u \in \mathbb{Z}_p^\times$. We can think of this minimal Weierstrass equation as defining a curve over $\mathbb{Z}_p$, which will be the best "model" of E over $\mathbb{Z}_p$ among plane projective curves. We call it the ***Weierstrass minimal model*** of E.

Néron showed that if we allow our models to be curves over $\mathbb{Z}_p$ that are not necessarily embeddable in $\mathbb{P}^2$, then we may obtain a model that is better in some respects than any plane model. I shall attempt to explain what these Néron models are in this section. Unfortunately, this is a difficult topic, which requires the theory of schemes for a satisfactory explanation[5] and so the exposition will be very superficial. For a detailed account, see Silverman 1994, Chap. IV.

In order to be able to state Néron's results, and the earlier results of Kodaira, we need to expand our notion of a curve to allow "multiple components". For an affine plane curve, this means simply that we allow curves to be defined by polynomials f with repeated factors. For example, the equation

$$(Y - X)(Y - pX)(Y - p^2X) = 0$$

defines an affine plane curve in the sense of Chapter I. It consists of three irreducible components, namely, three lines through the origin. Modulo p,

[5] Néron himself did not use schemes. For a long period, the only rigorous foundations for algebraic geometry were provided by Weil 1946, 1962, which did not allow mixed characteristic. Consequently, those working in mixed characteristic were forced to devise their own extension of Weil's foundations, which makes their work difficult to understand by the modern reader. This, of course, all changed in the early 1960s with Grothendieck's schemes, but some authors, Néron included, continued with the old way.

the equation becomes

$$(Y - X)Y^2 = 0,$$

which is the union of two lines $Y - X = 0$ and $Y = 0$, the second of which has multiplicity 2.

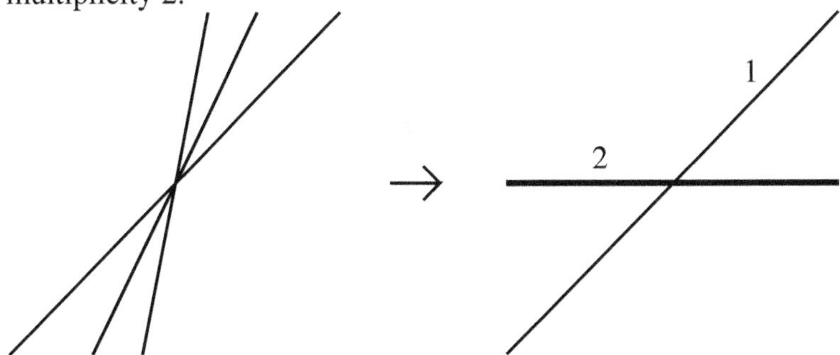

For more general curves, the idea is the same.

The work of Kodaira

Before considering Néron models, we look at an analogous situation, which was its precursor. Consider an equation

$$Y^2 Z = X^3 + a(T)XZ^2 + b(T)Z^3, \quad \Delta(T) \stackrel{\text{def}}{=} 4a(T)^3 + 27b(T)^2 \neq 0$$

with $a(T), b(T) \in \mathbb{C}[T]$. We can view this as defining three different objects:

(a) an elliptic curve E over the field $\mathbb{C}(T)$;

(b) a surface S in $\mathbb{P}^2(\mathbb{C}) \times \mathbb{A}^1(\mathbb{C})$ whose points are the pairs $((x{:}y{:}z), t)$ satisfying the equation;

(c) a family of (possibly degenerate) elliptic curves $E(T)$ parameterized by T.

By (c) we mean the following: for each $t_0 \in \mathbb{C}$ we have a curve

$$E(t_0) : Y^2 Z = X^3 + a(t_0)XZ^2 + b(t_0)Z^3, \quad a(t_0), b(t_0) \in \mathbb{C},$$

with discriminant $\Delta(t_0)$. This is nonsingular, and hence an elliptic curve, if and only if t_0 is not a root of the polynomial $\Delta(T)$; otherwise, it will have a singularity, and we view it as a degenerate elliptic curve. Note that the projection $\mathbb{P}^2(\mathbb{C}) \times \mathbb{A}^1(\mathbb{C}) \to \mathbb{A}^1(\mathbb{C})$ induces a map $S \to \mathbb{A}^1(\mathbb{C})$ whose fibres are the curves $E(t)$. We can view S as a "model" of E over $\mathbb{C}[T]$

(or over $\mathbb{A}^1(\mathbb{C})$). We should choose the equation of E so that $\Delta(T)$ has the smallest possible degree so that there are as few singular fibres as possible.

For convenience, we now drop the Z, and consider the equation

$$S : Y^2 = X^3 + a(T)X + b(T), \quad a(T), b(T) \in \mathbb{C}[T],$$

— strictly, we should work with the family of projective curves.

Let $P = (x, y, t) \in S(\mathbb{C})$, and let $f(X,Y,T) = X^3 + a(T)X + b(T) - Y^2$. Then P is singular on the curve $E(t)$ if and only if it satisfies the following equations,

$$\frac{\partial f}{\partial Y} = -2Y = 0$$
$$\frac{\partial f}{\partial X} = 3X^2 + a(T) = 0.$$

It is singular on the surface S if in addition it satisfies the equation,

$$\frac{\partial f}{\partial T} = \frac{da}{dT}X + \frac{db}{dT} = 0.$$

Thus, P might be singular in its fibre $E(t)$ without being singular on S.

EXAMPLE 7.1 (a) Consider the equation

$$Y^2 = X^3 - T, \quad \Delta(T) = 27T^2.$$

The origin is singular (in fact, it is a cusp) when regarded as a point on the curve $E(0) : Y^2 = X^3$, but not when regarded as a point on the surface $S : Y^2 = X^3 - T$. In fact, the tangent plane to S at the origin is the (X,Y)-plane, $T = 0$.

(b) Consider the equation

$$Y^2 = X^3 - T^2, \quad \Delta(T) = 27T^4.$$

In this case, the origin is singular when regarded as a point on $E(0)$ *and* when regarded as a point on S.

(c) Consider the equation

$$\begin{aligned} Y^2 &= (X - 1 + T)(X - 1 - T)(X + 2) \\ &= X^3 - (3 + T^2)X + 2 - 2T^2. \end{aligned}$$

The discriminant is

$$\Delta(T) = -324T^2 + 72T^4 - 4T^6.$$

The curve $E(0)$ is

$$Y^2 = X^3 - 3X + 2 = (X-1)^2(X+2),$$

which has a node at $(1,0)$. When we replace $X-1$ in the original equation with X in order to translate $(1,0,0)$ to the origin, the equation becomes

$$\begin{aligned} Y^2 &= (X+T)(X-T)(X+3) \\ &= (X^2 - T^2)(X+3) \\ &= X^3 + 3X^2 - T^2X - 3T^2. \end{aligned}$$

This surface has a singularity at the origin because its equation has no linear term.

Kodaira (1960) showed that, by blowing up points, and blowing down curves, etc., it is possible to obtain from the surface

$$S: Y^2Z = X^3 + a(T)XZ^2 + b(T)Z^3, \quad a(T), b(T) \in \mathbb{C}[T], \quad \Delta[T] \neq 0$$

a new surface S' and a regular map $S' \to \mathbb{A}^1$ having the following properties:

(a) S' is nonsingular;

(b) S' regarded as a curve over $\mathbb{C}(T)$ is equal to S regarded as a curve over $\mathbb{C}(T)$ (for the experts, the maps $S \to \mathbb{A}^1$ and $S' \to \mathbb{A}^1$ have the same generic fibres);

(c) the fibres $E'(t_0)$ of S' over $\mathbb{A}^1(\mathbb{C})$ are projective curves; moreover $E'(t_0) = E(t_0)$ if the points of $E(t_0)$ are nonsingular when regarded as points on S (for example, if $E(t_0)$ itself is nonsingular);

(d) S' is minimal with the above properties: if $S'' \to \mathbb{A}^1$ is a second map with properties (a,b,c), then every regular map $S' \to S''$ over $\mathbb{A}^1$ that is an isomorphism on the generic fibres is an isomorphism.

The map $S' \to \mathbb{A}^1$ is uniquely determined by these properties up to a unique isomorphism. Kodaira classified the possible fibres of $S' \to \mathbb{A}^1$.

"Blowing up" a nonsingular point P in a surface S leaves the surface unchanged except that it replaces the point P with the projective space of lines through the origin in the tangent plane to S at P. For a curve C in S through P, the inverse image of $C \smallsetminus P$ is a curve in the blown up variety whose closure meets the projective space at the point corresponding to the tangent line of C at P. Even when $S \subset \mathbb{P}^m$, the blown-up surface does not have a natural embedding into a projective space.

EXAMPLE 7.2 To illustrate the phenomenon of "blowing up", consider the map

$$\sigma : k^2 \to k^2, \quad (x,y) \mapsto (x,xy).$$

Its image omits only the points on the Y-axis where $Y \neq 0$. A point in the image is the image of a unique point in k^2 except for $(0,0)$, which has been "blown up" to the whole of the Y-axis. In other words, the map is one-to-one, except that the Y-axis has been "blown down" to a point. The line $L : Y = \alpha X$ has inverse image $XY = \alpha X$, which is the union of the Y-axis and the line $Y = \alpha$; the closure of the inverse image of $L \smallsetminus \{(0,0)\}$ is the line $Y = \alpha$. The singular curve

$$C : Y^2 = X^3 + \alpha X^2$$

has as inverse image the curve

$$Y^2 X^2 = X^3 + \alpha X^2,$$

which is the union of the curve $X^2 = 0$ (the Y-axis with "multiplicity" 2) and the nonsingular curve $Y^2 = X + \alpha$. Note that the latter meets the Y-axis at the points $(0, \pm\sqrt{\alpha})$, i.e., at the points corresponding to the slopes of the tangents of C at $(0,0)$.

In this example, $(0,0)$ in $\mathbb{A}^2(k)$ is blown up to an affine line. In a true blowing-up, it would be replaced by a projective line, and the description of the map would be more complicated. (See Fulton 1969, Chap. 7, for blow-ups of points in $\mathbb{P}^2$.)

The work of Néron

Néron proved an analogue of Kodaira's result for elliptic curves over $\mathbb{Q}_p$.[6] To explain his result, we need to talk about schemes. For the nonexperts, a ***scheme*** $\mathcal{E}$ over $\mathbb{Z}_p$ is simply the object defined by a collection of polynomial

[6] More accurately, he showed that Kodaira's results for curves over the discrete valuation ring $\mathbb{C}[[T]]$ also apply to curves over any complete valuation rings with perfect residue field.

equations with coefficients in $\mathbb{Z}_p$. The object defined by the same equations regarded as having coefficients in $\mathbb{Q}_p$ is a variety E over $\mathbb{Q}_p$ called the ***generic fibre*** of $\mathcal{E}/\mathbb{Z}_p$, and the object defined by the equations with the coefficients reduced modulo p is a variety $\bar{E}$ over $\mathbb{F}_p$ called the ***special fibre*** of $\mathcal{E}/\mathbb{Z}_p$. For example, if $\mathcal{E}$ is the scheme defined by the equation

$$Y^2Z + a_1XYZ + a_3YZ^2 = X^3 + a_2X^2Z + a_4XZ^2 + a_6Z^3, \quad a_i \in \mathbb{Z}_p,$$

then E is the elliptic curve over $\mathbb{Q}_p$ defined by the same equation, and $\bar{E}$ is the elliptic curve over $\mathbb{F}_p$ by the equation

$$Y^2Z + \bar{a}_1XYZ + \bar{a}_3YZ^2 = X^3 + \bar{a}_2X^2Z + \bar{a}_4XZ^2 + \bar{a}_6Z^3,$$

where $\bar{a}_i$ is the image of a_i in $\mathbb{Z}/p\mathbb{Z} = \mathbb{F}_p$.

Given an elliptic curve $E/\mathbb{Q}_p$, Néron constructs a scheme $\mathcal{E}$ over $\mathbb{Z}_p$ having the following properties:

(a) $\mathcal{E}$ is regular; this means that all the associated local rings are regular local rings, i.e., if the local ring has dimension d, then its maximal ideal can be generated by d elements. (For an algebraic variety over an algebraically closed field, "regular" is equivalent to "nonsingular".)

(b) the generic fibre of $\mathcal{E}$ is the original curve E;

(c) $\mathcal{E}$ is proper over $\mathbb{Z}_p$; this simply means that $\bar{E}$ is a projective curve;

(d) $\mathcal{E}$ is minimal with the above properties: if $\mathcal{E}'$ is a second scheme over $\mathbb{Z}_p$ having the properties (a,b,c), then any regular map $\mathcal{E} \to \mathcal{E}'$ over $\mathbb{Z}_p$ giving an isomorphism on the generic fibres is an isomorphism.

Moreover, Néron classified the possible special fibres over $\mathbb{F}_p^{\text{al}}$ and obtained essentially the same list as Kodaira.

The curve $\mathcal{E}$ over $\mathbb{Z}_p$ is called the ***complete Néron (minimal) model***. It has some defects: we need not have $\mathcal{E}(\mathbb{Z}_p) \simeq E(\mathbb{Q}_p)$; it does not have a group structure; its special fibre $\bar{E}$ may be singular. All three defects are eliminated by simply removing all singular points and multiple curves in the special fibre. One then obtains the ***smooth Néron (minimal) model***, which however has the defect that it not proper.

Given an elliptic curve E over $\mathbb{Q}_p$ we now have three models over $\mathbb{Z}_p$:

(a) $\mathcal{E}^w$, the Weierstrass minimal model of E;

(b) $\mathcal{E}$, the complete Néron model of E;

(c) $\mathcal{E}'$, the smooth Néron model of E.

They are related as follows: to get $\mathcal{E}'$ from $\mathcal{E}$, remove all multiple curves and singular points; to get the Weierstrass model with the singular point in the closed fibre removed from $\mathcal{E}'$, remove all connected components of the special fibre except that containing O.

ASIDE 7.3 By definition, an ***abelian variety*** is a projective algebraic variety with a group structure defined by regular maps. Thus, an abelian variety of dimension 1 is an elliptic curve. Much (but not all) of the theory of elliptic curves extends to all abelian varieties, but, because it is no longer possible to write equations, the proofs require much more of the machinery of algebraic geometry. For an abelian variety of dimension > 1, only the smooth Néron minimal model exists.

The different types

We describe three of the possible ten (or eleven, depending how one counts) different types of models using both Kodaira's numbering $(I_0, I_n, II, \ldots)$ and Néron's numbering $(a, b_n, c1, \ldots)$. We describe the special fibre over the algebraic closure of $\mathbb{F}_p$ rather than $\mathbb{F}_p$ itself. For example, in the case of nodal reduction (type $(I_n,\ b_n)$), the identity connected component of the special fibre of the smooth Néron model will be a twisted $\mathbb{G}_m$ over $\mathbb{F}_p$ unless the tangents are rational, and some of the components of the special fibre might not be rational over $\mathbb{F}_p$.

(I_0, a) In this case E has good reduction and all three models are the same.

$(I_n, b_n), n > 1$ In this case E has nodal reduction; let $n = \mathrm{ord}_p(\Delta)$; the special fibres for the three models are the following:

(a) a cubic curve with a node;

(b) n curves, each of genus 0, each intersecting exactly two other of the curves;

(c) an algebraic group G such that the identity component G° of G is $\mathbb{G}_m$, and such that G/G° is a cyclic group of order n.

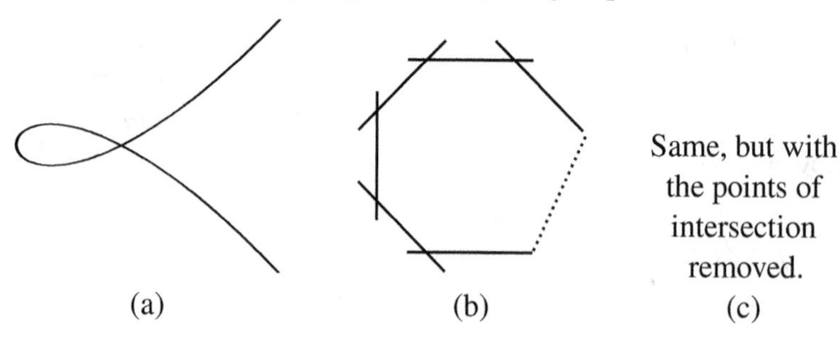

(I_0^*, c_4) In this case E has cuspidal reduction and $\mathrm{ord}_p(\Delta) = 6$ $(p \neq 2)$; the special fibres for the three models are the following:

(a) a cubic curve with a cusp;

(b) four disjoint curves of genus 0 and multiplicity 1, together with one curve of genus 0 and multiplicity 2 crossing each of the four curves transversally;

(c) an algebraic group G whose identity component is $\mathbb{G}_a$ and such that G/G° is a group of order 4 isomorphic to $\mathbb{Z}/2\mathbb{Z} \times \mathbb{Z}/2\mathbb{Z}$.

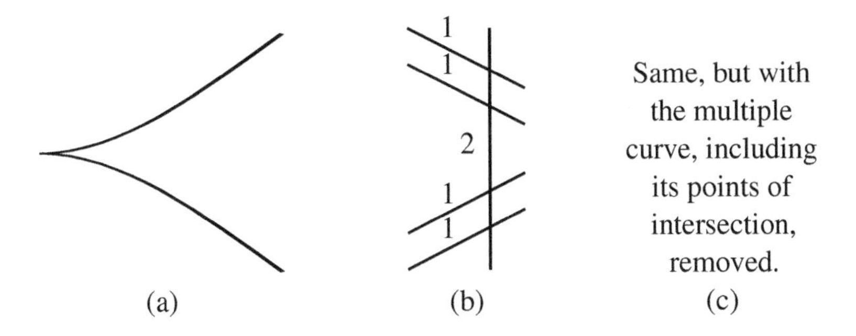

When the minimal equation is used, the mysterious quotient $\frac{E(\mathbb{Q}_p)}{E^0(\mathbb{Q}_p)}$ of §4 is equal to $\frac{G(\mathbb{F}_p)}{G^\circ(\mathbb{F}_p)}$, where G is the special fibre of the smooth Néron model and G° is the identity component of G. In the above three examples, it is (a) the trivial group; (b) a subgroup of a cyclic group of order n (and equal to a cyclic group of order n if E has split nodal reduction); (c) a subgroup of $(\mathbb{Z}/2\mathbb{Z})^2$.

Summary of minimal models $(E, \mathcal{E}, \bar{E})$.

	Weierstrass	complete Néron	smooth Néron
$\mathcal{E}$ a plane curve	Yes	Not always	Not always
$\mathcal{E}$ regular	Not always	Yes	Yes
$\bar{E}$ projective	Yes	Yes	Not always
$\bar{E}$ nonsingular	Not always	Not always	Yes
$\bar{E}$ a group	Not always	Not always	Yes

8 Algorithms for elliptic curves

About 1965, Weil suggested that there is a precise correspondence between elliptic curves over $\mathbb{Q}$ with a fixed conductor N and cusp forms of dimension -2 for $\Gamma_0(N)$ (Weil 1967; see V, §5). This prompted Tate[7] to find algorithms for computing the minimal Weierstrass equation, discriminant, conductor, j-invariant, fibres of the Néron model, etc., of an elliptic curve over $\mathbb{Q}$ so that Weil's idea could be tested. He included them in the letter to Cassels mentioned earlier (p. 68), which was published as Tate 1975. These are painful for humans to use, but easy for computers. Fortunately, they have been implemented in computer programs, for example, in the program Pari/gp, which is specifically designed for calculations in algebraic number theory, including elliptic curves. In the following, I explain how to use Pari as a supercalculator. You can also program it, but for that you will have to read the manual.

Recall (p. 67) that the general Weierstrass equation of an elliptic curve E over a field k is

$$Y^2Z + a_1XYZ + a_3YZ^2 = X^3 + a_2X^2Z + a_4XZ^2 + a_6Z^3,$$

and that we attach to the curve quantities $b_2, b_4, b_6, b_8, c_4, c_6$ and

$$\Delta = -b_2^2b_8 - 8b_4^3 - 27b_6^2 + 9b_2b_4b_6$$
$$j = c_4^3/\Delta.$$

The curve is nonsingular if and only if $\Delta \neq 0$. The differential one-form $\omega = \frac{dx}{2y+a_1x+a_3}$ is invariant under translation. A Weierstrass equation for an elliptic curve E is unique up to a coordinate transformation of the form

$$X = u^2X' + r \qquad Y = u^3Y' + su^2X' + t\ , \quad u, r, s, t \in k, \quad u \neq 0.$$

The quantities Δ, j, ω transform according to the rules:

$$u^{12}\Delta' = \Delta, \quad j' = j, \quad \omega' = u\omega.$$

[7] "I wrote to Cassels from Paris that same year 65–66. I believe it was in the fall. Early that summer, Weil had told me of the idea that all elliptic curves over $\mathbb{Q}$ are modular. That motivated Swinnerton-Dyer to make a big computer search for elliptic curves over $\mathbb{Q}$ with not too big discriminant, in order to test Weil's idea. But of course it was necessary to be able to compute the conductor to do that test. That was my main motivation. I don't know why I wrote to Cassels rather than Swinnerton-Dyer. Maybe because I knew him much better." Tate, personal communication.

A minimal Weierstrass equation for an elliptic curve E over $\mathbb{Q}$ is an equation of the above form with the $a_i \in \mathbb{Z}$ and Δ minimal. It is unique up to a coordinate transformation of the above form with $r,s,t,u \in \mathbb{Z}$ and $u \in \mathbb{Z}^{\times} = \{\pm 1\}$.

In gp (**g**rande calculatrice **p**ari), an elliptic curve is specified by giving a vector

`v=[a1,a2,a3,a4,a6]`.

`e=ellinit(v,1)` Defines `e` to be the elliptic curve $Y^2 + a_1XYZ + \cdots$ and computes the 13-component vector

$$[a_1,a_2,a_3,a_4,a_6,b_2,b_4,b_6,b_8,c_4,c_6,\Delta,j]$$

`e=ellinit(v)` Defines `e` to be the elliptic curve $Y^2 + a_1XYZ + \cdots$ and computes the above 13-component vector plus some other information useful for other computations.

`elladd(e,z1,z2)` Computes the sum of the points `z1=[x1,y2]` and `z2=[x2,y2]` on `e`.

`elltors(e)` Computes `[t,v1,v2]`, where `t` is the order of $E(\mathbb{Q})_{\text{tors}}$, `v1` gives the structure of $E(\mathbb{Q})_{\text{tors}}$ as a product of cyclic groups, and `v2` gives the generators of the cyclic groups.

`ellglobalred(e)` Computes the vector `[N,v,c]`, where `N` is the conductor of the curve and `v=[u,r,s,t]` is the coordinate transformation giving the Weierstrass minimal model with $a_1 = 0$ or 1, $a_2 = 0,1,-1$, and $a_3 = 0,1$. Such a model is unique.

`e'=ellchangecurve(e,v)` Changes `e` to `e'`, where `e'` is the 13+-component vector corresponding to the curve obtained by the change of coordinates `v=[u,r,s,t]`.

`ellocalred(e',p)` Computes the type of the reduction at p using Kodaira's notation. It produces `[f,n,...]`, where `f` is the exponent of p in the conductor of e, $n = 1$ means good reduction (type I_0), $n = 2,3,4$ means reduction of type II,III,IV, $n = 4+\nu$ means type I_ν, and -1, -2 etc. mean I^* II^* etc.

`ellap(e,p)` Computes $a_p \stackrel{\text{def}}{=} p + 1 - \#E(\mathbb{F}_p)$. Requires e to be minimal at p.

`ellan(e,n)` Computes the first n a_k. Requires e to be minimal at p.

`ellgenerators(e)` Computes a basis for $E(\mathbb{Q})/\text{tors}$ (requires John Cremona's elliptic curve data to be available).

Example.
Start gp
gp> `v=[0,-4,0,0,16]` Sets v equal to the vector $(0,-4,\ldots)$.
`%1=[0,-4,0,0,16]`
gp> `ellinit(v,1)` Computes the vector $(a_1,\ldots,j)$.
`%2=[0,-4,0,0,16,-16,0,64,-256,256,-9728,-45056,-4096/11]`
For example, $\Delta = -45056$ and $j = -4096/11$.
gp> `e=ellinit(v)` Computes the vector $(a_1,\ldots,j,\ldots)$ and sets e equal to the elliptic curve $Y^2 = X^3 - 4X^2 + 16$.
`%3=[0,-4,0,0,16,...]`
gp> `elltors(e)` Computes the torsion subgroup (it is cyclic of order 5).
`%4=[5,[5],[0,4]]`
gp> `ellglobalred(e)` Computes the minimum conductor and the change of coordinates required to give the minimal equation.
`%5=[11, [2,0,0,4],1]`
gp> `ellchangecurve(e,[2,0,0,4])`
`%6=[0,-1,1,0,0,-4,0,1,-1,16,-152,-11,-4096/11,...]`
Computes the minimal equation for E, $Y^2 + Y = X^3 - X^2$, which now has discriminant -11 but (of course) the same j-invariant.
gp> `elllocalred(%6,2)`
`%7 = [0,1,[1,0,0,0],1]` So E now has good reduction at 2.
gp> `elllocalred(%6,11)`
`%8 = [1,5,[1,0,0,0],1]` So E has bad reduction at 11, with conductor 11^1 (hence the singularity is a node), and the Kodaira type of the special fibre of the Néron model is I_1.
gp> `ellap(%6,13)` Computes a_{13}.
`%9=4` So $\#E(\mathbb{F}_{13}) = 13 + 1 - 4 = 10$.
gp> `ellgenerators(e)`
`%10=[]` So $E(\mathbb{Q})$ is finite.
gp> `ellinit([6,-3,9,-16,-14])`
`%11=[6,-3,9,-16,-14,24,22,25,29,48,-216,37,110592/37,...]`
gp> `ellgenerators(%11)`
`%12=[[-2,2]]` So $E(\mathbb{Q})/\mathrm{tors}$ is free of rank one generated by $(-2,2)$.
Pari/gp is freely available from `https://pari.math.u-bordeaux.fr/`. The following sources are recommended: the books Cohen 1993, 2000, and Cremona 1997; the database *The L-functions and Modular Forms Database* (`https://www.lmfdb.org/`); the article Best et al. 2020.

Chapter III

Elliptic Curves over the Complex Numbers

The case of the ground field $\mathbb{C}$ is just the classical theory of elliptic functions which can be found in any sufficiently old-fashioned text-book of analysis.
Cassels 1966.

The theory of elliptic curves simplifies when the base field is the complex numbers.

1 Lattices and bases

A ***lattice*** in $\mathbb{C}$ is any subgroup generated by two complex numbers that are linearly independent over $\mathbb{R}$. Thus

$$\Lambda = \mathbb{Z}\omega_1 + \mathbb{Z}\omega_2, \quad \text{some } \omega_1, \omega_2 \in \mathbb{C},$$

and since neither ω_1 nor ω_2 is a real multiple of the other, we can order them so that $\Im(\omega_1/\omega_2) > 0$. If $\{\omega_1', \omega_2'\}$ is a second pair of elements of Λ, then

$$\omega_1' = a\omega_1 + b\omega_2, \quad \omega_2' = c\omega_1 + d\omega_2, \quad \text{some } a, b, c, d \in \mathbb{Z},$$

i.e.,

$$\begin{pmatrix} \omega_1' \\ \omega_2' \end{pmatrix} = A \begin{pmatrix} \omega_1 \\ \omega_2 \end{pmatrix},$$

with A a 2×2 matrix with integer coefficients. The pair (ω_1',ω_2') is a $\mathbb{Z}$-basis for Λ if and only if A is invertible and so has determinant ± 1. Let $z=\omega_1/\omega_2$ and $z'=\omega_1'/\omega_2'$; then

$$\Im(z')=\Im\left(\frac{az+b}{cz+d}\right)=\frac{\Im(adz+bc\bar{z})}{|cz+d|^2}=\frac{(ad-bc)\Im(z)}{|cz+d|^2}$$

and so $\Im(\omega_1'/\omega_2')>0$ if and only if $\det A>0$. Therefore, the group $\mathrm{SL}_2(\mathbb{Z})$ of matrices with integer coefficients and determinant 1 acts transitively on the set of bases (ω_1,ω_2) for Λ with $\Im(\omega_1/\omega_2)>0$. We have proved the following statement.

PROPOSITION 1.1 *Let M be the set of pairs of complex numbers (ω_1,ω_2) such that $\Im(\omega_1/\omega_2)>0$, and let $\mathcal{L}$ be the set of lattices in $\mathbb{C}$. Then the map $(\omega_1,\omega_2)\mapsto\mathbb{Z}\omega_1+\mathbb{Z}\omega_2$ induces a bijection*

$$\mathrm{SL}_2(\mathbb{Z})\backslash M\to\mathcal{L}. \tag{19}$$

Here $\mathrm{SL}_2(\mathbb{Z})\backslash M$ denotes the set of orbits in M for the action

$$\begin{pmatrix}a&b\\c&d\end{pmatrix}\begin{pmatrix}\omega_1\\\omega_2\end{pmatrix}=\begin{pmatrix}a\omega_1+b\omega_2\\c\omega_1+d\omega_2\end{pmatrix}.$$

Let $\mathbb{H}$ be the complex upper half-plane,

$$\mathbb{H}=\{z\in\mathbb{C}\mid\Im(z)>0\}.$$

Let $z\in\mathbb{C}^\times$ act on M by the rule $z(\omega_1,\omega_2)=(z\omega_1,z\omega_2)$ and on $\mathcal{L}$ by the rule $z\Lambda=\{z\lambda\mid\lambda\in\Lambda\}$. The map $(\omega_1,\omega_2)\mapsto\omega_1/\omega_2$ induces a bijection

$$M/\mathbb{C}^\times\to\mathbb{H}. \tag{20}$$

The action of $\mathrm{SL}_2(\mathbb{Z})$ on M corresponds to the action

$$\begin{pmatrix}a&b\\c&d\end{pmatrix}\tau=\frac{a\tau+b}{c\tau+d}$$

on $\mathbb{H}$. On combining (19) and (20), we obtains bijections

$$\begin{array}{ccccc}\mathcal{L}/\mathbb{C}^\times & \xleftrightarrow{1:1} & \mathrm{SL}_2(\mathbb{Z})\backslash M/\mathbb{C}^\times & \xleftrightarrow{1:1} & \mathrm{SL}_2(\mathbb{Z})\backslash\mathbb{H}\\[2ex] (\mathbb{Z}\omega_1+\mathbb{Z}\omega_2)\cdot\mathbb{C}^\times & \leftrightarrow & \mathrm{SL}_2(\mathbb{Z})\cdot(\omega_1,\omega_2)\cdot\mathbb{C}^\times & \leftrightarrow & \mathrm{SL}_2(\mathbb{Z})\cdot\dfrac{\omega_1}{\omega_2}.\end{array}$$

For a lattice Λ with basis $\{\omega_1,\omega_2\}$, the interior of any parallelogram with vertices z_0, $z_0+\omega_1$, $z_0+\omega_2$, $z_0+\omega_1+\omega_2$ is called a ***fundamental domain*** or ***period parallelogram*** for Λ. We usually choose z_0 so that the domain contains 0.

2 Doubly periodic functions

Let Λ be a lattice in $\mathbb{C}$. To give a function on $\mathbb{C}/\Lambda$ is the same as giving a function on $\mathbb{C}$ such that

$$f(z+\omega) = f(z) \qquad \text{(as functions on } \mathbb{C}) \tag{21}$$

for all $\omega \in \Lambda$. If $\{\omega_1, \omega_2\}$ is a basis for Λ, then this condition is equivalent to

$$\begin{cases} f(z+\omega_1) = f(z) \\ f(z+\omega_2) = f(z) \end{cases}.$$

For this reason, functions satisfying (21) are said to be ***doubly periodic***. In this section, we study meromorphic functions on $\mathbb{C}$ that are doubly periodic meromorphic for a lattice Λ, and in the next section we interpret these functions as meromorphic functions on the quotient Riemann surface $\mathbb{C}/\Lambda$.

PROPOSITION 2.1 *Let $f(z)$ be a doubly periodic meromorphic function for Λ, not identically zero, and let D be a fundamental domain for Λ such that f has no zeros or poles on the boundary of D. Then*

(a) $\sum_{P \in D} \operatorname{Res}_P(f) = 0$;

(b) $\sum_{P \in D} \operatorname{ord}_P(f) = 0$;

(c) $\sum_{P \in D} \operatorname{ord}_P(f) \cdot P \equiv 0 \mod \Lambda$.

The first sum is over the points in D where f has a pole, and the other sums are over the points where it has a zero or pole (and $\operatorname{ord}_P(f)$ is the order of the zero or the negative of the order of the pole). Each sum is finite.

PROOF. According to the residue theorem (Cartan 1963, III, 5.2),

$$\int_\Gamma f(z)dz = 2\pi i \sum_{P \in D} \operatorname{Res}_P(f),$$

where Γ is the boundary of D. Because f is periodic, the integrals over opposite sides of D cancel, and so the integral is zero. This gives (a). For (b) one applies the residue theorem to f'/f, noting that this is again doubly periodic and that $\operatorname{Res}_P(f'/f) = \operatorname{ord}_P(f)$. For (c) one applies the residue theorem to $z \cdot f'(z)/f(z)$. This is no longer doubly periodic, but the integral of it around Γ lies in Λ. □

COROLLARY 2.2 *A nonconstant doubly periodic meromorphic function has at least two poles (or one double pole).*

PROOF. A holomorphic doubly periodic function is bounded on the closure of any fundamental domain (by compactness), and hence on the entire plane (by periodicity). It is therefore constant by Liouville's theorem (Cartan 1963, III, 1.2). It is impossible for a doubly periodic meromorphic function to have a single simple pole in a period parallelogram because, by (a) of the proposition, the residue at the pole would have to be zero there, and so the pole would not be simple. □

The Weierstrass $\wp$ function

Let Λ be a lattice in $\mathbb{C}$. The Riemann–Roch theorem applied to the quotient $\mathbb{C}/\Lambda$ proves the existence of nonconstant doubly periodic meromorphic functions for Λ, but here we shall construct them explicitly. When G is a *finite* group acting on a set S, it is easy to construct functions invariant under the action of G: take f to be any function $f: S \to \mathbb{C}$, and define

$$F(s) = \sum_{g \in G} f(gs);$$

then $F(g's) = \sum_{g \in G} f(g'gs) = F(s)$ because, as g runs over G, so does $g'g$; thus F is invariant, and (obviously) all invariant functions are of this form. When G is not finite, we have to verify that the series converges — in fact, in order for the result to be independent of the order of summation, we need (at least) absolute convergence.

Let D be an open subset of $\mathbb{C}$, and let $f_0, f_1, \ldots$ be a sequence of holomorphic functions on D. Recall (Cartan 1963, I, 2) that the series $\sum_n f_n$ is said to ***converge normally*** on a subset A of D if the series of positive terms $\sum_n \|f_n\|$ converges, where $\|f_n\| = \sup_{z \in A} |f_n(z)|$. The series $\sum_n f_n$ is then both uniformly convergent and absolutely convergent on A. When $f_0, f_1, \ldots$ is a sequence of *meromorphic* functions, the series is said to ***converges normally*** on A if, after a finite number of terms f_n have been removed, it becomes a normally convergent series of holomorphic functions (ibid., V, 2). If a series $\sum_n f_n$ of meromorphic functions is normally convergent on compact subsets of D, then the sum f of the series is a meromorphic function on D; moreover, the series of derivatives converges normally on compact subsets of D, and its sum is the derivative of f (ibid., V, 2).

Now let $\varphi(z)$ be a meromorphic function $\mathbb{C}$ and write

$$\Phi(z) = \sum_{\omega \in \Lambda} \varphi(z + \omega).$$

Assume that as $|z| \to \infty$, $\varphi(z) \to 0$ so fast that the series for $\Phi(z)$ is normally convergent on compact subsets. Then $\Phi(z)$ is doubly periodic with respect to Λ, because replacing z by $z+\omega_0$ for some $\omega_0 \in \Lambda$ merely rearranges the terms in the sum.

To prove the normal convergence for the functions we are interested in, we shall need the following result.

LEMMA 2.3 *For every lattice Λ in $\mathbb{C}$, the series $\sum_{\omega\in\Lambda,\omega\neq 0} 1/|\omega|^3$ converges.*

PROOF. Let ω_1, ω_2 be a basis for Λ, and, for each integer $n \geq 1$, consider the parallelogram

$$P(n) = \{a_1\omega_1 + a_2\omega_2 \mid a_1, a_2 \in \mathbb{R},\ \max(|a_1|, |a_2|) = n\}.$$

There are $8n$ points of Λ on $P(n)$, and the distance between 0 and any of them is at least kn, where k is the shortest distance from 0 to a point of $P(1) \cap \Lambda$. Therefore, the contribution of the points on $P(n)$ to the sum is bounded by $8n/k^3n^3$, and so

$$\sum_{\omega\in\Lambda,\omega\neq 0} \frac{1}{|\omega|^3} \leq \frac{8}{k^3}\sum_n \frac{1}{n^2} < \infty.$$

□

We know from 2.2 that the simplest possible nonconstant doubly periodic meromorphic function is one with a double pole at each point of Λ and no other poles. Suppose that $f(z)$ is such a function. Then $f(z) - f(-z)$ is doubly periodic with no poles except perhaps simple ones at the points of Λ. Hence it must be constant, and since it is an odd function it must vanish. Thus $f(z)$ is even, and we can make it unique by imposing the normalization condition

$$f(z) = z^{-2} + 0 + z^2 g(z)$$

with $g(z)$ holomorphic near $z = 0$. There is such a function, namely, the Weierstrass function $\wp(z)$, but we cannot define it directly from $1/z^2$ by the method at the start of this subsection because $\sum_{\omega\in\Lambda} 1/(z+\omega)^2$ is not normally convergent. Instead, we define

$$\wp(z) = \frac{1}{z^2} + \sum_{\omega\in\Lambda,\ \omega\neq 0}\left(\frac{1}{(z-\omega)^2} - \frac{1}{\omega^2}\right)$$

$$\wp'(z) = \sum_{\omega\in\Lambda} \frac{-2}{(z-\omega)^{-3}}.$$

PROPOSITION 2.4 *The two series above converge normally on compact subsets of $\mathbb{C}$, and their sums $\wp$ and $\wp'$ are doubly periodic meromorphic functions on $\mathbb{C}$ with $\wp' = \frac{d\wp}{dz}$.*

PROOF. Note that $\wp'(z) = \sum_{\omega\in\Lambda}\varphi(z)$ with $\varphi(z) = \frac{-2}{z^3} = \frac{d}{dz}\left(\frac{1}{z^2}\right)$, and that $\sum_{\omega\in\Lambda}\varphi(z+\omega)$ converges normally on any compact disk $|z|\leq r$ by comparison with $\sum\frac{1}{|\omega|^3}$. Thus, $\wp'(z)$ is a doubly periodic meromorphic function on $\mathbb{C}$ by the above remarks.

For $|z|\leq r$, and for all but the finitely many ω with $|\omega|\leq 2r$, we have that

$$\left|\frac{1}{(z-\omega)^2}-\frac{1}{\omega^2}\right| = \left|\frac{-z^2+2\omega z}{\omega^2(z-\omega)^2}\right| = \frac{\left|z\left(2-\frac{z}{\omega}\right)\right|}{\left|\omega^3\right|\left|1-\frac{z}{\omega}\right|^2} \leq \frac{r\frac{5}{2}}{\left|\omega^3\right|\cdot\frac{1}{4}} = \frac{10r}{|\omega|^3}$$

and so $\wp(z)$ also converges normally on the compact disk $|z|\leq r$. Because its derivative is doubly periodic, so also is $\wp(z)$. □

Eisenstein series

Let Λ be a lattice in $\mathbb{C}$, and consider the sum

$$\sum_{\omega\in\Lambda,\,\omega\neq 0}\frac{1}{\omega^n}.$$

The map $\omega\mapsto -\omega\colon\Lambda\to\Lambda$ has order 2, and its only fixed point is 0. Therefore $\Lambda\smallsetminus\{0\}$ is a disjoint union of its orbits, and it follows that the sum is zero if n is odd. We write

$$G_{2k}(\Lambda) = \sum_{\omega\in\Lambda,\,\omega\neq 0}\frac{1}{\omega^{2k}},$$

and we let $G_{2k}(z) = G_{2k}(\mathbb{Z}z+\mathbb{Z})$, $z\in\mathbb{H}$. Note that

$$G_{2k}(c\Lambda) = c^{-2k}G_{2k}(\Lambda) \text{ for } c\in\mathbb{C}^\times,$$

and so $G_{2k}(\mathbb{Z}\omega_1+\mathbb{Z}\omega_2) = \omega_2^{-2k}G_{2k}(\mathbb{Z}\frac{\omega_1}{\omega_2}+\mathbb{Z})$.

PROPOSITION 2.5 *For all integers $k\geq 2$, $G_{2k}(z)$ converges to a holomorphic function on $\mathbb{H}$.*

PROOF. Let

$$D = \{z \in \mathbb{C} \mid |z| \geq 1, \quad |\Re(z)| \leq 1/2\}.$$

For $z \in D$,

$$|mz+n|^2 = m^2 z\bar{z} + 2mn\Re(z) + n^2 \geq m^2 - mn + n^2 = |m\rho - n|^2,$$

where $\rho = e^{2\pi i/3}$. Therefore, Lemma 2.3 shows that, for $k \geq 2$, $G_{2k}(z)$ converges normally on D. For all $A \in \mathrm{SL}_2(\mathbb{Z})$, $G_{2k}(A^{-1}z)$ also converges normally on D, which shows that $G_{2k}(z)$ converges normally on AD. In V, 1.3, we shall see that the sets AD cover $\mathbb{H}$, and so this shows that $G_{2k}(z)$ is holomorphic on the whole of $\mathbb{H}$. □

The functions $G_{2k}(\Lambda)$ and $G_{2k}(z)$ are called ***Eisenstein series.***

The field of doubly periodic meromorphic functions

Fix a lattice Λ in $\mathbb{C}$, and let $\wp(z)$ be the Weierstrass $\wp$ function for Λ. The doubly periodic functions for Λ form a subfield of the field $M(\mathbb{C})$ of meromorphic functions on $\mathbb{C}$, which we now determine.

PROPOSITION 2.6 *There is the following relation between $\wp$ and $\wp'$,*

$$\wp'(z)^2 = 4\wp(z)^3 - g_4\wp(z) - g_6,$$

where $g_4 = 60G_4(\Lambda)$ and $g_6 = 140G_6(\Lambda)$.

PROOF. We compute the Laurent expansion of $\wp(z)$ near 0. Recall that for $|t| < 1$,

$$\frac{1}{1-t} = 1 + t + t^2 + \cdots.$$

On differentiating this, we find that

$$\frac{1}{(1-t)^2} = \sum_{n\geq 1} nt^{n-1} = \sum_{n\geq 0}(n+1)t^n.$$

Hence, for $|z| < |\omega|$,

$$\frac{1}{(z-\omega)^2} - \frac{1}{\omega^2} = \frac{1}{\omega^2}\left(\frac{1}{\left(1-\frac{z}{\omega}\right)^2} - 1\right) = \sum_{n\geq 1}(n+1)\frac{z^n}{\omega^{n+2}}.$$

On putting this into the definition of $\wp(z)$ and changing the order of summation, we find that for $|z| < |\omega|$,

$$\begin{aligned}\wp(z) &= \frac{1}{z^2} + \sum_{n\geq 1}\sum_{\omega\neq 0}(n+1)\frac{z^n}{\omega^{n+2}} \\ &= \frac{1}{z^2} + \sum_{k\geq 1}(2k+1)G_{2k+2}(\Lambda)z^{2k} \\ &= \frac{1}{z^2} + 3G_4z^2 + 5G_6z^4 + \cdots .\end{aligned}$$

Therefore,

$$\wp'(z) = -\frac{2}{z^3} + 6G_4z + 20G_6z^3 + \cdots .$$

These expressions contain enough terms to show that the Laurent expansion of

$$\wp'(z)^2 - 4\wp(z)^3 + 60G_4(\Lambda)\wp(z) + 140G_6(\Lambda)$$

has no nonzero term in z^n with $n \leq 0$. Therefore this function is holomorphic at 0 and takes the value 0 there. Since it is doubly periodic and has no poles in a suitable fundamental domain containing 0, we see that it is constant, and in fact zero. □

PROPOSITION 2.7 *The field of doubly periodic meromorphic functions is the subfield $\mathbb{C}(\wp,\wp')$ of $M(\mathbb{C})$ generated by $\wp$ and $\wp'$, i.e., every doubly periodic meromorphic function can be expressed as a rational function of $\wp$ and $\wp'$.*

PROOF. We begin by showing that every even doubly periodic meromorphic function f lies in $\mathbb{C}(\wp)$.

Observe that, because $f(z) = f(-z)$, the kth derivative of f,

$$f^{(k)}(z) = (-1)^k f^{(k)}(-z).$$

Therefore, if f has a zero of order m at z_0, then it has a zero of order m at $-z_0$. On applying this remark to $1/f$, we obtain the same statement with "zero" replaced by "pole".

Similarly, because $f^{(2k+1)}(z_0) = -f^{(2k+1)}(-z_0)$, if $z_0 \equiv -z_0 \bmod \Lambda$, then the order of zero (or pole) of f at z_0 is even.

Choose a set of representatives mod Λ for the zeros and poles of f not in Λ and number them $z_1, \ldots, z_m, -z_1, \ldots, -z_m, z_{m+1}, \ldots, z_n$ so that (modulo Λ)

$$\begin{aligned} z_i &\not\equiv -z_i, & 1 \le i \le m \\ z_i &\equiv -z_i \not\equiv 0, & m < i \le n. \end{aligned}$$

Let m_i be the order of f at z_i; according to the second observation, m_i is even for $i > m$.

Now $\wp(z) - \wp(z_i)$ is also an even doubly periodic function. Since it has exactly two poles in a fundamental domain, it must have exactly two zeros there. When $i \le m$, it has simple zeros at $\pm z_i$; when $i > m$, it has a double zero at z_i (by the second observation). Define

$$g(z) = \prod_{i=1}^{m} (\wp(z) - \wp(z_i))^{m_i} \cdot \prod_{i=m+1}^{n} (\wp(z) - \wp(z_i))^{m_i/2}.$$

Then $f(z)$ and $g(z)$ have exactly the same zeros and poles at points z not on Λ. We deduce from 2.1b that they also have the same order at $z = 0$, and so f/g, being holomorphic and doubly periodic, is constant: $f = cg \in \mathbb{C}(\wp)$.

Now consider an arbitrary doubly periodic meromorphic function f. We can write f as the sum of an even and of an odd doubly periodic function,

$$f(z) = \frac{f(z) + f(-z)}{2} + \frac{f(z) - f(-z)}{2}.$$

We know the even doubly periodic meromorphic functions lie in $\mathbb{C}(\wp)$, and clearly the odd doubly periodic meromorphic functions lie in $\wp' \cdot \mathbb{C}(\wp)$. □

Let

$$\mathbb{C}[x, y] = \frac{\mathbb{C}[X, Y]}{(Y^2 - 4X^3 + g_4 X + g_6)},$$

and let $\mathbb{C}[\wp, \wp']$ be the $\mathbb{C}$-algebra of meromorphic functions on $\mathbb{C}$ generated by $\wp$ and $\wp'$. Proposition 2.6 shows that the map $(X, Y) \mapsto (\wp(z), \wp'(z))$ defines a homomorphism

$$\mathbb{C}[x, y] \to \mathbb{C}[\wp, \wp'],$$

which we shall show to be an isomorphism. For this, we have to show that the only polynomials $g(X, Y) \in \mathbb{C}[X, Y]$ such that $g(\wp, \wp') = 0$ are those divisible by $f(X, Y) \stackrel{\text{def}}{=} Y^2 - 4X^3 + g_4 X + g_6$. The theory of resultants

(I, 1.24) shows that, for any polynomial $g(X,Y)$, there exist polynomials $a(X,Y)$ and $b(X,Y)$ such that

$$a(X,Y)f(X,Y)+b(X,Y)g(X,Y)=r(X)\in\mathbb{C}[X]$$

with $\deg_Y(b)<\deg_Y(f)$. If $g(\wp,\wp')=0$, then $r(\wp)=0$, but it is easy to see that $\wp$ is transcendental over $\mathbb{C}$ (for example, it has infinitely many poles). Therefore $r=0$, and so $f(X,Y)$ divides $b(X,Y)g(X,Y)$. As $f(X,Y)$ is irreducible, it must divide either $b(X,Y)$ or $g(X,Y)$. Because of the degrees, it cannot divide b, and so it must divide g.

The isomorphism $\mathbb{C}[x,y]\to\mathbb{C}[\wp,\wp']$ induces an isomorphism on the fields of fractions

$$\mathbb{C}(x,y)\to\mathbb{C}(\wp,\wp').$$

Thus, $\mathbb{C}(x,y)$ is canonically isomorphic to the field of all doubly periodic meromorphic functions for Λ.

3 Elliptic curves as Riemann surfaces

The notion of a Riemann surface

Let X be a connected Hausdorff topological space admitting a countable base for its open sets. A ***coordinate neighbourhood*** for X is a pair (U,z) with U an open subset of X and z a homeomorphism of U onto an open subset of the complex plane $\mathbb{C}$. Two coordinate neighbourhoods (U_1,z_1) and (U_2,z_2) are ***compatible*** if the function

$$z_1\circ z_2^{-1}\colon z_2(U_1\cap U_2)\to z_1(U_1\cap U_2)$$

and its inverse are holomorphic. A family of coordinate neighbourhoods $(U_i,z_i)_{i\in I}$ is a ***coordinate covering*** if $X=\bigcup_i U_i$ and (U_i,z_i) is compatible with (U_j,z_j) for all pairs $(i,j)\in I\times I$. Two coordinate coverings are said to be ***equivalent*** if the their union is also a coordinate covering. This defines an equivalence relation on the set of coordinate coverings, and an equivalence class is called a ***complex structure*** on X. The space X equipped with a complex structure is a ***Riemann surface.***

Let X be a Riemann surface. A function $f\colon U\to\mathbb{C}$ on an open subset U of X is ***holomorphic*** if it satisfies the following condition for one (hence every) coordinate covering $(U_i,z_i)_{i\in I}$ in the equivalence class defining the complex structure: $f\circ z_i^{-1}\colon z_i(U\cap U_i)\to\mathbb{C}$ is holomorphic for all $i\in I$.

A map $f: X \to X'$ from one Riemann surface to a second is ***holomorphic*** if $g \circ f$ is holomorphic whenever g is a holomorphic function on an open subset of X'. For this, it suffices to check, for some coordinate coverings of X and X', that for every point P in X, there are coordinate neighbourhoods (U, z) of P and (U', z') of $f(P)$ such that $z' \circ f \circ z^{-1}: z(U) \to z'(U')$ is holomorphic. An ***isomorphism*** of Riemann surfaces is a bijective holomorphic map.

Recall that a ***meromorphic function*** on an open subset U of $\mathbb{C}$ is a holomorphic function f on $U \smallsetminus \Xi$ for some discrete subset $\Xi \subset U$ having at worst a pole at each point of Ξ, i.e., such that for every $a \in \Xi$, there exists an m for which $(z-a)^m f(z)$ is holomorphic in a neighbourhood of a. A ***meromorphic function*** on an open subset of a Riemann surface is defined similarly. Equivalently it is a holomorphic map from the Riemann surface to the Riemann sphere (Cartan 1963, VI, 4.5).

EXAMPLE 3.1 Every open subset U of $\mathbb{C}$ is a Riemann surface with a single coordinate neighbourhood, namely, U itself with the identity map z.

EXAMPLE 3.2 Consider the unit sphere

$$S_2 : X^2 + Y^2 + Z^2 = 1$$

in $\mathbb{R}^3$, and let P be the north pole $(0,0,1)$. Stereographic projection from P is the map

$$(x, y, z) \mapsto \frac{x+iy}{1-z}: S_2 \smallsetminus P \to \mathbb{C}.$$

Take this to be a coordinate neighbourhood for S_2. Stereographic projection from the south pole gives a second coordinate neighbourhood. These two coordinate neighbourhoods define a complex structure on S_2, and S_2 together with this complex structure is called the ***Riemann sphere***. On the other hand, $\mathbb{P}^1(\mathbb{C})$ acquires a complex structure from the coordinate covering $\mathbb{P}^1(\mathbb{C}) = U_0(\mathbb{C}) \cup U_1(\mathbb{C})$,

$$U_0(\mathbb{C}) \stackrel{\text{def}}{=} \{(x{:}\, y) \mid x \neq 0\} \xrightarrow[\simeq]{(x:y)\mapsto y/x} \mathbb{C}$$

$$U_1(\mathbb{C}) \stackrel{\text{def}}{=} \{(x{:}\, y) \mid y \neq 0\} \xrightarrow[\simeq]{(x:y)\mapsto x/y} \mathbb{C}.$$

The map $(x, y, z) \mapsto (x{:}\, y{:}\, z): S_2 \to \mathbb{P}^1(\mathbb{C})$ is an isomorphism of Riemann surfaces.

Differential one-forms on Riemann surfaces

A differential one-form on an open subset of $\mathbb{C}$ is simply an expression $\omega = f dz$, with f a meromorphic function. Given a smooth curve γ

$$t \mapsto z(t): [a,b] \to \mathbb{C}, \quad [a,b] = \{t \in \mathbb{R} \mid a \leq t \leq b\},$$

we can form the integral

$$\int_\gamma \omega = \int_a^b f(z(t)) \cdot z'(t) \cdot dt \in \mathbb{C}.$$

Now consider a compact Riemann surface X. If ω is a differential one-form on X and (U_i, z_i) is a coordinate neighbourhood for X, then $\omega|U_i = f_i(z_i)dz_i$. If (U_j, z_j) is a second coordinate neighbourhood, so that $z_j = w(z_i)$ on $U_i \cap U_j$, then

$$f_i(z_i)dz_i = f_j(w(z_i))w'(z_i)dz_i$$

on $U_i \cap U_j$. Thus, to give a differential one-form on X is the same as giving a differential one-form $f_i dz_i$ on each U_i satisfying the above equation on the overlaps. For any (real) curve $\gamma: I \to X$ and differential one-form ω on X, the integral $\int_\gamma \omega$ makes sense.

A differential one-form is ***holomorphic*** if it is represented on the coordinate neighbourhoods by forms $f dz$ with f holomorphic.

It is an important fact that the holomorphic differential one-forms on a Riemann surface of genus g form a complex vector space $\Omega^1(X)$ of dimension g.

For example, the Riemann sphere S has genus 0 and so should have no nonzero holomorphic differential one-forms. Note that dz is holomorphic on $\mathbb{C} = S \smallsetminus \{\text{ north pole}\}$, but that $z = 1/z'$ on $S \smallsetminus \{\text{ poles}\}$, and so $dz = -\frac{1}{z'^2}dz'$, which has a pole at the north pole. Hence dz does not extend to a holomorphic differential one-form on the whole of S.

Quotients of $\mathbb{C}$ by lattices

Let Λ be a lattice in $\mathbb{C}$. Topologically the quotient $\mathbb{C}/\Lambda$ is isomorphic to $\mathbb{R}^2/\mathbb{Z}^2$, which is a one-holed torus (the surface of a donut). Write $\pi: \mathbb{C} \to \mathbb{C}/\Lambda$ for the quotient map. Then $\mathbb{C}/\Lambda$ has a unique complex structure for which π is a local isomorphism of Riemann surfaces. It can be described as follows: for all $P \in \mathbb{C}/\Lambda$ and $Q \in f^{-1}(P)$, there exist open neighbourhoods U of P and V of Q such that $\pi|V: V \to U$ is a homeomorphism; the coordinate neighbourhoods $(U, (\pi|V)^{-1}: U \to V)$ form a coordinate covering of $\mathbb{C}/\Lambda$, and so define a complex structure on $\mathbb{C}/\Lambda$. Relative to this structure, a function $\varphi: U \to \mathbb{C}$ on an open subset U of $\mathbb{C}/\Lambda$ is holomorphic (resp. meromorphic) if and only if the composite $\varphi \circ \pi: \pi^{-1}(U) \to \mathbb{C}$ is holomorphic (resp. meromorphic) in the usual sense.

We shall see that, although any two quotients $\mathbb{C}/\Lambda$, $\mathbb{C}/\Lambda'$ are homeomorphic, they will be isomorphic as Riemann surfaces only if $\Lambda' = \alpha\Lambda$ for some $\alpha \in \mathbb{C}^{\times}$.

The holomorphic maps $\mathbb{C}/\Lambda \to \mathbb{C}/\Lambda'$

Let Λ and Λ' be lattices in $\mathbb{C}$. The map $\pi: \mathbb{C} \to \mathbb{C}/\Lambda$ realizes $\mathbb{C}$ as the universal covering space of $\mathbb{C}/\Lambda$. As the same is true of $\pi': \mathbb{C} \to \mathbb{C}/\Lambda'$, every continuous map $\varphi: \mathbb{C}/\Lambda \to \mathbb{C}/\Lambda'$ such that $\varphi(0) = 0$ lifts uniquely to a continuous map $\tilde{\varphi}: \mathbb{C} \to \mathbb{C}$ such that $\tilde{\varphi}(0) = 0$,

$$\begin{array}{ccc} \mathbb{C} & \overset{\tilde{\varphi}}{\dashrightarrow} & \mathbb{C} \\ \downarrow{\scriptstyle \pi} & & \downarrow{\scriptstyle \pi'} \\ \mathbb{C}/\Lambda & \xrightarrow{\varphi} & \mathbb{C}/\Lambda'. \end{array}$$

Because π and π' are local isomorphisms of Riemann surfaces, the map φ is holomorphic if and only if $\tilde{\varphi}$ is holomorphic.

PROPOSITION 3.3 *Let Λ and Λ' be lattices in $\mathbb{C}$. A complex number α such that $\alpha\Lambda \subset \Lambda'$ defines a holomorphic map*

$$[z] \mapsto [\alpha z]: \mathbb{C}/\Lambda \to \mathbb{C}/\Lambda'$$

sending 0 to 0, and every holomorphic map $\mathbb{C}/\Lambda \to \mathbb{C}/\Lambda'$ sending 0 to 0 is of this form (for a unique α).

PROOF. It is obvious from the above remarks that α defines a holomorphic map $\mathbb{C}/\Lambda \to \mathbb{C}/\Lambda'$. Conversely, let $\varphi: \mathbb{C}/\Lambda \to \mathbb{C}/\Lambda'$ be a holomorphic map such that $\varphi(0) = 0$, and let $\tilde{\varphi}$ be its unique lifting to a holomorphic map $\mathbb{C} \to \mathbb{C}$ sending 0 to 0. For every $\omega \in \Lambda$, the map $z \mapsto \tilde{\varphi}(z+\omega) - \tilde{\varphi}(z)$ is continuous and takes values in $\Lambda' \subset \mathbb{C}$; because $\mathbb{C}$ is connected and Λ' is discrete, the map must be constant, and so its derivative is zero,

$$\tilde{\varphi}'(z+\omega) = \tilde{\varphi}'(z).$$

Therefore $\tilde{\varphi}'(z)$ is doubly periodic. As it is holomorphic, it must be constant, say, $\tilde{\varphi}'(z) = \alpha$ for all z. On integrating, we find that $\tilde{\varphi}(z) = \alpha z + \beta$, and $\beta = \tilde{\varphi}(0) = 0$. □

COROLLARY 3.4 *Riemann surfaces* $\mathbb{C}/\Lambda$ *and* $\mathbb{C}/\Lambda'$ *are isomorphic if and only if* $\Lambda' = \alpha\Lambda$ *for some* $\alpha \in \mathbb{C}^{\times}$.

PROOF. If $\mathbb{C}/\Lambda$ and $\mathbb{C}/\Lambda'$ are isomorphic, then every isomorphism $\mathbb{C}/\Lambda \to \mathbb{C}/\Lambda'$ is of the form $[z] \mapsto [\alpha z]$ for some $\alpha \in \mathbb{C}$, and so $\Lambda' = \alpha\Lambda$. Conversely, if $\Lambda' = \alpha\Lambda$, then $[z] \mapsto [\alpha z]$ is an isomorphism $\mathbb{C}/\Lambda \to \mathbb{C}/\Lambda'$. □

COROLLARY 3.5 *Every holomorphic map* $\mathbb{C}/\Lambda \to \mathbb{C}/\Lambda'$ *sending* 0 *to* 0 *is a homomorphism of groups.*

PROOF. Clearly $[z] \mapsto [\alpha z]$ is a homomorphism of groups. □

The proposition shows that

$$\mathrm{Hom}(\mathbb{C}/\Lambda, \mathbb{C}/\Lambda') \simeq \{\alpha \in \mathbb{C} \mid \alpha\Lambda \subset \Lambda'\},$$

and the first corollary shows that there is a one-to-one correspondence

$$\{\mathbb{C}/\Lambda \text{ modulo isomorphism}\} \longleftrightarrow \mathcal{L}/\mathbb{C}^{\times}.$$

The elliptic curve $E(\Lambda)$

Let Λ be a lattice in $\mathbb{C}$.

LEMMA 3.6 *The polynomial* $f(X) = 4X^3 - g_4(\Lambda)X - g_6(\Lambda)$ *has distinct roots.*

PROOF. The function $\wp'(z)$ is odd, so $\wp'(\omega_1/2) = -\wp'(-\omega_1/2)$, and doubly periodic, so $\wp'(\omega_1/2) = \wp'(-\omega_1/2)$. Thus, $\wp'(\omega_1/2) = 0$, and Proposition 2.6 shows that $\wp(\omega_1/2)$ is a root of $f(X)$. The same argument shows that $\wp(\omega_2/2)$ and $\wp((\omega_1+\omega_2)/2)$ are also roots of $f(X)$. It remains to prove that these three numbers are distinct.

The function $\wp(z) - \wp(\omega_1/2)$ has a zero at $\omega_1/2$, which must be a double zero because its derivative is also 0 there. Since $\wp(z) - \wp(\omega_1/2)$ has only one (double) pole in a fundamental domain D containing 0, Proposition 2.1 shows that $\omega_1/2$ is the only zero of $\wp(z) - \wp(\omega_1/2)$ in D, i.e., that $\wp(z)$ takes the value $\wp(\omega_1/2)$ only at $z = \omega_1/2$ within D. In particular, $\wp(\omega_1/2)$ is not equal to $\wp(\omega_2/2)$ or $\wp((\omega_1+\omega_2)/2)$. Similarly, $\wp(\omega_2/2)$ is not equal to $\wp((\omega_1+\omega_2)/2)$. □

The lemma shows that

$$E(\Lambda): Y^2Z = 4X^3 - g_4(\Lambda)XZ^2 - g_6(\Lambda)Z^3$$

is an elliptic curve. Recall that $c^4 g_4(c\Lambda) = g_4(\Lambda)$ and $c^6 g_6(c\Lambda) = g_6(\Lambda)$ for any $c \in \mathbb{C}^\times$, and so $c\Lambda$ defines essentially the same elliptic curve as Λ.

For an elliptic curve

$$E: Y^2Z = X^3 + a_4XZ^2 + a_6Z^3,$$

the closed subset $E(\mathbb{C})$ of $\mathbb{P}^2(\mathbb{C})$ has a natural complex structure — for example, in a neighbourhood of a point $P \in E(\mathbb{C})$ such that $y(P) \neq 0 \neq z(P)$, the function x/z provides a local coordinate.

PROPOSITION 3.7 *The map*

$$\begin{cases} z \mapsto (\wp(z) : \wp'(z) : 1), & z \neq 0 \\ 0 \mapsto (0:1:0) \end{cases}$$

is an isomorphism of Riemann surfaces $\mathbb{C}/\Lambda \to E(\Lambda)(\mathbb{C})$.

PROOF. It is certainly a well-defined map. The function $\wp(z)\colon \mathbb{C}/\Lambda \to \mathbb{P}^1(\mathbb{C})$ is $2:1$ in a fundamental domain containing 0, except at the points $\frac{\omega_1}{2}, \frac{\omega_2}{2}, \frac{\omega_1+\omega_2}{2}$, where it is one-to-one. Therefore, $\wp$ realizes $\mathbb{C}/\Lambda$ as a covering of degree 2 of the Riemann sphere, and it is a local isomorphism except at the four listed points. Similarly, x/z realizes $E(\Lambda)(\mathbb{C})$ as a covering of degree 2 of the Riemann sphere, and it is a local isomorphism except at $(0:1:0)$ and the three points where $y = 0$. It follows that $\mathbb{C}/\Lambda \to E(\Lambda)(\mathbb{C})$ is an isomorphism outside the two sets of four points, and hence an isomorphism. □

THE ADDITION FORMULA

For a fixed z', $\wp(z + z')$ is a doubly periodic function of z, and therefore a rational function of $\wp$ and $\wp'$. The next result exhibits the rational function.

PROPOSITION 3.8 *The following formula holds,*

$$\wp(z+z') = \frac{1}{4}\left(\frac{\wp'(z) - \wp'(z')}{\wp(z) - \wp(z')}\right)^2 - \wp(z) - \wp(z').$$

PROOF. Let $f(z)$ denote the difference of the left and the right sides. Its only possible poles (in a fundamental domain for Λ) are at 0 or $\pm z'$, and by examining the Laurent expansion of $f(z)$ near these points one sees that it has no pole at 0 or $-z'$, and at worst a simple pole at z'. Since it is doubly periodic, it must be constant, and since $f(0) = 0$, it must be identically zero. □

COROLLARY 3.9 *The map* $z \mapsto (\wp(z):\wp'(z):1) : \mathbb{C}/\Lambda \to E(\Lambda)$ *is a homomorphism of groups.*

PROOF. Let $Y = mX + c$ be the line through the points $P = (x, y)$ and $P' = (x', y')$ on the curve $Y^2 = 4X^3 - g_4 X - g_6$. Then the x, x', and $x(P + P')$ are the roots of the polynomial

$$(mX + c)^2 - 4X^3 + g_4 X + g_6,$$

and so

$$x(P + P') + x + x' = \frac{m^2}{4} = \frac{1}{4}\left(\frac{y - y'}{x - x'}\right)^2.$$

This agrees with the formula in the proposition. □

Classification of elliptic curves over $\mathbb{C}$

THEOREM 3.10 *Every elliptic curve* E *over* $\mathbb{C}$ *is isomorphic to* $E(\Lambda)$ *for some lattice* Λ.

PROOF. Recall (II, 2.1c) that, over an algebraically closed field, the elliptic curves are classified up to isomorphism by their j-invariants. For a lattice Λ in $\mathbb{C}$, the curve

$$E(\Lambda) : Y^2 Z = 4X^3 - g_4(\Lambda) X Z^2 - g_4(\Lambda) Z^3$$

has discriminant $\Delta(\Lambda) = g_4(\Lambda)^3 - 27 g_6(\Lambda)^2$ and j-invariant

$$j(\Lambda) = \frac{1728 g_4(\Lambda)^3}{g_4(\Lambda)^3 - 27 g_6(\Lambda)^2}.$$

For $c \in \mathbb{C}^\times$, $g_4(c\Lambda) = c^{-4} g_4(\Lambda)$ and $g_6(c\Lambda) = c^{-6} g_6(\Lambda)$, and so the isomorphism class of $E(\Lambda)$ depends only on Λ up to scaling. Define

$$j(\tau) = j(\mathbb{Z}\tau + \mathbb{Z}).$$

Then, for all $\begin{pmatrix} a & b \\ c & d \end{pmatrix} \in \mathrm{SL}_2(\mathbb{Z})$,

$$j\left(\frac{a\tau+b}{c\tau+d}\right) = j(\tau).$$

In V, 2.2, we shall show that $j : \mathbb{H} \to \mathbb{C}$ is surjective, which completes the proof. $\square$

Let k be a subfield of $\mathbb{C}$. The following diagram summarizes the situation,

$$\begin{array}{ccccccc}
\dfrac{\{\text{Elliptic curves}/\mathbb{C}\}}{\approx} & \xleftrightarrow{1:1} & \mathcal{L}/\mathbb{C}^{\times} & \xleftrightarrow{1:1} & \mathrm{SL}_2(\mathbb{Z})\backslash\mathbb{H} & \xrightarrow{j} & \mathbb{C} \\
\uparrow & & & & & & \uparrow \\
\dfrac{\{\text{Elliptic curves}/k\}}{\approx} & & & \xrightarrow{\qquad\qquad j \qquad\qquad} & & & k.
\end{array}$$

The upper j is a bijection (V, 2.2) and the lower j is surjective (II, 2.3). If k is algebraically closed, then both maps j are bijective.

REMARK 3.11 The above picture can be made more precise. Consider the isomorphism

$$z \mapsto (\wp(z) : \wp'(z) : 1) : \mathbb{C}/\Lambda \to E(\mathbb{C}).$$

Since $x = \wp(z)$ and $y = \wp'(z)$,

$$\frac{dx}{y} = \frac{\wp'(z)dz}{\wp'(z)} = dz.$$

Thus the differential one-form dz on $\mathbb{C}$ corresponds to the differential one-form $\frac{dx}{y}$ on $E(\mathbb{C})$. Conversely, from a holomorphic differential one-form ω on $E(\mathbb{C})$ we can obtain an realization of E as a quotient $\mathbb{C}/\Lambda$ as follows. For $P \in E(\mathbb{C})$, consider $\varphi(P) = \int_O^P \omega \in \mathbb{C}$. This is a not well defined because it depends on the choice of a path from O to P. However, if we choose a $\mathbb{Z}$-basis (γ_1, γ_2) for $H_1(E(\mathbb{C}), \mathbb{Z})$, and set $\omega_1 = \int_{\gamma_1} \omega$, $\omega_2 = \int_{\gamma_2} \omega$, then $\Lambda = \mathbb{Z}\omega_1 + \mathbb{Z}\omega_2$ is a lattice in $\mathbb{C}$, and $P \mapsto \varphi(P)$ is an isomorphism $E(\mathbb{C}) \to \mathbb{C}/\Lambda$. In this way, we obtain a natural one-to-one correspondence between $\mathcal{L}$ and the set of isomorphism classes of pairs (E, ω) consisting of an elliptic curve E over $\mathbb{C}$ and a holomorphic differential one-form ω on E,

$$\begin{array}{ccc}\{(E,\omega) \text{ up to isomorphism}\} & \xleftrightarrow{1:1} & \mathcal{L} \\ \downarrow & & \downarrow \\ \{E \text{ up to isomorphism}\} & \xleftrightarrow{1:1} & \mathcal{L}/\mathbb{C}^{\times}.\end{array}$$

Torsion points

Let E be an elliptic curve over $\mathbb{C}$. From an isomorphism $E(\mathbb{C}) \simeq \mathbb{C}/\Lambda$, we obtain an isomorphism

$$E(\mathbb{C})_n \simeq \operatorname{Ker}(n\colon \mathbb{C}/\Lambda \to \mathbb{C}/\Lambda) = \tfrac{1}{n}\Lambda/\Lambda.$$

Because of this description, torsion points on elliptic curves are often called ***division points.*** If $\Lambda = \mathbb{Z}\omega_1 + \mathbb{Z}\omega_2$, then

$$\begin{aligned}\tfrac{1}{n}\Lambda/\Lambda &= \frac{\{\frac{a}{n}\omega_1 + \frac{b}{n}\omega_2 \mid a,b \in \mathbb{Z}\}}{\mathbb{Z}\omega_1 + \mathbb{Z}\omega_2} \\ &\simeq \tfrac{1}{n}\mathbb{Z}/\mathbb{Z} \oplus \tfrac{1}{n}\mathbb{Z}/\mathbb{Z} \\ &\simeq \mathbb{Z}/n\mathbb{Z} \oplus \mathbb{Z}/n\mathbb{Z},\end{aligned}$$

and so $E(\mathbb{C})_n$ is a free $\mathbb{Z}/n\mathbb{Z}$-module of rank 2.

THEOREM 3.12 *Let E be an elliptic curve over an algebraically closed field k of characteristic zero. For all $n \geq 1$, $E(k)_n$ is a free $\mathbb{Z}/n\mathbb{Z}$-module of rank 2.*

PROOF. If there exists an embedding of k into $\mathbb{C}$, this follows from the next proposition. In general, there exists an algebraically closed subfield k_0 of finite transcendence degree over $\mathbb{Q}$ such that E arises from a curve E_0 over k_0 (e.g., take k_0 to be the algebraic closure in k of the subfield generated over $\mathbb{Q}$ by the coefficients of the equation defining E). Now k_0 can be embedded into $\mathbb{C}$, and we can apply the next proposition (twice). □

PROPOSITION 3.13 *Let $k \subset \Omega$ be algebraically closed fields, and let E be an elliptic curve over k. The map $E(k) \to E(\Omega)$ induces an isomorphism on the torsion subgroups.*

PROOF. The first lemma below shows that the elements of $E(k)_n$ are the common solutions of equations

$$\begin{cases} Y^2 = X^3 + aX + b \\ \psi_n(X,Y) = 0 \end{cases}$$

where ψ_n is a polynomial not divisible by $Y^2 - X^3 - aX - b$ and so we can apply the second lemma below. □

LEMMA 3.14 *Let E be the elliptic curve $Y^2Z = X^3 + aXZ^2 + bZ^3$ over a field of characteristic zero. Let*

$$\begin{aligned} &\psi_1 = 1, \quad \psi_2 = 2Y, \\ &\psi_3 = 3X^4 + 6aX^2 + 12bX - a^2 \\ &\psi_4 = 4Y(X^6 + 5aX^4 + 20bX^3 - 5a^2X^2 - 4abX - 8b^2 - a^3) \end{aligned}$$

and, inductively,

$$\begin{aligned} Y\psi_{2n} &= \psi_n\left(\psi_{n-1}^2\psi_{n+2} - \psi_{n+1}^2\psi_{n-2}\right) \\ \psi_{2n+1} &= \psi_n^3\psi_{n+2} - \psi_n^3\psi_{n-1}. \end{aligned}$$

Then, for any point $P = (x\colon y\colon 1)$ of E,

$$nP = (2X\psi_n^4 - 2\psi_{n-1}\psi_n^2\psi_{n+1} : \psi_{2n} : 2\psi_n^4).$$

PROOF. When $k = \mathbb{C}$, these are standard formulas from the theory of the Weierstrass $\wp$ function (Weber, H., Lehrbuch der Algebra, III, §58, 1908). For an elegant algebraic proof, see Cassels 1991, Formulary. □

LEMMA 3.15 *Let $k \subset \Omega$ be algebraically closed fields. If $F, G \in k[X,Y]$ have no common factor, then every common solution of the equations*

$$\begin{cases} F(X,Y) = 0 \\ G(X,Y) = 0 \end{cases}$$

with coordinates in Ω has coordinates in k.

PROOF. From the theory of resultants (I, 1.22, 1.24), we know that there exist polynomials $a(X,Y)$, $b(X,Y)$, and $r(X)$ with coefficients in k such that

$$a(X,Y)F(X,Y) + b(X,Y)G(X,Y) = r(X)$$

and $r(x_0) = 0$ if and only if $F(x_0, Y)$ and $G(x_0, Y)$ have a common zero. In other words, the roots of r are the x-coordinates of the common zeros of $F(X,Y)$ and $G(X,Y)$. Since $r(X)$ is a polynomial in one variable, its roots all lie in k. Similarly, for a given $x_0 \in k$, all the common roots of $F(x_0, Y)$ and $G(x_0, Y)$ lie in k. □

REMARK 3.16 Theorem 3.12 holds also when k has characteristic $p \neq 0$ provided p does not divide n (II, 6.3). For a power p^m of p, either $E(k)_{p^m}$ is a free $\mathbb{Z}/p^m\mathbb{Z}$-module of rank 1 (the "ordinary" case) or zero (the "supersingular" case).

Endomorphisms

Let K be a number field. Every $\alpha \in K$ satisfies a monic equation,

$$\alpha^m + a_1\alpha^{m-1} + \cdots + a_m = 0, \quad a_i \in \mathbb{Q}, \quad m = [K:\mathbb{Q}]. \tag{22}$$

If it satisfies such an equation with the $a_i \in \mathbb{Z}$, then α is said to be an ***algebraic integer*** of K. The algebraic integers form a subring $\mathcal{O}_K$ of K. For example, if $K = \mathbb{Q}[\sqrt{d}]$ with $d \in \mathbb{Z}$ and square free, then

$$\mathcal{O}_K = \begin{cases} \mathbb{Z} + \mathbb{Z}\sqrt{d} & d \not\equiv 1 \mod 4 \\ \mathbb{Z} + \mathbb{Z}\frac{1+\sqrt{d}}{2} & d \equiv 1 \mod 4. \end{cases}$$

A subring R of K is an ***order*** in K if it is generated as a $\mathbb{Z}$-module by a $\mathbb{Q}$-basis for K. For example, $\mathcal{O}_K$ is an order, and contains all other orders. The orders in $K = \mathbb{Q}[\sqrt{d}]$ are the subrings $\mathbb{Z}1 + c\mathcal{O}_K$ with c a positive integer.

PROPOSITION 3.17 *Let $\Lambda = \mathbb{Z}\omega_1 + \mathbb{Z}\omega_2$ be a lattice in $\mathbb{C}$ with $\tau = \omega_1/\omega_2$ in $\mathbb{H}$. The ring of endomorphisms of the Riemann surface $\mathbb{C}/\Lambda$ is $\mathbb{Z}$ unless $[\mathbb{Q}[\tau]:\mathbb{Q}] = 2$, in which case it is an order in $\mathbb{Q}[\tau]$.*

PROOF. Suppose that there exists an $\alpha \in \mathbb{C}$, $\alpha \notin \mathbb{Z}$, such that $\alpha\Lambda \subset \Lambda$. Then

$$\begin{aligned} \alpha\omega_1 &= a\omega_1 + b\omega_2 \\ \alpha\omega_2 &= c\omega_1 + d\omega_2, \end{aligned}$$

with $a, b, c, d \in \mathbb{Z}$. On dividing through by ω_2 we obtain the equations

$$\begin{aligned} \alpha\tau &= a\tau + b \\ \alpha &= c\tau + d. \end{aligned}$$

Note that $c \neq 0$.

On eliminating α from the two equations, we find that

$$c\tau^2 + (d-a)\tau + b = 0.$$

Therefore $\mathbb{Q}[\tau]$ is of degree 2 over $\mathbb{Q}$.

On eliminating τ from the two equations, we find that

$$\alpha^2 - (a+d)\alpha - bc = 0.$$

Therefore α is integral over $\mathbb{Z}$, and hence is contained in the ring of integers of $\mathbb{Q}[\tau]$. Hence $\mathbb{Z} \neq \mathrm{End}(\mathbb{C}/\Lambda) \subset \mathcal{O}_{\mathbb{Q}[\tau]}$, which implies that $\mathrm{End}(\mathbb{C}/\Lambda)$ is an order in $\mathcal{O}_{\mathbb{Q}[\tau]}$. □

Note that if $\mathbb{Q}[\tau]$ is a quadratic extension of $\mathbb{Q}$, then it is imaginary quadratic, and so it equals $\mathbb{Q}[\sqrt{-d}]$ for some integer $d > 0$. The endomorphisms of $E(\Lambda) = \mathbb{C}/\Lambda$ are the same whether we regard it as an elliptic curve or a Riemann surface. Therefore, the endomorphism ring of an elliptic curve E over $\mathbb{C}$ is either $\mathbb{Z}$ or an order in an imaginary quadratic field $\mathbb{Q}[\sqrt{-d}]$. In the second case, we say that the curve E has ***complex multiplication.***

EXAMPLE 3.18 We exhibit some endomorphisms not in $\mathbb{Z}$.

(a) Consider

$$E : Y^2Z = X^3 + aXZ^2,$$

and let $i = \sqrt{-1}$ be a primitive 4th root of 1. Then $(x{:}\,y{:}\,z) \mapsto (-x{:}\,iy{:}\,z)$ is an endomorphism of E of order 4, and $\mathrm{End}(E) = \mathbb{Z}[i]$. Note that E has j-invariant 1728.

(b) Consider

$$E : Y^2Z = X^3 + bZ^3,$$

and let $\rho = \frac{-1+\sqrt{-3}}{2}$ be a primitive cube root of 1. Then $(x{:}\,y{:}\,z) \mapsto (\rho x{:}\,y{:}\,z)$ is an endomorphism of E of order 3 of E, and $\mathrm{End}(E) = \mathbb{Z}[\rho]$. Note that E has j-invariant 0.

ASIDE 3.19 A complex number α is said to be ***algebraic*** if it is algebraic over $\mathbb{Q}$, and is otherwise said to be ***transcendental***. There is a general philosophy that a transcendental meromorphic function f should take transcendental values at the algebraic points in $\mathbb{C}$, except at some "special" points where it has interesting "special values". We illustrate this for two functions.

(a) Define $e(z) = e^{2\pi i z}$. If z is algebraic but not rational, then $e(z)$ is transcendental. On the other hand, if $z \in \mathbb{Q}$, then $e(z)$ is algebraic — in fact, it is a root of 1, and $\mathbb{Q}[e(z)]$ is a finite extension of $\mathbb{Q}$ with abelian Galois group. The famous Kronecker–Weber theorem states that every abelian extension of $\mathbb{Q}$ is contained in $\mathbb{Q}[e(\frac{1}{m})]$ for some m.

(b) Let $\tau \in \mathbb{H}$ be algebraic. If τ generates a quadratic extension of $\mathbb{Q}$, then $j(\tau)$ is algebraic, and otherwise $j(\tau)$ is transcendental (the second statement was proved by Siegel in 1949). In fact, when $[\mathbb{Q}[\tau]:\mathbb{Q}] = 2$, one can say much more. Assume that $\mathbb{Z}[\tau]$ is the ring of integers in $K \stackrel{\text{def}}{=} \mathbb{Q}[\tau]$. Then $j(\tau)$ is an algebraic integer, and

$$[\mathbb{Q}[j(\tau)] : \mathbb{Q}] = [K[j(\tau)] : K] = h_K,$$

where h_K is the class number of K. Moreover, $K[j(\tau)]$ is the Hilbert class field of K (the largest abelian extension of K unramified at all primes of K including the infinite primes).

The characteristic polynomial of an endomorphism

For an R-linear endomorphism α of a free module Λ of finite rank over a ring R, we let $\det(\alpha)$ denote the determinant of the matrix of α relative to some basis for Λ. It is independent of the choice of the basis.

LEMMA 3.20 *Let Λ be a free $\mathbb{Z}$-module of finite rank, and let $\alpha\colon \Lambda \to \Lambda$ be a $\mathbb{Z}$-linear map with nonzero determinant.*

(a) *The cokernel of α is finite, with order equal to $|\det(\alpha)|$.*

(b) *The kernel of the map*

$$\tilde{\alpha}\colon (\Lambda \otimes \mathbb{Q})/\Lambda \to (\Lambda \otimes \mathbb{Q})/\Lambda$$

defined by α is finite with order equal to $|\det(\alpha)|$.

PROOF. (a) As α has nonzero determinant, it maps Λ isomorphically onto a subgroup $\alpha(\Lambda)$ of Λ. According to a basic structure theorem, there exist bases $e_1,\ldots,e_m$ and $e'_1,\ldots,e'_m$ for Λ such that

$$\alpha(e_i) = n_i e'_i \quad n_i \in \mathbb{Z}, \quad n_i > 0, \quad i = 1,\ldots,n.$$

Therefore the cokernel of α is finite, with order equal to $n_1 \cdots n_m$. The matrix of α with respect to the bases $e_1,\ldots e_m$ and $e'_1,\ldots e'_m$ is $\mathrm{diag}(n_1,\ldots,n_m)$. As the transition matrix from one basis to the second has determinant ± 1, we see that the

$$\det(\alpha) = \pm\det(\mathrm{diag}(n_1,\ldots,n_m)) = \pm n_1 \cdots n_m.$$

(b) Consider the commutative diagram,

$$\begin{array}{ccccccccc} 0 & \longrightarrow & \Lambda & \longrightarrow & \Lambda\otimes\mathbb{Q} & \longrightarrow & (\Lambda\otimes\mathbb{Q})/\Lambda & \longrightarrow & 0 \\ & & \downarrow{\scriptstyle\alpha} & & \downarrow{\scriptstyle\alpha\otimes 1} & & \downarrow{\scriptstyle\tilde{\alpha}} & & \\ 0 & \longrightarrow & \Lambda & \longrightarrow & \Lambda\otimes\mathbb{Q} & \longrightarrow & (\Lambda\otimes\mathbb{Q})/\Lambda & \longrightarrow & 0. \end{array}$$

Because $\det(\alpha) \neq 0$, the map $\alpha \otimes 1$ is an isomorphism. Therefore the snake lemma gives an isomorphism

$$\operatorname{Ker}(\tilde{\alpha}) \to \operatorname{Coker}(\alpha),$$

and so (b) follows from (a). □

PROPOSITION 3.21 *The degree of a nonzero endomorphism α of an elliptic curve E with $E(\mathbb{C}) = \mathbb{C}/\Lambda$ is the determinant of α acting on Λ.*

PROOF. We have $E(\mathbb{C})_{\mathrm{tors}} = \mathbb{Q}\Lambda/\Lambda$, where

$$\begin{aligned} \mathbb{Q}\Lambda &= \{r\lambda \in \mathbb{C} \mid r \in \mathbb{Q}, \lambda \in \Lambda\} \\ &= \{z \in \mathbb{C} \mid mz \in \Lambda \text{ some } m \in \mathbb{Z}\} \\ &\simeq \mathbb{Q} \otimes_{\mathbb{Z}} \Lambda. \end{aligned}$$

Let α be a nonzero endomorphism of E, and let n be the order of its kernel on $E(\mathbb{C})$ (which equals its kernel on $E(\mathbb{C})_{\mathrm{tors}}$). Because $\alpha(\mathbb{C})$ is a homomorphism of groups, it is $n:1$, and so α has degree n (see I, 4.24a). Thus the statement follows from the lemma. □

THEOREM 3.22 *Let E be an elliptic curve over $\mathbb{C}$, and let α and β be endomorphisms of E. There exist $r, s, t \in \mathbb{Z}$, depending only on α, β, such that*

$$\deg(m\alpha + n\beta) = rm^2 + smn + tn^2$$

for all $m, n \in \mathbb{Z}$.

PROOF. Choose an isomorphism $E(\mathbb{C}) \approx \mathbb{C}/\Lambda$, and let $A = (a_{ij})$ and $B = (b_{ij})$ be the matrices of α and β acting on Λ relative to some basis. Then

$$\begin{aligned} \deg(m\alpha + n\beta) &= \det(mA + nB) \\ &= m^2 \det(A) + smn + n^2 \det(B) \\ &= m^2 \deg(\alpha) + smn + n^2 \deg(\beta) \end{aligned}$$

where $s = a_{11}b_{22} + a_{22}b_{11} - a_{12}b_{21} - a_{21}b_{12}$. □

Let α be an endomorphism of an elliptic curve E over $\mathbb{C}$. We define the ***characteristic polynomial*** of α to be its characteristic polynomial as an endomorphism of $\Lambda \otimes \mathbb{Q}$. It is the unique polynomial

$$f(T) = T^2 - sT + t \in \mathbb{Z}[T]$$

such that $f(n) = \deg(n - \alpha)$ for all $n \in \mathbb{Z}$. Note that $f(\alpha) = 0$.

Recall (p. 91) that the Tate module $T_\ell E$ of an elliptic curve E over a field k is defined to be $T_\ell E = \varprojlim E(k^{\mathrm{al}})_{\ell^n}$. It is a free $\mathbb{Z}_\ell$-module of rank 2 such that $T_\ell E/\ell^n T_\ell E \simeq E(k)_{\ell^n}$ for all n. If $k = \mathbb{C}$ and $E(\mathbb{C}) = \mathbb{C}/\Lambda$, then

$$E(\mathbb{C})_{\ell^n} = \tfrac{1}{\ell^n}\Lambda/\Lambda \simeq \Lambda/\ell^n\Lambda \simeq \Lambda \otimes (\mathbb{Z}/\ell^n\mathbb{Z}),$$

and so

$$T_\ell E \simeq \Lambda \otimes_{\mathbb{Z}} \mathbb{Z}_\ell.$$

Therefore the characteristic polynomial of α is equal to its characteristic polynomial as an endomorphism of $T_\ell E \otimes_{\mathbb{Z}_\ell} \mathbb{Q}_\ell$. In particular,

$$\deg(\alpha) = \det(T_\ell \alpha).$$

EXERCISE 3.23 (a) Prove that, for all z_1, z_2,

$$\begin{vmatrix} \wp(z_1) & \wp'(z_1) & 1 \\ \wp(z_2) & \wp'(z_2) & 1 \\ \wp(z_1+z_2) & -\wp'(z_1+z_2) & 1 \end{vmatrix} = 0.$$

(b) Compute sufficiently many initial terms for the Laurent expansions of $\wp'(z)$, $\wp'(z)^2$, etc., to verify the equation in Proposition 2.6.

A p-adic analogue: the Tate curve

For an elliptic curve over $\mathbb{C}$, we saw that the choice of a differential one-form ω on E realizes $E(\mathbb{C})$ as the quotient $\mathbb{C}/\Lambda \simeq E(\mathbb{C})$ of $\mathbb{C}$ by the lattice of periods of ω. More precisely, it realizes the Riemann surface E as the quotient of the Riemann surface $\mathbb{C}$ by the action of the discrete group Λ.

For an elliptic curve E over $\mathbb{Q}_p$, there is no similar description of $E(\mathbb{Q}_p)$ because there are no nonzero discrete subgroups of $\mathbb{Q}_p$ (if $\lambda \in \mathbb{Q}_p$, then $p^n\lambda \to 0$ as $n \to \infty$). However, there is an alternative uniformization of elliptic curves over $\mathbb{C}$. Let Λ be the lattice $\mathbb{Z} + \mathbb{Z}\tau$ in $\mathbb{C}$. Then the exponential map $\underline{e}\colon \mathbb{C} \to \mathbb{C}^\times$ sends $\mathbb{C}/\Lambda$ isomorphically onto $\mathbb{C}^\times/q^{\mathbb{Z}}$ where

$q = \underline{e}(\tau)$, and so $\mathbb{C}^{\times}/q^{\mathbb{Z}} \simeq E$ (as Riemann surfaces). If $\mathrm{Im}(\tau) > 0$, then $|q| < 1$, and the elliptic curve E_q is given by the equation

$$Y^2Z + XYZ = X^3 - b_2XZ^2 - b_3Z^3, \tag{23}$$

where

$$\begin{cases} b_2 = 5\sum_{n=1}^{\infty} \dfrac{n^3q^n}{1-q^n} = 5q + 45q^2 + 140q^3 + \cdots \\ b_3 = \sum_{n=1}^{\infty} \dfrac{7n^5+5n^3}{12} \dfrac{q^n}{1-q^n} = q + 23q^2 + 154q^3 + \cdots \end{cases} \tag{24}$$

are power series with integer coefficients. The discriminant and j-invariant of E_q are given by the usual formulas

$$\Delta = q\prod_{n\geq 1}(1-q^n)^{24} \tag{25}$$

$$j(E_q) = \frac{(1+48b_2)^3}{q\prod_{n\geq 1}(1-q^n)^{24}} = \frac{1}{q} + 744 + 196884q + \cdots. \tag{26}$$

Now let q be an element of $\mathbb{Q}_p^{\times}$ with $|q| < 1$. The series (24) converge in $\mathbb{Q}_p$, and Tate showed (in 1959) that (23) is an elliptic curve E_q such that $K^{\times}/q^{\mathbb{Z}} \simeq E_q(K)$ for all finite extensions K of $\mathbb{Q}_p$. It follows from certain power series identities, valid over $\mathbb{Z}$, that the discriminant and j-invariant of E_q are given by (25) and (26). Every $j \in \mathbb{Q}_p^{\times}$ with $|j| > 1$ arises from a q ((26) can be used to express q as a power series in $1/j$ with integer coefficients). The function field of E_q consists of the quotients F/G of Laurent series

$$F = \sum_{-\infty}^{\infty} a_n z^n, \quad G = \sum_{-\infty}^{\infty} b_n z^n, \quad a_n, b_n \in \mathbb{Q}_p,$$

converging for all nonzero z, such that

$$F(qz)/G(qz) = F(z)/G(z).$$

The elliptic curves E over $\mathbb{Q}_p$ with $|j(E)| > 1$ that arise in this way are exactly those whose reduction over $\mathbb{F}_p$ has a node with rational tangents. They are called ***Tate curves***. See Silverman 1994, Chap. V.

NOTES After Hensel introduced the p-adic number field $\mathbb{Q}_p$ in the 1890s, there were attempts to develop a theory of analytic functions over $\mathbb{Q}_p$, the most prominent being that of Krasner. The problem is that every disk D in $\mathbb{Q}_p$ can be written as a disjoint union of arbitrarily many open-closed smaller disks, and so there are too many functions on D that can be represented locally by power series. Once Tate had discovered the uniformization $\mathbb{Q}_p^\times/q^{\mathbb{Z}} = E_q(\mathbb{Q}_p)$, he became persuaded that there should exist a category in which E itself, not just its points, is a quotient; in other words, that there exists a category in which E, as an "analytic space", is the quotient of $\mathbb{Q}_p^\times$, as an "analytic space", by the discrete group $q^{\mathbb{Z}}$. Two years later, Tate constructed the correct category of "rigid analytic spaces", thereby founding a new subject in mathematics "rigid analytic geometry".

Chapter IV

The Arithmetic of Elliptic Curves

The fundamental theorem proved in this chapter is the finite basis theorem.

THEOREM (FINITE BASIS) *For an elliptic curve E over a number field K, the group $E(K)$ is finitely generated.*

In other words, from a finite set of points on E, it is possible to construct all points by drawing tangents and chords.

The theorem was proved by Mordell (1922) when $K = \mathbb{Q}$, and for all number fields by Weil in his thesis (1928). The theorem over $\mathbb{Q}$ is also called Mordell's theorem and over general number fields the Mordell–Weil theorem. The group $E(K)$ is called the ***Mordell–Weil group*** of E and its rank the ***rank*** of E.

The first step in proving the theorem is to prove a weaker result.

THEOREM (WEAK FINITE BASIS) *For every elliptic curve E over a number field K and integer $n \geq 1$, $E(K)/nE(K)$ is finite.*

Clearly, for an abelian group M,

$$M \text{ finitely generated} \implies M/nM \text{ finite for all } n \geq 1,$$

but the converse is false. For example, $(\mathbb{Q}, +)$ has the property that $\mathbb{Q} = n\mathbb{Q}$ for all $n \geq 1$, but it is not finitely generated because the elements of any finitely generated subgroup have a common denominator.

We sketch a proof, in the case $K = \mathbb{Q}$, that the weak theorem implies the full theorem. We define the height of a point P of $E(\mathbb{Q})$ to be its height as an element of $\mathbb{P}^2(\mathbb{Q})$ (see I, 2.3). Assume that $E(\mathbb{Q})/2E(\mathbb{Q})$ is finite, and let $P_1,\ldots,P_s \in E(\mathbb{Q})$ be a set of representatives for the elements of $E(\mathbb{Q})/2E(\mathbb{Q})$. Then each point $Q \in E(\mathbb{Q})$ can be written

$$Q = P_i + 2Q'$$

with $Q' \in E(\mathbb{Q})$. The height h has the property that there exists a constant C such that, if $h(Q) > C$, then $h(Q') < h(Q)$.

Now consider a $Q \in E(\mathbb{Q})$. If $h(Q) > C$, then $h(Q') < h(Q)$. If also $h(Q') > C$, then we repeat the argument with Q'. Continuing in this way, we obtain

$$Q = P_i + 2Q' = P_i + 2(P_{i'} + 2Q'') = \cdots,$$
$$h(Q) > h(Q') > h(Q'') > \cdots.$$

Eventually, $h(Q^{(n)}) \leq C$, and Q will be the sum of a linear combination of the P_i and of $2^n Q^{(n)}$. Thus the P_i, together with the finite set of points of height $\leq C$, generate $E(\mathbb{Q})$.

NOTES (a) The argument in the last paragraph is called "proof by descent". Fermat used it to answer (in the margin of his copy of Diophantus) an old question of Leonardo Pisano: find all rational squares x^2, y^2 such that $x^2 + y^2$ and $x^2 - y^2$ are both squares. If (x,y) is one solution with $y \neq 0$, then Fermat showed that there exists another (x',y') with $0 < |y'| < |y|$, which yields a contradiction if (x,y) is chosen to have $|y|$ as small as possible (Cassels 1966, p. 207). Fermat is usually credited with originating proof by descent, but, in some sense, it goes back to the ancient Greeks. Consider the proof that $Y^2 = 2X^2$ has no solution in integers. Define the height of a pair (m,n) of integers to be $\max(|m|,|n|)$. The usual argument shows that if (m,n) is one solution to the equation, then there exists another of smaller height, which again gives a contradiction.

(b) In his thesis, Weil proved[1] that, for a curve C of *arbitrary* genus over a number field K, the group $\mathrm{Pic}^0(C)$ of divisors of degree 0 modulo linear equivalence is finitely generated. For an elliptic curve, $\mathrm{Pic}^0(C) = C(K)$ (see I, 4.10), and, for a general curve $\mathrm{Pic}^0(C) = J_C(K)$, where J_C is the jacobian variety of C. Thus, once the jacobian variety J_C of a curve C over a number field had been defined, Weil's result showed that $J_C(K)$ is finitely generated. For an arbitrary

[1] Weil's exposition is sometimes sketchy, and he does not use heights At one point, where he writes that "il est aisé de combler cette lacune" (it is easy to fill this gap) and then proposes an inadequate argument, it would today be most natural to use the theory of heights. See Weil's comments on the article in his Collected Papers. Hadamard famously discouraged Weil from publishing his thesis until he had "completed" the proof of Mordell's conjecture.

abelian variety A over a number field K, every curve C on A defines a homomorphism $J_C \to A$, which is surjective if C is sufficiently general (e.g., Milne 1986b, 10.1). It follows that A is a direct factor of J_C (e.g., Milne 1986a, 12.1), and so $A(K)$ is finitely generated. This also is called the Mordell–Weil theorem.

1 Group cohomology

In proving the weak finite basis theorem, and also later in the study of the Tate–Shafarevich group, we shall use a little group cohomology.

Cohomology of groups

Let G be a group, and let M be an abelian group. An ***action*** of G on M is a map $G \times M \to M$ such that

(a) $\sigma(m+m') = \sigma m + \sigma m'$ for all $\sigma \in G$, $m, m' \in M$;

(b) $(\sigma\tau)(m) = \sigma(\tau m)$ for all $\sigma, \tau \in G$, $m \in M$;

(c) $1_G m = m$ for all $m \in M$.

Thus, to give an action of G on M is the same as giving a homomorphism $G \to \mathrm{Aut}(M)$. A ***G-module*** is an abelian group together with an action of G.

EXAMPLE 1.1 Let L be a finite Galois extension of a field K with Galois group G, and let E be an elliptic curve over K. Then L, $L^\times$, and $E(L)$ all have obvious G-actions.

For a G-module M, we define

$$H^0(G, M) = M^G = \{m \in M \mid \sigma m = m, \text{ all } \sigma \in G\}.$$

In the above examples,

$$H^0(G, L) = K, \quad H^0(G, L^\times) = K^\times, H^0(G, E(L)) = E(K).$$

A ***crossed homomorphism*** is a map $f: G \to M$ such that

$$f(\sigma\tau) = f(\sigma) + \sigma f(\tau), \quad \text{all } \sigma, \tau \in G.$$

Note that the condition implies that $f(1) = f(1 \cdot 1) = f(1) + f(1)$, and so $f(1) = 0$. For every $m \in M$, we obtain a crossed homomorphism by putting

$$f(\sigma) = \sigma m - m, \qquad \text{all } \sigma \in G.$$

Such crossed homomorphisms are said to be ***principal***. The sum and difference of two crossed homomorphisms is again a crossed homomorphism, and the sum and difference of two principal crossed homomorphisms is again principal. Thus we can define

$$H^1(G,M) = \frac{\{\text{crossed homomorphisms}\}}{\{\text{principal crossed homomorphisms}\}}$$

(quotient abelian group). There are also cohomology groups $H^n(G,M)$ for $n > 1$, but we will not need them.

EXAMPLE 1.2 When G acts trivially on M, i.e., $\sigma m = m$ for all $\sigma \in G$ and $m \in M$, a crossed homomorphism is simply a homomorphism, and every principal crossed homomorphism is zero. Hence $H^1(G,M) = \mathrm{Hom}(G,M)$.

PROPOSITION 1.3 *Let L be a finite Galois extension of K with Galois group G; then $H^1(G,L^\times) = 0$, i.e., every crossed homomorphism $G \to L^\times$ is principal.*

PROOF. Let f be a crossed homomorphism $G \to L^\times$. In multiplicative notation, this means that

$$f(\sigma\tau) = f(\sigma)\cdot\sigma(f(\tau)), \quad \sigma,\tau \in G,$$

and we have to find a $\gamma \in L^\times$ such that $f(\sigma) = \sigma\gamma/\gamma$ for all $\sigma \in G$. Because the $f(\tau)$ are nonzero, Dedekind's theorem on the independence of characters (FT, 5.14) shows that

$$\sum_{\tau\in G} f(\tau)\tau\colon L \to L$$

is not the zero map, i.e., that there exists an $\alpha \in L$ such that

$$\beta \overset{\text{def}}{=} \sum_{\tau\in G} f(\tau)\tau\alpha \neq 0.$$

But then, for $\sigma \in G$,

$$\begin{aligned}\sigma\beta = \sum_{\tau\in G}\sigma(f(\tau))\cdot\sigma\tau(\alpha) &= \sum_{\tau\in G} f(\sigma)^{-1}\cdot f(\sigma\tau)\cdot\sigma\tau(\alpha)\\ &= f(\sigma)^{-1}\sum_{\tau\in G} f(\sigma\tau)\cdot\sigma\tau(\alpha)\\ &= f(\sigma)^{-1}\beta,\end{aligned}$$

and so $f(\sigma) = \beta/\sigma\beta = \sigma(\beta^{-1})/\beta^{-1}$. □

COROLLARY 1.4 *A point $P = (x_0{:}\cdots{:}x_n) \in \mathbb{P}^n(L)$ is fixed by G if and only if it is represented by an $n+1$-tuple in K.*

PROOF. If $\sigma P = P$ for $\sigma \in G$, then

$$\sigma(x_0,\ldots,x_n) = c(\sigma)(x_0,\ldots,x_n)$$

for some $c(\sigma) \in L^\times$. The map $\sigma \mapsto c(\sigma)$ is a crossed homomorphism, and so $c(\sigma) = \sigma c^{-1}/c^{-1}$ for some $c \in L^\times$. Now

$$\sigma(cx_0,\ldots,cx_n) = (cx_0,\ldots,cx_n),$$

and so the cx_i lie in K. □

PROPOSITION 1.5 *For any exact sequence of G-modules*

$$0 \to M \to N \to P \to 0,$$

there is a canonical exact sequence

$$0 \to H^0(G,M) \to H^0(G,N) \to H^0(G,P) \xrightarrow{\delta} H^1(G,M) \to H^1(G,N) \to H^1(G,P)$$

PROOF. The map δ is defined as follows: let $p \in P^G$; there exists an $n \in N$ mapping to p, and $\sigma n - n \in M$ for all $\sigma \in G$; the map $\sigma \mapsto \sigma n - n{:}\, G \to M$ is a crossed homomorphism, whose class we define to be $\delta(p)$. Another n' mapping to p gives rise to a crossed homomorphism differing from the first by the principal crossed homomorphism $\sigma \mapsto \sigma(n'-n) - (n'-n)$, and so $\delta(p)$ is well-defined. The rest of the proof is routine (and should be written out by all readers unfamiliar with group cohomology). □

Let H be a subgroup of G and M a G-module. A crossed homomorphism f for G restricts to a crossed homomorphism for H, and in this way we get a "restriction" homomorphism

$$\text{Res}\colon H^1(G,M) \to H^1(H,M).$$

Now assume that H has finite index in G, and let S be a set of representatives for G/H, so $G = \bigsqcup_{s \in S} sH$. Let f be a crossed homomorphism for H. Then $\sigma \mapsto \sum_{s\in S} t \cdot f(t^{-1} \cdot \sigma s)$, where t is the unique element of S such that $tH = \sigma s H$, is a crossed homomorphism for G, and in this way we get a "corestriction" homomorphism

$$\text{Cor}\colon H^1(H,M) \to H^1(G,M).$$

The composite $\text{Cor} \circ \text{Res} = (G{:}\,H)$. See Serre 1979, VIII, §2.

PROPOSITION 1.6 *If G has order m, then $mH^1(G,M)=0$.*

PROOF. On taking $H=1$ in the above discussion, we see that multiplication by m on $H^1(G,M)$ factors through $H^1(1,M)=0$. . □

REMARK 1.7 Let H be a normal subgroup of a group G, and let M be a G-module. Then M^H is a G/H-module, and a crossed homomorphism $f\colon G/H\to M^H$ defines a crossed homomorphism $G\to M$ by composition,

$$\begin{array}{ccc} G & \dashrightarrow & M \\ \downarrow & & \uparrow \\ G/H & \xrightarrow{f} & M^H. \end{array}$$

In this way we obtain an "inflation" homomorphism

$$\text{Inf: } H^1(G/H, M^H) \to H^1(G,M),$$

and one verifies easily that the sequence

$$0 \to H^1(G/H, M^H) \overset{\text{Inf}}{\to} H^1(G,M) \overset{\text{Res}}{\to} H^1(H,M)$$

is exact.

Cohomology of infinite Galois groups

Let k be a perfect field and k^{al} an algebraic closure of k. The group G of automorphisms of k^{al} fixing the elements of k has a natural topology, called the ***Krull topology***, for which a subgroup is open if and only if it consists of the automorphisms fixing the elements of some finite extension of k in k^{al}. When endowed with its Krull topology, G is called the ***Galois group*** $\text{Gal}(k^{\text{al}}/k)$ of k^{al} over k. The open subgroups of G form a neighbourhood base for the identity element. As always, open subgroups are closed, and so every intersection of open subgroups is closed; conversely, every closed subgroup is an intersection of open subgroups. The group G is compact, and so open subgroups of G have finite index, and continuous maps from G are uniformly continuous. The usual Galois theory extends to give a one-to-one correspondence between the intermediate fields K, $k\subset K\subset k^{\text{al}}$, and the *closed* subgroups of G, under which the fields of finite degree over k correspond to open subgroups of G. For any finite Galois extension K of

k in k^{al}, the map $\text{Gal}(k^{\text{al}}/k) \to \text{Gal}(K/k)$ is surjective with open normal kernel, and these maps realize $\text{Gal}(k^{\text{al}}/k)$ as an inverse limit

$$\text{Gal}(k^{\text{al}}/k) = \varprojlim \text{Gal}(K/k).$$

See FT, §7, for the details.

A G-module M is said to be ***discrete*** if the map $G \times M \to M$ is continuous relative to the discrete topology on M and the Krull topology on G. This is equivalent to requiring that

$$M = \bigcup\nolimits_H M^H, \quad H \text{ open in } G,$$

i.e., to requiring that every element of M is fixed by the subgroup of G fixing some finite extension of k. For example, k^{al}, $k^{\text{al}\times}$, and $E(k^{\text{al}})$ are all discrete G-modules because

$$k^{\text{al}} = \bigcup K, \quad k^{\text{al}\times} = \bigcup K^\times, \; E(k^{\text{al}}) = \bigcup E(K),$$

where, in each case, the union runs over the finite extensions K of k contained in k^{al}.

When M is discrete, a crossed homomorphism $f: G \to M$ is continuous if and only if it is constant on the cosets of some open normal subgroup H of G, and so arises by inflation from a crossed homomorphism $G/H \to M^H$. Every principal crossed homomorphism is continuous because every element of M is fixed by an open normal subgroup of G.

For an infinite Galois group G and a discrete G-module M, we define $H^1(G, M)$ to be the group of continuous crossed homomorphisms $f: G \to M$ modulo the subgroup of principal crossed homomorphisms. With this definition

$$H^1(G, M) = \varinjlim\nolimits_H H^1(G/H, M^H),$$

where H runs through the open normal subgroups of G. Explicitly, this means that

(a) $H^1(G, M)$ is the union of the images of the inflation maps

$$\text{Inf}: H^1(G/H, M^H) \to H^1(G, M),$$

where H runs over the open normal subgroup of G, and

(b) an element $\gamma \in H^1(G/H, M^H)$ maps to zero in $H^1(G, M)$ if and only if it maps to zero in $H^1(G/H', M^{H'})$ for some open normal subgroup H' of G contained in H.

In particular, Proposition 1.6 shows that the group $H^1(G,M)$ is torsion.

The proofs of the statements in the last three paragraphs are left as an exercise.

EXAMPLE 1.8 (a) Proposition 1.3 shows that

$$H^1(G,k^{\mathrm{al}\times}) = \varinjlim_K H^1(\mathrm{Gal}(K/k),K^{\times}) = 0.$$

(b) For a field L and an integer $n > 1$, let

$$\mu_n(L) = \{\zeta \in L^{\times} \mid \zeta^n = 1\}.$$

From the exact sequence

$$1 \to \mu_n(k^{\mathrm{al}}) \to k^{\mathrm{al}\times} \xrightarrow{n} k^{\mathrm{al}\times} \to 1,$$

we get an exact sequence of cohomology groups

$$1 \to \mu_n(k) \to k^{\times} \xrightarrow{n} k^{\times} \to H^1(G,\mu_n(k^{\mathrm{al}})) \to 1,$$

and hence a canonical isomorphism

$$H^1(G,\mu_n(k^{\mathrm{al}})) \simeq k^{\times}/k^{\times n}.$$

When k is a number field, this group is infinite. For example, the numbers

$$(-1)^{\varepsilon(\infty)} \prod_{p \text{ prime}} p^{\varepsilon(p)},$$

where each exponent is 0 or 1 and all but finitely many are zero, form a set of representatives for the elements of $\mathbb{Q}^{\times}/\mathbb{Q}^{\times 2}$, which is therefore an infinite-dimensional vector space over $\mathbb{F}_2$.

(c) If G acts trivially on M, then $H^1(G,M)$ is the set of continuous homomorphisms $\alpha\colon G \to M$. Because M is discrete, the kernel H of such an α is open, and α defines an injective homomorphism $\mathrm{Gal}(K/k) \to M$, where K is the fixed field of H.

For an elliptic curve E over k, we shorten $H^i(\mathrm{Gal}(k^{\mathrm{al}}/k),E(k^{\mathrm{al}}))$ to $H^i(k,E)$.

1.9 Let E be an elliptic curve over $\mathbb{Q}$, and let $\mathbb{Q}^{\mathrm{al}}$ be the algebraic closure of $\mathbb{Q}$ in $\mathbb{C}$. Choose an extension of the p-adic valuation of $\mathbb{Q}$ to $\mathbb{Q}^{\mathrm{al}}$, and let $\mathbb{Q}_p^{\mathrm{al}}$ be the algebraic closure of $\mathbb{Q}_p$ in the completion of $\mathbb{Q}$ defined by the valuation. Then $\mathbb{Q}_p^{\mathrm{al}}$ is an algebraic closure of $\mathbb{Q}_p$, and we have a commutative diagram

$$\begin{array}{ccc} \mathbb{Q}^{\mathrm{al}} & \hookrightarrow & \mathbb{Q}_p^{\mathrm{al}} \\ \uparrow & & \uparrow \\ \mathbb{Q} & \hookrightarrow & \mathbb{Q}_p. \end{array} \tag{27}$$

The action of $\mathrm{Gal}(\mathbb{Q}_p^{\mathrm{al}}/\mathbb{Q}_p)$ on $\mathbb{Q}^{\mathrm{al}} \subset \mathbb{Q}_p^{\mathrm{al}}$ defines a homomorphism

$$\mathrm{Gal}(\mathbb{Q}_p^{\mathrm{al}}/\mathbb{Q}_p) \to \mathrm{Gal}(\mathbb{Q}^{\mathrm{al}}/\mathbb{Q}).$$

Hence every crossed homomorphism $\mathrm{Gal}(\mathbb{Q}^{\mathrm{al}}/\mathbb{Q}) \to E(\mathbb{Q}^{\mathrm{al}})$ defines (by composition) a crossed homomorphism $\mathrm{Gal}(\mathbb{Q}_p^{\mathrm{al}}/\mathbb{Q}_p) \to E(\mathbb{Q}_p^{\mathrm{al}})$. In this way, we obtain a homomorphism

$$H^1(\mathbb{Q}, E) \to H^1(\mathbb{Q}_p, E),$$

which is independent of the choice of the diagram (27). A similar remark applies to the cohomology groups of μ_n and E_n. See Serre 1979, VII, §5, Proposition 3. In §7 below, we shall give a more natural geometric interpretation of these "localization" homomorphisms.

EXAMPLE 1.10 Let k be a perfect field, and let Γ denote its Galois group. Let C be a curve over k, and let $\bar{C}$ be the corresponding curve over k^{al}. We examine whether $\mathrm{Pic}(C) \stackrel{\text{def}}{=} \mathrm{Pic}(\bar{C})^{\Gamma}$ is the group of divisors on C modulo principal divisors. From the exact sequence

$$0 \to k^{\mathrm{al}}(\bar{C})^{\times}/k^{\mathrm{al}\times} \to \mathrm{Div}(\bar{C}) \to \mathrm{Pic}(\bar{C}) \to 0$$

of Γ-modules, we get an exact cohomology sequence

$$\left(k^{\mathrm{al}}(\bar{C})^{\times}/k^{\mathrm{al}\times}\right)^{\Gamma} \to \mathrm{Div}(\bar{C})^{\Gamma} \to \left(\mathrm{Pic}(\bar{C})\right)^{\Gamma} \to H^1(\Gamma, k^{\mathrm{al}}(\bar{C})^{\times}/k^{\mathrm{al}\times}),$$

and from the exact sequence

$$0 \to k^{\mathrm{al}\times} \to k^{\mathrm{al}}(\bar{C})^{\times} \to k^{\mathrm{al}}(\bar{C})^{\times}/k^{\mathrm{al}\times} \to 0,$$

we get (using 1.8) exact cohomology sequences

$$0 \to k^{\times} \to k(C)^{\times} \to \left(k^{\mathrm{al}}(\bar{C})^{\times}/k^{\mathrm{al}\times}\right)^{\Gamma} \to 0$$
$$0 \to H^{1}(\Gamma, k^{\mathrm{al}}(\bar{C})^{\times}/k^{\mathrm{al}\times}) \to H^{2}(\Gamma, k^{\mathrm{al}\times}).$$

When we use these last sequences to replace terms in the earlier cohomology sequence, and then use that $\mathrm{Div}(\bar{C})^{\Gamma} = \mathrm{Div}(C)$ (see p. 46), we obtain an exact sequence

$$0 \to k^{\times} \to k(C)^{\times} \to \mathrm{Div}(C) \to \mathrm{Pic}(C) \to H^{2}(\Gamma, k^{\mathrm{al}\times}).$$

Now $H^{2}(\Gamma, k^{\mathrm{al}\times})$ is the Brauer group of k (Serre 1979, X, §5). Thus, $\mathrm{Pic}(C)$ is the group of divisor classes on C if the Brauer group of k is zero, for example, if k is finite (Wedderburn's theorem; ibid., §7).

REMARK 1.11 If k is not perfect, then k^{al} is not a Galois extension of k; in particular, k is not the fixed field of the group of automorphisms of k^{al}/k. However, when we replace k^{al} with a separable closure k^{sep} of k, the above discussion is essentially unchanged, except that in 1.8b, we must take n prime to the characteristic of k in order for $k^{\mathrm{sep}\times} \xrightarrow{n} k^{\mathrm{sep}\times}$ to be surjective.

2 The Selmer and Tate–Shafarevich groups

We now set $\mathbb{Q}_{\infty} = \mathbb{R}$.

LEMMA 2.1 *For every elliptic curve E over an algebraically closed field k and integer $n \geq 1$, the map $P \mapsto nP\colon E(k) \to E(k)$ is surjective.*

PROOF. When $k = \mathbb{C}$, this follows from III, 3.10. If k has characteristic zero, then, as in the proof of III, 3.12, we may assume that $k \subset \mathbb{C}$. Given a point $P \in E(k)$, to show that there exists a point Q such that $nQ = P$ amounts to showing that certain polynomials in X and Y have a common zero (see III, 3.14). Because these equations have a solution in $\mathbb{C}$, the polynomials generate a *proper* ideal in $k[X,Y]$, and hence have a solution in k by the Nullstellensatz.

For an arbitrary algebraically closed k, we use that, for every $m \neq \mathrm{char}(k)$, there exists a nonzero point $P \in E(k)$ of order m (see II, 6.3). In particular, there exists a nonzero point $P \in E(k)$ of order m prime to n. Then $nP \neq 0$, and so the regular map $n\colon E \to E$ is nonconstant, hence dominant (I, 4.23a), and hence surjective (I, 4.23b). □

Let E be an elliptic curve over $\mathbb{Q}$. For any integer $n > 1$, we let E_n denote the algebraic subvariety of E such that

$$E_n(\mathbb{Q}^{\mathrm{al}}) = E(\mathbb{Q}^{\mathrm{al}})_n \stackrel{\text{def}}{=} \operatorname{Ker}(E(\mathbb{Q}^{\mathrm{al}}) \xrightarrow{n} \mathbb{Q}^{\mathrm{al}}).$$

From the lemma, we obtain an exact sequence

$$0 \to E_n(\mathbb{Q}^{\mathrm{al}}) \to E(\mathbb{Q}^{\mathrm{al}}) \xrightarrow{n} E(\mathbb{Q}^{\mathrm{al}}) \to 0$$

and an exact cohomology sequence

$$0 \to E_n(\mathbb{Q}) \to E(\mathbb{Q}) \xrightarrow{n} E(\mathbb{Q}) \to H^1(\mathbb{Q}, E_n) \to H^1(\mathbb{Q}, E) \xrightarrow{n} H^1(\mathbb{Q}, E),$$

from which we extract the sequence

$$\boxed{0 \to E(\mathbb{Q})/nE(\mathbb{Q}) \to H^1(\mathbb{Q}, E_n) \to H^1(\mathbb{Q}, E)_n \to 0.} \tag{28}$$

Here, as usual, $H^1(\mathbb{Q}, E)_n$ is the subgroup of $H^1(\mathbb{Q}, E)$ killed by n. If $H^1(\mathbb{Q}, E_n)$ were finite, then we could deduce that $E(\mathbb{Q})/nE(\mathbb{Q})$ is finite, but it need not be. For example, if all the points of order 2 on E have coordinates in $\mathbb{Q}$, so that $\operatorname{Gal}(\mathbb{Q}^{\mathrm{al}}/\mathbb{Q})$ acts trivially on $E_2(\mathbb{Q}^{\mathrm{al}}) \approx (\mathbb{Z}/2\mathbb{Z})^2$, then

$$H^1(\mathbb{Q}, E_2) \approx H^1(\mathbb{Q}, \mu_2 \times \mu_2) \simeq \mathbb{Q}^\times/\mathbb{Q}^{\times 2} \times \mathbb{Q}^\times/\mathbb{Q}^{\times 2},$$

which we saw to be infinite in 1.8b. Instead, we replace $H^1(\mathbb{Q}, E_n)$ with a subgroup (the Selmer group S) that is large enough to contain the image of $E(\mathbb{Q})/nE(\mathbb{Q})$ but which we can show to be finite.

When we regard E as an elliptic curve over $\mathbb{Q}_p$, we obtain a similar exact sequence to (28), and a commutative diagram (see 1.9),

$$\begin{array}{ccccccccc}
0 & \longrightarrow & E(\mathbb{Q})/nE(\mathbb{Q}) & \longrightarrow & H^1(\mathbb{Q}, E_n) & \longrightarrow & H^1(\mathbb{Q}, E)_n & \longrightarrow & 0 \\
 & & \downarrow & & \downarrow & & \downarrow & & \\
0 & \longrightarrow & E(\mathbb{Q}_p)/nE(\mathbb{Q}_p) & \longrightarrow & H^1(\mathbb{Q}_p, E_n) & \longrightarrow & H^1(\mathbb{Q}_p, E)_n & \longrightarrow & 0.
\end{array}$$

If $\gamma \in H^1(\mathbb{Q}, E_n)$ comes from an element of $E(\mathbb{Q})$, then certainly its image γ_p in $H^1(\mathbb{Q}_p, E_n)$ comes from an element of $E(\mathbb{Q}_p)$. This suggests defining

$$\begin{aligned}
S^{(n)}(E/\mathbb{Q}) &= \{\gamma \in H^1(\mathbb{Q}, E_n) \mid \text{for all } p,\ \gamma_p \text{ comes from } E(\mathbb{Q}_p)\} \\
&= \operatorname{Ker}\left(H^1(\mathbb{Q}, E_n) \to \prod_{p=2,3,5,\ldots,\infty} H^1(\mathbb{Q}_p, E) \right).
\end{aligned}$$

The group $S^{(n)}(E/\mathbb{Q})$ is called the ***Selmer group***. In the same spirit, we define the ***Tate–Shafarevich group*** to be

$$Ш(E/\mathbb{Q}) = \operatorname{Ker}\left(H^1(\mathbb{Q}, E) \to \prod_{p=2,3,5,\ldots,\infty} H^1(\mathbb{Q}_p, E)\right).$$

We shall need an elementary lemma.

LEMMA 2.2 *Any pair of homomorphisms of abelian groups*

$$A \xrightarrow{\alpha} B \xrightarrow{\beta} C$$

gives rise to an exact (kernel-cokernel) *sequence*

$$0 \longrightarrow \operatorname{Ker}(\alpha) \longrightarrow \operatorname{Ker}(\beta \circ \alpha) \xrightarrow{\alpha} \operatorname{Ker}(\beta)$$
$$\longrightarrow \operatorname{Coker}(\alpha) \xrightarrow{\beta} \operatorname{Coker}(\beta \circ \alpha) \longrightarrow \operatorname{Coker}(\beta) \longrightarrow 0.$$

PROOF. Prove directly, or apply the snake lemma to the diagram

$$\begin{array}{ccccccc} & & A & \xrightarrow{\alpha} & B & \longrightarrow & \operatorname{Coker}(\alpha) & \longrightarrow & 0 \\ & & \downarrow{\scriptstyle \beta\circ\alpha} & & \downarrow{\scriptstyle \beta} & & \downarrow & & \\ 0 & \longrightarrow & C & \xrightarrow{\text{id}} & C & \longrightarrow & 0 & & \end{array}$$

□

When we apply the lemma to the maps

$$H^1(\mathbb{Q}, E_n) \to H^1(\mathbb{Q}, E)_n \to \prod_{p=2,3,\ldots,\infty} H^1(\mathbb{Q}_p, E)_n,$$

we obtain the fundamental exact sequence

$$\boxed{0 \to E(\mathbb{Q})/nE(\mathbb{Q}) \to S^{(n)}(E/\mathbb{Q}) \to Ш(E/\mathbb{Q})_n \to 0.} \tag{29}$$

We shall prove that $E(\mathbb{Q})/nE(\mathbb{Q})$ is finite by showing that $S^{(n)}(E/\mathbb{Q})$ is finite.

Note that $Ш(E/\mathbb{Q})$ is a torsion group (see p. 142). Later we shall give a geometric interpretation of $Ш(E/\mathbb{Q})$ which shows that it provides a measure of the failure of the Hasse principle for curves of genus 1 over $\mathbb{Q}$. Also, we shall see that these definitions extend to elliptic curves over all number fields.

NOTES The group $H^1(k,E)$ is also called the ***Weil–Châtelet group*** and denoted by $\mathrm{WC}(E/k)$ (Lang and Tate 1958; Tate 1958). The Selmer group was named by Cassels (1962a), who also admits responsibility for denoting the Tate–Shafarevich group by the Cyrillic letter Ш (Cassels 1991, p. 109).[2]

3 The finiteness of the Selmer group

In this section, we prove the following theorem.

THEOREM 3.1 *For every elliptic curve E over $\mathbb{Q}$ and integer $n \geq 1$, the Selmer group $S^{(n)}(E/\mathbb{Q})$ is finite.*

In fact, we prove the statement over an arbitrary number field L.

Preliminaries

LEMMA 3.2 *Let E be an elliptic curve over $\mathbb{Q}_p$ with good reduction, and let n be an integer prime to p. A point P in $E(\mathbb{Q}_p)$ is of the form nQ for some $Q \in E(\mathbb{Q}_p)$ if and only if its image $\bar{P}$ in $\bar{E}(\mathbb{F}_p)$ is of the form $n\bar{Q}$ for some $\bar{Q} \in \bar{E}(\mathbb{F}_p)$.*

PROOF. As $P \mapsto \bar{P}$ is a homomorphism, the necessity is obvious, and the sufficiency follows from a diagram chase in

$$\begin{array}{ccccccccc}
0 & \longrightarrow & E^1(\mathbb{Q}_p) & \longrightarrow & E(\mathbb{Q}_p) & \longrightarrow & \bar{E}(\mathbb{F}_p) & \longrightarrow & 0 \\
 & & \simeq\downarrow n & & \downarrow n & & \downarrow n & & \\
0 & \longrightarrow & E^1(\mathbb{Q}_p) & \longrightarrow & E(\mathbb{Q}_p) & \longrightarrow & \bar{E}(\mathbb{F}_p) & \longrightarrow & 0
\end{array}$$

using that the first vertical arrow is an isomorphism (II, 4.2). In detail, let $P \in E(\mathbb{Q}_p)$ be such that $\bar{P} = n\bar{Q}$ for some $Q \in E(\mathbb{Q}_p)$; then $P - nQ$ maps to zero in $\bar{E}(\mathbb{F}_p)$, and so lies in $E^1(\mathbb{Q}_p)$. Therefore, $P - nQ = nQ'$ for some $Q' \in E^1(\mathbb{Q}_p)$, and so $P = n(Q + Q')$. □

We shall need to use some local algebraic number theory. For a finite extension K of $\mathbb{Q}_p$, the integral closure $\mathcal{O}_K$ of $\mathbb{Z}_p$ in K is a principal ideal domain with a single maximal ideal (π). Thus, $p = \text{unit} \times \pi^e$ for some e, called the ***ramification index*** of K over $\mathbb{Q}_p$. When $e = 1$, so that the maximal ideal in $\mathcal{O}_K$ is (p), then K is said to be ***unramified*** over $\mathbb{Q}_p$.

[2]Which he claims to be his "most lasting contribution to the subject".

LEMMA 3.3 *For any finite extension k of $\mathbb{F}_p$, there exists an unramified extension K of $\mathbb{Q}_p$ of degree $[k:\mathbb{F}_p]$ such that $\mathcal{O}_K/p\mathcal{O}_K = k$.*

PROOF. Let a be a primitive element for k over $\mathbb{F}_p$, and let $f_0(X)$ be the minimum polynomial for a over $\mathbb{F}_p$, so that

$$k = \mathbb{F}_p[a] \simeq F_p[X]/(f_0(X))$$

(FT, 5.1). Let $f(X)$ be any lifting of $f_0(X)$ to a monic polynomial in $\mathbb{Z}_p[X]$. As a is a simple root of f_0, it lifts to a root α of $f(X)$ (see I, 2.13), which lies in $\mathcal{O}_K$. Now $K = \mathbb{Q}_p[\alpha]$ has the required properties. □

REMARK 3.4 Let $K \supset \mathcal{O}_K \to k$ be as in the lemma. Let q be the order of k, so that the elements of k are the roots of $X^q - X$. Then Hensel's lemma (I, 2.12) holds for $\mathcal{O}_K$, and so all the roots of $X^q - X$ in k lift to $\mathcal{O}_K$. Therefore K contains the splitting field of $X^q - X$, and equals it.

Let K be as in the Lemma 3.3. Because $\mathcal{O}_K$ is a principal ideal domain with p as its only prime element (up to units), every element α of $K^\times$ can be written uniquely in the form up^m with $u \in \mathcal{O}_K^\times$ and $m \in \mathbb{Z}$. Define $\mathrm{ord}_p(\alpha) = m$. Then ord_p is a homomorphism $K^\times \to \mathbb{Z}$ extending $\mathrm{ord}_p\colon \mathbb{Q}_p^\times \to \mathbb{Z}$.

Chapter II, §4, holds word-for-word with $\mathbb{Q}_p$ replaced by an unramified extension K, except that now

$$E^0(K)/E^1(K) \simeq \bar{E}^{\mathrm{ns}}(k), \quad E^n(K)/E^{n+1}(K) \simeq k.$$

Therefore, Lemma 3.2 remains valid with $\mathbb{Q}_p$ replaced by K and $\mathbb{F}_p$ by k.

LEMMA 3.5 *Let E be an elliptic curve over $\mathbb{Q}_p$ with good reduction, and let n be an integer prime to p. For each $P \in E(\mathbb{Q}_p)$, there exists a finite unramified extension K of $\mathbb{Q}_p$ such that $P \in nE(K)$.*

PROOF. Let $P \in E(\mathbb{Q}_p)$. Then $\bar{P} \in n\bar{E}(k)$ for some finite extension k of $\mathbb{F}_p$ (see 2.1). Let K be the unramified extension of $\mathbb{Q}_p$ with residue field k (see 3.3). Then $P \in nE(K)$ by Lemma 3.2 with K for $\mathbb{Q}_p$. □

PROPOSITION 3.6 *Let E be an elliptic curve over $\mathbb{Q}$. For every $\gamma \in S^{(n)}(E/\mathbb{Q})$ and prime p of good reduction not dividing n, there exists a finite unramified extension K of $\mathbb{Q}_p$ such that γ maps to zero in $H^1(K, E_n)$.*

PROOF. From the definition of the Selmer group, we know that there exists a $P \in E(\mathbb{Q}_p)$ mapping to the image γ_p of γ in $H^1(\mathbb{Q}_p, E_n)$. As E has good reduction at p, there exists an unramified extension K of $\mathbb{Q}_p$ such that $P \in nE(K)$. It follows that γ_p maps to zero in $H^1(K, E_n)$,

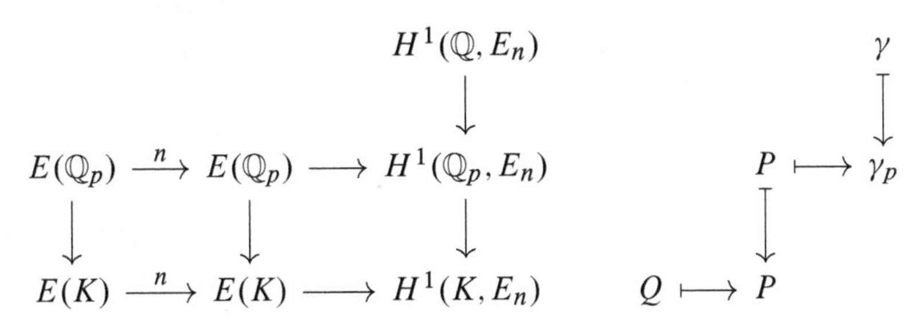

□

Proof of the finiteness in a special case

We prove that $S^{(2)}(E/\mathbb{Q})$ is finite in the case that the points of order 2 on E have coordinates in $\mathbb{Q}$. This condition means that the equation for E has the form

$$Y^2Z = (X - \alpha Z)(X - \beta Z)(X - \gamma Z), \quad \alpha, \beta, \gamma \in \mathbb{Q}$$

(I, 3.4). It implies that

$$E_2(\mathbb{Q}^{\mathrm{al}}) \simeq E_2(\mathbb{Q}) \approx (\mathbb{Z}/2\mathbb{Z})^2 \simeq (\mu_2)^2,$$

all with the trivial action of $\mathrm{Gal}(\mathbb{Q}^{\mathrm{al}}/\mathbb{Q})$, and so

$$H^1(\mathbb{Q}, E_2) \approx H^1(\mathbb{Q}, \mu_2)^2 \simeq (\mathbb{Q}^\times/\mathbb{Q}^{\times 2})^2$$

(1.8b). The isomorphisms depend only on the choice of the isomorphism $E_2(\mathbb{Q}) \approx (\mathbb{Z}/2\mathbb{Z})^2$, which we fix.

Let $\gamma \in S^{(2)}(E/\mathbb{Q}) \subset H^1(\mathbb{Q}, E_2)$. For each prime p_0 not dividing 2Δ, there exists by 3.6 a finite unramified extension K of $\mathbb{Q}_{p_0}$ such that γ_{p_0} maps to zero in $H^1(K, E_2)$,

$$\begin{array}{ccccc}
\gamma & & \gamma_{p_0} & & 0 \\
H^1(\mathbb{Q}, E_2) & \longrightarrow & H^1(\mathbb{Q}_{p_0}, E_2) & \longrightarrow & H^1(K, E_2) \\
{\scriptstyle 1.8}\downarrow\simeq & & {\scriptstyle 1.8}\downarrow\simeq & & {\scriptstyle 1.8}\downarrow\simeq \\
(\mathbb{Q}^\times/\mathbb{Q}^{\times 2})^2 & \longrightarrow & (\mathbb{Q}_{p_0}^\times/\mathbb{Q}_{p_0}^{\times 2}) & \longrightarrow & (K^\times/K^{\times 2})^2.
\end{array}$$

Suppose that γ corresponds to

$$\left((-1)^{\varepsilon(\infty)}\prod_p p^{\varepsilon(p)}, (-1)^{\varepsilon'(\infty)}\prod_p p^{\varepsilon'(p)}\right), \quad 0 \le \varepsilon(p), \varepsilon'(p) \le 1,$$

under the first isomorphism. Then γ_{p_0} corresponds to $(p_0^{\varepsilon(p_0)}, p_0^{\varepsilon'(p_0)})$ under the second isomorphism, and this maps to zero in $(K^\times/K^{\times 2})^2$. As p_0 is a prime element of the ring of integers of K, this implies that $\varepsilon(p_0) = \varepsilon'(p_0) = 0$. Since this holds for all p_0 not dividing 2Δ, there are only finitely many possibilities for γ.

REMARK 3.7 It is possible to prove that $E(\mathbb{Q})/2E(\mathbb{Q})$ is finite in this case without mentioning cohomology groups. Consider an elliptic curve

$$Y^2Z = (X - \alpha Z)(X - \beta Z)(X - \gamma Z), \quad \alpha, \beta, \gamma \in \mathbb{Z}.$$

Define $\varphi_\alpha\colon E(\mathbb{Q})/2E(\mathbb{Q}) \to \mathbb{Q}^\times/\mathbb{Q}^{\times 2}$ by

$$\varphi_\alpha((x{:}y{:}z)) = \begin{cases} (x/z - \alpha)\mathbb{Q}^{\times 2} & z \neq 0, \quad x \neq \alpha z; \\ (\alpha - \beta)(\alpha - \gamma)\mathbb{Q}^{\times 2} & z \neq 0, \quad x = \alpha z \\ \mathbb{Q}^\times & (x{:}y{:}z) = (0{:}1{:}0). \end{cases}$$

Define φ_β similarly. One can prove directly that φ_α and φ_β are homomorphisms, that the kernel of $(\varphi_\alpha, \varphi_\beta)\colon E(\mathbb{Q}) \to (\mathbb{Q}^\times/\mathbb{Q}^{\times 2})^2$ is $2E(\mathbb{Q})$, and that $\varphi_\alpha(P)$ and $\varphi_\beta(P)$ are represented by $\pm$ a product of primes dividing 2Δ (see Cassels 1991, Chap. 15).

Proof of the finiteness in the general case: overview

We saw in Chapter I that $\mathbb{Q}$ has one valuation, hence one completion $\mathbb{Q}_p$, for each nonzero prime ideal (p) in $\mathbb{Z}$ and one other completion $\mathbb{R}$, which it is convenient to denote $\mathbb{Q}_\infty$. Similarly, a number field L has one valuation, hence completion, for each prime ideal of $\mathcal{O}_L$ and one valuation for each embedding of L into $\mathbb{R}$ or complex-conjugate pair of embeddings of L into $\mathbb{C}$. Write $\mathcal{P}(p)$ for the set of valuations of L extending $|\cdot|_p$. Then (ANT, 8.2),

$$L \otimes_{\mathbb{Q}} \mathbb{Q}_p \simeq \prod_{v \in \mathcal{P}(p)} L_v,$$

where L_v is the completion of L for v. Let $\mathcal{P} = \bigcup_{p=2,3,\ldots,\infty} \mathcal{P}(p)$.

For an elliptic curve E over a number field L, we define

$$S^{(n)}(E/L) = \operatorname{Ker}\left(H^1(L, E_n) \to \prod_{v \in \mathcal{P}} H^1(L_v, E)\right).$$

The next lemma shows that, in proving $S^{(n)}(E/L)$ finite, we may replace L with a larger field.

LEMMA 3.8 *Let E be an elliptic curve over a number field L, and let L' be a finite Galois extension of L. For all $n \geq 1$, the kernel of*

$$S^{(n)}(E/L) \to S^{(n)}(E/L')$$

is finite.

PROOF. As $S^{(n)}(E/L)$ and $S^{(n)}(E/L')$ are subgroups of $H^1(L, E_n)$ and $H^1(L', E_n)$ respectively, it suffices to prove that the kernel of

$$H^1(L, E_n) \to H^1(L', E_n)$$

is finite. But (cf. 1.7), this kernel is $H^1(\mathrm{Gal}(L'/\mathbb{Q}), E_n(L'))$, which is finite because both $\mathrm{Gal}(L'/\mathbb{Q})$ and $E_n(L')$ are finite. □

In the proof of the finiteness in the special case, we used the following facts:

(A) $\mathbb{Q}$ contains a primitive square root of 1;

(B) $E(\mathbb{Q})_2 = E(\mathbb{Q}^{\mathrm{al}})_2$ (assumption on E);

(C) for any finite set T of prime numbers, the kernel of

$$r \mapsto (\mathrm{ord}_p(r) \bmod 2) \colon \mathbb{Q}^\times/\mathbb{Q}^{\times 2} \to \bigoplus_{p \notin T} \mathbb{Z}/2\mathbb{Z}$$

is finite.

Let E be an elliptic curve over a number field L. In proving that $S^{(n)}(E/L)$ is finite, Lemma 3.8 allows us to assume that L contains a primitive nth root of 1 and that $E(L)$ contains all points of order n in $E(\mathbb{Q}^{\mathrm{al}})$. In Proposition 3.13 below, we show that the analogue of (C) for L and n follows from the three basic theorems proved in every course on algebraic number theory. Now, the proof of finiteness in the special case carries over to the case of E/L.

Review of algebraic number theory

In this section, L is a finite extension of $\mathbb{Q}$ and $\mathcal{O}_L$ is the ring of all algebraic integers in L (integral closure of $\mathbb{Z}$ in L); see p. 128.

Every element of $\mathcal{O}_L$ is a product of irreducible (i.e., "unfactorable") elements, but this factorization may not be unique. For example, in $\mathbb{Z}[\sqrt{-5}]$ we have

$$6 = 2\cdot 3 = (1+\sqrt{-5})(1-\sqrt{-5})$$

and 2, 3, $1+\sqrt{-5}$, $1-\sqrt{-5}$ are irreducible with no two differing by a unit. The idea of Kummer and Dedekind to remedy this problem was to enlarge the set of numbers with "ideal numbers", now called ideals, to recover unique factorization. For ideals $\mathfrak{a}$ and $\mathfrak{b}$, the set of finite sums $\sum a_i b_i$ with $a_i \in \mathfrak{a}$ and $b_i \in \mathfrak{b}$ is an ideal, denoted $\mathfrak{ab}$. The ring $\mathcal{O}_L$ is a unique factorization domain if and only if it is a principal ideal domain.

THEOREM 3.9 (DEDEKIND) *Every ideal in $\mathcal{O}_L$ can be written uniquely as a product of prime ideals.*

PROOF. See ANT, 3.7, 3.29, or Samuel 2008, §3.4. □

For example, in $\mathbb{Z}[\sqrt{-5}]$,

$$(6) = (2, 1+\sqrt{-5})^2(3, 1+\sqrt{-5})(3, 1-\sqrt{-5}).$$

For an element $a \in \mathcal{O}_L$ and a prime ideal $\mathfrak{p}$ in $\mathcal{O}_L$, let $\mathrm{ord}_{\mathfrak{p}}(a)$ denote the exponent of $\mathfrak{p}$ in the unique factorization of the ideal (a), so that

$$(a) = \prod_{\mathfrak{p}} \mathfrak{p}^{\mathrm{ord}_{\mathfrak{p}}(a)}.$$

For $x = \frac{a}{b} \in L$, define $\mathrm{ord}_{\mathfrak{p}}(x) = \mathrm{ord}_{\mathfrak{p}}(a) - \mathrm{ord}_{\mathfrak{p}}(b)$. The ***ideal class group*** C of $\mathcal{O}_L$ (or L) is defined to be the cokernel of the homomorphism

$$\begin{array}{ccccccc} L^{\times} & \longrightarrow & \bigoplus_{\mathfrak{p}\subset\mathcal{O}_L,\,\mathfrak{p}\text{ prime}} \mathbb{Z} & \longrightarrow & C & \longrightarrow & 0 \\ x & \longmapsto & (\mathrm{ord}_{\mathfrak{p}}(x)). & & & & \end{array}$$

Theorem 3.9 allows us to identify the monoid of ideals in $\mathcal{O}_L$ with $\bigoplus_{\mathfrak{p}} \mathbb{N} \subset \bigoplus_{\mathfrak{p}} \mathbb{Z}$. Every element of C is represented by a nonzero ideal in $\mathcal{O}_L$ and two nonzero ideals $\mathfrak{a}, \mathfrak{b}$ in $\mathcal{O}_L$ represent the same element of C if and only if $(a)\mathfrak{a} = (b)\mathfrak{b}$ for some $a, b \in \mathcal{O}_L$. The group $C = 0$ if and only if $\mathcal{O}_L$ is a principal ideal domain, and so C measures the failure of factorization in $\mathcal{O}_L$ to be unique.

THEOREM 3.10 (FINITENESS OF THE CLASS NUMBER) *The ideal class group C is finite.*

PROOF. The numerical norm $\mathbb{N}\mathfrak{a}$ of an ideal $\mathfrak{a}$ of $\mathcal{O}_L$ is defined to be the index $(\mathcal{O}_L:\mathfrak{a})$. If $\mathfrak{a}=\prod \mathfrak{p}_i^{r_i}$, then $\mathbb{N}\mathfrak{a}=\prod(\mathbb{N}\mathfrak{p})^{r_i}$. From this, and the fact that only finitely many prime ideals $\mathfrak{p}$ of $\mathcal{O}_L$ lie over each prime ideal (p) of $\mathbb{Z}$, we see that, for any real number B, there exist only finitely ideals $\mathfrak{a}$ with $\mathbb{N}\mathfrak{a}\leq B$. To prove the finiteness of C, one shows (more precisely) that every element of C is represented by an ideal with

$$\mathbb{N}\mathfrak{a}\leq \frac{n!}{n^n}\left(\frac{4}{\pi}\right)^s |\Delta_L|^{\frac{1}{2}},$$

where $n=[L:\mathbb{Q}]$, $2s$ is the number of nonreal complex embeddings of L, and Δ_L is the discriminant of $\mathcal{O}_L$ over $\mathbb{Z}$. See ANT, 4.3, or Samuel 2008, §4.3. □

We next need to understand the group of units $\mathcal{O}_L^\times$ of $\mathcal{O}_L$. The group $\mathbb{Z}^\times=\{\pm 1\}$, but already for the ring of integers $\mathbb{Z}[\sqrt{2}]$ in $\mathbb{Q}[\sqrt{2}]$, the group is infinite because $\sqrt{2}+1$ is a unit in $\mathbb{Z}[\sqrt{2}]$,

$$(\sqrt{2}+1)(\sqrt{2}-1)=1.$$

In fact

$$\mathbb{Z}[\sqrt{2}]^\times=\{\pm(1+\sqrt{2})^n \mid n\in\mathbb{Z}\}\simeq \mathbb{Z}/2\mathbb{Z}\oplus\mathbb{Z}.$$

THEOREM 3.11 (DEDEKIND UNIT THEOREM) *The group $\mathcal{O}_L^\times$ of units of $\mathcal{O}_L$ is finitely generated.*

PROOF. The group $\mu(L)$ of roots of 1 in L is finite (because, for all $n\geq 1$, the nth cyclotomic polynomial is irreducible over $\mathbb{Q}$; FT, 5.9) and cyclic (because, for all $n\geq 1$, it contains at most n elements of order dividing n).

Choose an isomorphism

$$\sigma: L\otimes_{\mathbb{Q}}\mathbb{R}\to\mathbb{R}^r\times\mathbb{C}^s,\quad \alpha\otimes 1\mapsto(\sigma_1\alpha,\ldots,\sigma_r\alpha,\sigma_{r+1}\alpha,\ldots,\sigma_{r+s}\alpha).$$

This gives a homomorphism

$$\mathcal{O}_L^\times\to\mathbb{R}^{r+s},\quad \alpha\mapsto(\log|\sigma_1\alpha|,\ldots,\log|\sigma_r\alpha|,\log|\sigma_{r+1}\alpha|,\ldots,\log|\sigma_{r+s}\alpha|)$$

One proves the theorem by showing that this homomorphism has kernel the finite subgroup $\mu(L^\times)$ of $L^\times$ and image a full lattice in the hyperplane

$$H: x_1+\cdots+x_r+2x_{r+1}+\cdots+2x_{r+s}$$

of $\mathbb{R}^{r+s}$, so that

$$\mathcal{O}_L^\times \approx \mu(L) \times \mathbb{Z}^{r+s-1}.$$

See ANT, Chapter 5, or Samuel 2008, §4.4. □

An element a of $\mathcal{O}_L$ is a unit if and only if $(a) = \mathcal{O}_L$. This is equivalent to saying that $\mathrm{ord}_{\mathfrak{p}}(a) = 0$ for all prime ideals $\mathfrak{p}$, and so we have an exact sequence

$$0 \to U \to L^\times \to \bigoplus\nolimits_{\mathfrak{p}} \mathbb{Z} \to C \to 0$$

with U finitely generated and C finite. More generally, we have the following statement.

COROLLARY 3.12 *Let T be a finite set of prime ideals of $\mathcal{O}_L$, and define U_T and C_T by the exact sequence*

$$0 \longrightarrow U_T \longrightarrow L^\times \longrightarrow \bigoplus\nolimits_{\mathfrak{p}\notin T} \mathbb{Z} \longrightarrow C_T \longrightarrow 0$$

$$x \longmapsto (\mathrm{ord}_{\mathfrak{p}}(x)).$$

Then U_T is finitely generated and C_T is finite.

PROOF. The kernel-cokernel exact sequence (2.2) of

$$L^\times \longrightarrow \bigoplus\nolimits_{\text{all } \mathfrak{p}} \mathbb{Z} \xrightarrow{\text{project}} \bigoplus\nolimits_{\mathfrak{p}\notin T} \mathbb{Z}$$

is an exact sequence

$$0 \to U \to U_T \to \bigoplus\nolimits_{\mathfrak{p}\in T} \mathbb{Z} \to C \to C_T \to 0.$$

□

Proof of the finiteness in the general case: completion

We first generalize the statement (C), p. 151.

PROPOSITION 3.13 *For an algebraic number field L and $n \geq 1$, let N denote the kernel of*

$$a \mapsto (\mathrm{ord}_{\mathfrak{p}}(a) \bmod n) \colon L^\times / L^{\times n} \to \bigoplus\nolimits_{\mathfrak{p}\notin T} \mathbb{Z}/n\mathbb{Z}.$$

Then there is an exact sequence

$$0 \to U_T / U_T^n \to N \to (C_T)_n \to 0.$$

PROOF. Let $P = \operatorname{Ker}(\bigoplus_{\mathfrak{p}\notin T}\mathbb{Z} \to C_T)$. The required sequence is the kernel-cokernel sequence of the pair of maps

$$L^\times/L^{\times n} \xrightarrow{a} P/nP \xrightarrow{b} \bigoplus\nolimits_{\mathfrak{p}\notin T}\mathbb{Z}/n\mathbb{Z}.$$

To compute the kernels and cokernels of a and b, apply the snake lemma to the diagrams

$$\begin{array}{ccccccc} & U_T & \to & L^\times & \to & P & \to 0 \\ & \downarrow n & & \downarrow n & & \downarrow n & \\ 0 \to & U_T & \to & L^\times & \to & P & \end{array} \qquad \begin{array}{ccccccc} & P & \to & \bigoplus_{\mathfrak{p}\notin T}\mathbb{Z} & \to & C_T & \to 0 \\ & \downarrow n & & \downarrow n & & \downarrow n & \\ 0 \to & P & \to & \bigoplus_{\mathfrak{p}\notin T}\mathbb{Z} & \to & C_T. & \end{array}$$

□

COROLLARY 3.14 *The group N is finite.*

PROOF. This follows from 3.12 and 3.13. □

Now the same argument as in the special case proves that $S^{(n)}(E/L)$ is finite. After Lemma 3.8, we may suppose that L contains a primitive nth root of 1 and that $E(L)_n = E(L^{\mathrm{al}})_n$, so that

$$H^1(L, E_n) \simeq L^\times/L^{\times n} \times L^\times/L^{\times n}. \tag{30}$$

Let T be the (finite) set of prime ideals of $\mathcal{O}_L$ not prime to n or where E has bad reduction. Let $\gamma \in S^{(n)}(E/L)$. For each $\mathfrak{p} \notin T$, we have a commutative diagram

$$\begin{array}{ccccc} H^1(L, E_n) & \longrightarrow & H^1(L_\mathfrak{p}, E_n) & \longrightarrow & H^1(L'_\mathfrak{p}, E_n) \\ \downarrow \simeq & & \downarrow \simeq & & \downarrow \simeq \\ (L^\times/L^{\times n})^2 & \longrightarrow & (L_\mathfrak{p}^\times/L_\mathfrak{p}^{\times n})^2 & \longrightarrow & (L_\mathfrak{p}'^\times/L_\mathfrak{p}'^{\times n})^2 \\ & & \downarrow (\mathrm{ord}_\mathfrak{p})^2 & & \downarrow (\mathrm{ord}_\mathfrak{p})^2 \\ & & (\mathbb{Z}/n\mathbb{Z})^2 & = & (\mathbb{Z}/n\mathbb{Z})^2 \end{array}$$

where $L'_\mathfrak{p}$ is an unramified extension of $L_\mathfrak{p}$ such that the γ maps to zero in $H^1(L'_\mathfrak{p}, E_n)$ (cf. 3.6). It follows that, under the isomorphism (30), γ maps into the finite group $N \times N$.

ASIDE 3.15 The above proof of the finiteness of the Selmer group is simpler than the standard proof, which unnecessarily "translates the putative finiteness of $E(L)/nE(L)$ into a statement about certain field extensions of L." However, other arguments may be more obviously algorithmic.

4 Heights; proof of the finite basis theorem

Let $P = (a_0 : \ldots : a_n) \in \mathbb{P}^n(\mathbb{Q})$. We call $(a_0, \ldots, a_n)$ a ***primitive representative*** for P if

$$a_0, \ldots, a_n \in \mathbb{Z}, \quad \gcd(a_0, \ldots, a_n) = 1.$$

The ***height*** $H(P)$ of P is then defined to be

$$H(P) = \max_i |a_i|,$$

where $| * |$ is the usual absolute value. The ***logarithmic height*** $h(P)$ of P is defined to be $\log H(P)$.

Heights on $\mathbb{P}^1$

Let $F(X,Y)$ and $G(X,Y)$ be homogeneous polynomials of degree m in $\mathbb{Q}[X,Y]$ with no common zeros in $\mathbb{P}^1(\mathbb{Q}^{\text{al}})$. Then F and G define a map

$$\varphi : \mathbb{P}^1(\mathbb{Q}) \to \mathbb{P}^1(\mathbb{Q}), \quad (x : y) \mapsto (F(x,y) : G(x,y)).$$

LEMMA 4.1 *There exists a constant B such that*

$$|h(\varphi(P)) - mh(P)| \le B, \quad \text{all } P \in \mathbb{P}^1(\mathbb{Q}).$$

PROOF. Since multiplying F and G by a nonzero constant does not change φ, we may suppose that F and G have integer coefficients. Let $P \in \mathbb{P}^1(\mathbb{Q})$, and let $(a:b)$ be a primitive representative for it. For any monomial cX^iY^{m-i},

$$|ca^ib^j| \le |c| \max(|a|^m, |b|^m),$$

and so

$$|F(a,b)|, |G(a,b)| \le C\,(\max(|a|, |b|)^m$$

with

$$C = (m+1)\max(|\text{coefficient of } F \text{ or } G|).$$

As $F(a,b),\ G(a,b) \in \mathbb{Z}$,

$$H(\varphi(P)) \leq \max(|F(a,b)|, |G(a,b)|) \tag{31}$$
$$\leq C \cdot \max(|a|,|b|)^m = C \cdot H(P)^m. \tag{32}$$

On taking logs, we obtain the inequality

$$h(\varphi(P)) \leq mh(P) + \log C.$$

The problem with proving a reverse inequality is that $F(a,b)$ and $G(a,b)$ may have a large common factor, and so the inequality (31) may be strict. We use the hypothesis that F and G have no common zero in $\mathbb{Q}^{\mathrm{al}}$ to limit this problem.

The hypothesis says that the resultant R of F and G (as homogeneous polynomials) is nonzero (I, 1.26). Consider $Y^{-m}F(X,Y) = F\left(\frac{X}{Y},1\right)$ and $Y^{-m}G(X,Y) = G\left(\frac{X}{Y},1\right)$. When regarded as polynomials in the single variable $\frac{X}{Y}$, $F\left(\frac{X}{Y},1\right)$ and $G\left(\frac{X}{Y},1\right)$ have the same resultant as $F(X,Y)$ and $G(X,Y)$, and so (see I, §1) there are polynomials $U\left(\frac{X}{Y}\right)$, $V\left(\frac{X}{Y}\right) \in \mathbb{Z}\left[\frac{X}{Y}\right]$ of degree $m-1$ such that

$$U\left(\tfrac{X}{Y}\right)F\left(\tfrac{X}{Y},1\right) + V\left(\tfrac{X}{Y}\right)G\left(\tfrac{X}{Y},1\right) = R.$$

On multiplying through by Y^{2m-1} and renaming $Y^{m-1}U\left(\frac{X}{Y}\right)$ as $U(X,Y)$ and $Y^{m-1}V\left(\frac{X}{Y}\right)$ as $V(X,Y)$, we obtain the equation

$$U(X,Y)F(X,Y) + V(X,Y)G(X,Y) = RY^{2m-1}.$$

Similarly, there are homogenous polynomials $U'(X,Y)$ and $V'(X,Y)$ of degree $m-1$ such that

$$U'(X,Y)F(X,Y) + V'(X,Y)G(X,Y) = RX^{2m-1}.$$

Substitute (a,b) for (X,Y) to obtain the equations

$$\left.\begin{aligned} U(a,b)F(a,b) + V(a,b)G(a,b) &= Rb^{2m-1}, \\ U'(a,b)F(a,b) + V'(a,b)G(a,b) &= Ra^{2m-1}. \end{aligned}\right. \tag{33}$$

From these equations, we deduce that $\gcd(F(a,b), G(a,b))$ divides $\gcd(Ra^{2m-1}, Rb^{2m-1}) = R$. As in the first part of the proof, there is a $C > 0$ such that

$$U(a,b), U'(a,b), V(a,b), V'(a,b) \leq C\,(\max|a|,|b|)^{m-1}.$$

Therefore, the equations (33) show that

$$2C\,(\max|a|,|b|)^{m-1}\cdot\max\left(|F(a,b)|,|G(a,b)|\right)\geq |R||a|^{2m-1} \text{ and } |R||b|^{2m-1},$$

i.e., that

$$\max\left(|F(a,b)|,|G(a,b)|\right)\geq\frac{|R|\cdot H(P)^{2m-1}}{2C\cdot h(P)^{m-1}}=\frac{|R|}{2C}H(P)^{m-1}.$$

Because $\gcd(F(a,b),G(a,b))|R$,

$$H(\varphi(P))\geq\frac{1}{|R|}\max(|F(a,b)|,|G(a,b)|)\geq\frac{1}{2C}H(P)^{m}.$$

On taking logs, we obtain the inequality

$$h(\varphi(P))\geq mh(P)-\log 2C.$$

□

LEMMA 4.2 *Let R be the image of (P,Q) under the map*

$$(a{:}b),(c{:}d)\mapsto(ac{:}ad+bc{:}bd){:}\mathbb{P}^1(\mathbb{Q})\times\mathbb{P}^1(\mathbb{Q})\to\mathbb{P}^2(\mathbb{Q}).$$

Then

$$\frac{1}{2}\leq\frac{H(R)}{H(P)H(Q)}\leq 2.$$

PROOF. Choose $(a{:}b)$ and $(c{:}d)$ to be primitive representatives of P and Q. Then

$$\begin{aligned}H(R)&\leq\max(|ac|,|ad+bc|,|bd|)\\&\leq 2\max(|a|,|b|)\max(|c|,|d|)\\&=2H(P)H(Q).\end{aligned}$$

If a prime p divides both ac and bd, then either it divides a and d but not b or c, or the other way round. In either case, it does not divide $ad+bc$, and so $(ac,ad+bc,bd)$ is a primitive representative for R. It remains to show that

$$\max(|ac|,|ad+bc|,|bd|)\geq\frac{1}{2}(\max(|a|,|b|)(\max|c||d|),$$

but this is an elementary exercise (e.g., regard a,b,c,d as real numbers, and rescale the two pairs so that $a=1=c$).

□

Heights on E

Let E be the elliptic curve

$$E: Y^2Z = X^3 + aXZ^2 + bZ^3, \quad a, b \in \mathbb{Q}, \quad \Delta = 4a^3 + 27b^2 \neq 0.$$

For $P \in E(\mathbb{Q})$, define

$$H(P) = \begin{cases} H((x(P){:}z(P))) & \text{if } z(P) \neq 0 \\ 1 & \text{if } P = (0{:}1{:}0). \end{cases}$$

and

$$h(P) = \log H(P).$$

There are other definitions of h, but they differ by bounded amounts, and therefore they lead to the same canonical height (see below).

LEMMA 4.3 *For any constant B, the set of $P \in E(\mathbb{Q})$ such that $h(P) < B$ is finite.*

PROOF. Certainly, for any B, the set $\{P \in \mathbb{P}^1(\mathbb{Q}) \mid H(P) \leq B\}$ is finite, but for each point $(x_0{:}z_0) \in \mathbb{P}^1(\mathbb{Q})$, there are at most two points $(x_0{:}y{:}z_0) \in E(\mathbb{Q})$, and so $\{P \in E(\mathbb{Q}) \mid H(P) \leq B\}$ is finite. □

LEMMA 4.4 *There exists a constant A such that*

$$|h(2P) - 4h(P)| \leq A.$$

PROOF. Let $P = (x{:}y{:}z)$ and $2P = (x_2{:}y_2{:}z_2)$. According to the duplication formula (p. 66),

$$(x_2{:}z_2) = (F(x){:}G(x)),$$

where $F(X,Z)$ and $G(X,Z)$ are homogeneous polynomials of degree 4 such that

$$\begin{aligned} F(X,1) &= (3X^2 + a)^2 - 8X(X^3 + aX + b) \\ G(X,1) &= 4(X^3 + aX + b). \end{aligned}$$

Since $X^3 + aX + b$ and its derivative $3X^2 + a$ have no common root, neither do $F(X,1)$ and $G(X,1)$, and so Lemma 4.1 shows that

$$|h(2P) - 4h(P)| \leq A$$

for some constant A. □

LEMMA 4.5 *There exists at most one function $\hat{h}: E(\mathbb{Q}) \to \mathbb{R}$ satisfying the following conditions:*

(a) *$\hat{h}(P) - h(P)$ is bounded on $E(\mathbb{Q})$;*

(b) *$\hat{h}(2P) = 4\hat{h}(P)$.*

PROOF. If $\hat{h}$ satisfies (a) with bound B, then

$$\left|\hat{h}(2^n P) - h(2^n P)\right| \leq B.$$

If in addition it satisfies (b), then

$$\left|\hat{h}(P) - \frac{h(2^n P)}{4^n}\right| \leq \frac{B}{4^n},$$

and so

$$\hat{h}(P) = \lim_{n\to\infty} h(2^n P)/4^n.$$

□

Any function $\hat{h}: E(\mathbb{Q}) \to \mathbb{R}$ satisfying (a) and (b) of the lemma will be called the ***canonical***,[3] or ***Néron–Tate***, ***height function.*** According to the proof, if $\hat{h}$ exists then $\hat{h}(P)$ must be the limit of the sequence $h(2^n P)/4^n$.

LEMMA 4.6 *For all $P \in E(\mathbb{Q})$, the sequence $h(2^n P)/4^n$ is Cauchy in $\mathbb{R}$.*

PROOF. From Lemma 4.4, we know that there exists a constant A such that

$$|h(2P) - 4h(P)| \leq A$$

for all P. For $N \geq M \geq 0$ and $P \in E(\mathbb{Q})$,

$$\begin{aligned}
\left|\frac{h(2^N P)}{4^N} - \frac{h(2^M P)}{4^M}\right| &= \left|\sum_{n=M}^{N-1}\left(\frac{h(2^{n+1}P)}{4^{n+1}} - \frac{h(2^n P)}{4^n}\right)\right| \\
&\leq \sum_{n=M}^{N-1}\frac{1}{4^{n+1}}\left|h(2^{n+1}P) - 4h(2^n P)\right| \\
&\leq \sum_{n=M}^{N-1}\frac{1}{4^{n+1}}A \\
&\leq \frac{A}{4^{M+1}}\left(1 + \frac{1}{4} + \frac{1}{4^2} + \cdots\right) \\
&= \frac{A}{3\cdot 4^M}.
\end{aligned}$$

Therefore the sequence $h(2^n P)/4^n$ is Cauchy. □

[3] There are different normalizations of the canonical height in the literature, one of which is twice another (see Cremona 1997, 3.4).

The lemma allows us to define

$$\hat{h}(P) = \lim_{n\to\infty} \frac{h(2^n P)}{4^n}, \quad \text{all } P \in E(\mathbb{Q}).$$

THEOREM 4.7 *The function $\hat{h}: E(\mathbb{Q}) \to \mathbb{R}$ is a Néron–Tate height function. Moreover,*

(a) *for all $C \geq 0$, the set $\{P \in E(\mathbb{Q}) \mid \hat{h}(P) \leq C\}$ is finite;*

(b) *$\hat{h}(P) \geq 0$, with equality if and only if P has finite order.*

PROOF. When M is taken to be zero, the inequality in the proof of Lemma 4.6 becomes

$$\left| \frac{h(2^N P)}{4^N} - h(P) \right| \leq \frac{A}{3}.$$

On letting $N \to \infty$, we find that $\hat{h}$ satisfies condition (a) of Lemma 4.5. For condition (b), note that

$$\hat{h}(2P) = \lim_{n\to\infty} \frac{h(2^{n+1} P)}{4^n} = 4 \cdot \lim_{n\to\infty} \frac{h(2^{n+1} P)}{4^{n+1}} = 4 \cdot \hat{h}(P).$$

Thus, $\hat{h}$ is a Néron–Tate height function.

The set of P for which $\hat{h}(P) \leq C$ is finite because h has this property and the difference $\hat{h}(P) - h(P)$ is bounded.

Because $H(P)$ is an integer ≥ 1, both $h(P)$ and $\hat{h}(P)$ are ≥ 0. If P is torsion, then $\{2^n P \mid n \geq 0\}$ is finite, so $\hat{h}$ is bounded on it, say, by D, and then $\hat{h}(P) = \hat{h}(2^n P)/4^n \leq D/4^n$ for all n; therefore $\hat{h}(P) = 0$. On the other hand, if P has infinite order, then $\{2^n P \mid n \geq 0\}$ is infinite and $\hat{h}$ is unbounded on it (by (a)). Hence $\hat{h}(2^n P) > 1$ for some n, and so $\hat{h}(P) > 4^{-n} > 0$. □

Let $f: M \to K$ be a function from an abelian group M into a field K of characteristic $\neq 2$. Such an f is called a ***quadratic form*** if $f(2x) = 4f(x)$ and

$$B(x,y) \overset{\text{def}}{=} f(x+y) - f(x) - f(y)$$

is bi-additive. Then B is symmetric, and it is the only symmetric bi-additive form $B: M \times M \to K$ such that $f(x) = \frac{1}{2}B(x,x)$. We shall need the following criterion.

LEMMA 4.8 *A function $f: M \to K$ from an abelian group into a field K of characteristic $\neq 2$ is a quadratic form if it satisfies the parallelogram*[4] *law,*

$$f(x+y)+f(x-y)=2f(x)+2f(y) \quad \text{all } x,y\in M.$$

PROOF. Let f satisfy the parallelogram law. On taking $x=y=0$ in the parallelogram law, we find that $f(0)=0$; on taking $x=y$ we find that $f(2x)=4f(x)$; and on taking $x=0$ we find that $f(-y)=f(y)$. As $B(x,y)$ is symmetric, it remains to show that $B(x+y,z)=B(x,z)+B(y,z)$, i.e., that

$$f(x+y+z)-f(x+y)-f(x+z)-f(y+z)+f(x)+f(y)+f(z)=0.$$

Now four applications of the parallelogram law show that

$$\begin{aligned}
f(x+y+z)+f(x+y-z)-2f(x+y)-2f(z)&=0\\
f(x+y-z)+f(x-y+z)-2f(x)-2f(y-z)&=0\\
f(x+y+z)+f(x-y+z)-2f(x+z)-2f(y)&=0\\
2f(y+z)+2f(y-z)-4f(y)-4f(z)&=0.
\end{aligned}$$

The alternating sum of these equations is (double) the required equation. □

PROPOSITION 4.9 *The height function $\hat{h}: E(\mathbb{Q}) \to \mathbb{R}$ is a quadratic form.*

We have to prove the parallelogram law.

LEMMA 4.10 *There exists a constant C such that*

$$H(P_1+P_2)\cdot H(P_1-P_2)\leq C\cdot H(P_1)^2\cdot H(P_2)^2$$

for all $P_1, P_2 \in E(\mathbb{Q})$.

PROOF. Let $P_1+P_2=P_3$ and $P_1-P_2=P_4$, and let $P_i=(x_i:y_i:z_i)$. Then

$$(x_3x_4 : x_3z_4+x_4z_3 : z_3z_4)=(w_0 : w_1 : w_2),$$

[4] In elementary linear algebra, the parallelogram law says that, for vectors u and v in $\mathbb{R}^n$,

$$\|u+v\|^2+\|u-v\|^2=2\|u\|^2+2\|v\|^2.$$

where

$$\begin{aligned}
w_0 &= x_1^2x_2^2 - 2ax_1x_2z_1z_2 - 4b(x_1z_1z_2^2 + x_2z_1^2z_2) + a^2z_1^2z_2^2 \\
w_1 &= 2(x_1x_2 + az_1z_2)(x_1z_2 + x_2z_1) + 4bz_1^2z_2^2 \\
w_2 &= (x_2z_1 - x_1z_2)^2.
\end{aligned}$$

For example, by the addition formula (p. 66)

$$\begin{aligned}
x_3(x_1 - x_2)^2 &= x_1x_2^2 + x_1^2x_2 - 2y_1y_2 + a(x_1 + x_2) + 2b \\
x_4(x_1 - x_2)^2 &= x_1x_2^2 + x_1^2x_2 + 2y_1y_2 + a(x_1 + x_2) + 2b.
\end{aligned}$$

On adding these equations, and making them homogeneous, we obtain the equation

$$x_3z_4 + x_4z_3 = w_1/w_2.$$

Similarly,

$$x_3x_4 = w_0/w_2.$$

After choosing $(x_1{:}\,y_1{:}\,z_1)$ and $(x_2{:}\,y_2{:}\,z_2)$ to be primitive representatives for P_1 and P_2, it is clear that

$$H(w_0{:}\,w_1{:}\,w_2) \le C \cdot H(P_1)^2 \cdot H(P_2)^2$$

for some constant C. Here $H(w_0{:}\,w_1{:}\,w_2)$ is the height of $(w_0{:}\,w_1{:}\,w_2)$ as a point of $\mathbb{P}^2(\mathbb{Q})$. According to Lemma 4.2,

$$H(w_0{:}\,w_1{:}\,w_2) \ge \frac{1}{2}H(P_3)H(P_4).$$

On combining these two inequalities, we obtain the required inequality (with the constant $2C$). □

LEMMA 4.11 *The canonical height function $\hat{h}\colon E(\mathbb{Q}) \to \mathbb{R}$ satisfies the parallelogram law,*

$$\hat{h}(P + Q) + \hat{h}(P - Q) = 2\hat{h}(P) + 2\hat{h}(Q).$$

PROOF. On taking logs in the previous lemma, we find that

$$h(P + Q) + h(P - Q) \le 2h(P) + 2h(Q) + \log C.$$

On replacing P and Q with 2^nP and 2^nQ, dividing through by 4^n, and letting $n \to \infty$, we obtain the inequality

$$\hat{h}(P + Q) + \hat{h}(P - Q) \le 2\hat{h}(P) + 2\hat{h}(Q).$$

Putting $P' = P + Q$ and $Q' = P - Q$ in this gives the reverse inequality,

$$\hat{h}(P') + \hat{h}(Q') \leq 2\hat{h}\left(\frac{P'+Q'}{2}\right) + 2\hat{h}\left(\frac{P'-Q'}{2}\right)$$
$$= \frac{1}{2}\hat{h}(P'+Q') + \frac{1}{2}\hat{h}(P'-Q'). \qquad \square$$

REMARK 4.12 Let K be a number field. When $\mathcal{O}_K$ is not a principal ideal domain, there might not exist a primitive representative for a point[5] P of $\mathbb{P}^n(K)$, and so the definition we gave for the height of a point in $\mathbb{P}^n(\mathbb{Q})$ does not extend directly to number fields. Instead, we need a slightly different approach. Note that, for $c \in \mathbb{Q}^\times$,

$$\prod_{p=2,\ldots,\infty} |c|_p = 1 \quad \text{(product formula)}.$$

Here $|\cdot|_p$ is the usual absolute value when $p = \infty$ and otherwise is the p-adic valuation defined in I, §2. Hence, for $P = (a_0{:}a_1{:}\ldots{:}a_n) \in \mathbb{P}^n(\mathbb{Q})$,

$$H(P) \stackrel{\text{def}}{=} \prod_{p=2,\ldots,\infty} \max_i(|a_i|_p)$$

is independent of the choice of a representative for P. Moreover, when $(a_0,\ldots,a_n)$ is chosen to be a primitive representative, then $\max_i |a_i|_p = 1$ for all $p \neq \infty$, and so $H(P) = \max_i |a_i|_\infty$, which agrees with the earlier definition. For a number field K, it is possible to normalize the valuations so that the product formula holds (see ANT, 8.8), and then

$$H(P) \stackrel{\text{def}}{=} \prod_v \max_i(|a_i|_v), \quad P = (a_0{:}a_1{:}\ldots{:}a_n),$$

gives a good notion of a height on $\mathbb{P}^n(K)$. With this definition, all the results of this section extend to elliptic curves over number fields.

NOTES Every projective embedding $X \hookrightarrow \mathbb{P}^n$ of a projective variety over a number field K defines a height function on $X(K)$, and the height functions defined by two different embeddings differ by a bounded amount. In a Short Communication at the 1958 International Congress of Mathematicians, Néron conjectured that, when X is an abelian variety, among the possible height functions $X(K) \to \mathbb{R}$ there is one that is a quadratic form. Tate proved this in 1962 by the above argument. Néron later found another construction of the quadratic height, which is more complicated, but which has the advantage of expressing it as a sum of local heights.

[5] More precisely, there might not exist $a_0,\ldots,a_n \in \mathcal{O}_K$ such that $P = (a_0 : \ldots : a_n)$ and the ideal generated by the a_i is $\mathcal{O}_K$.

Proof of the finite basis theorem

We prove the following more precise statement.

PROPOSITION 4.13 *Let $C > 0$ be such that $S \stackrel{\text{def}}{=} \{P \in E(\mathbb{Q}) \mid \hat{h}(P) \leq C\}$ contains a set of coset representatives for $2E(\mathbb{Q})$ in $E(\mathbb{Q})$; then S generates $E(\mathbb{Q})$.*

PROOF. Suppose that there exists a $Q \in E(\mathbb{Q})$ not in the subgroup generated by S. Because $\hat{h}$ takes discrete values, we may choose Q so that $\hat{h}(Q)$ has the smallest possible value. From the definition of S, there exists a $P \in S$ such that $Q = P + 2R$ for some $R \in E(\mathbb{Q})$. Clearly, R cannot be in the subgroup generated by S, and so $\hat{h}(R) \geq \hat{h}(Q)$. Thus,

$$\begin{aligned} 2\hat{h}(P) &= \hat{h}(P+Q) + \hat{h}(P-Q) - 2\hat{h}(Q) \quad \text{(by 4.11)} \\ &\geq 0 + \hat{h}(2R) - 2\hat{h}(Q) \quad \text{(because } h(\cdot) \geq 0\text{)} \\ &= 4\hat{h}(R) - 2\hat{h}(Q) \quad \text{(see 4.5b)} \\ &\geq 2\hat{h}(Q), \end{aligned}$$

which is a contradiction because $\hat{h}(P) \leq C$ and $\hat{h}(Q) > C$. □

In view of Remark 4.12, this argument works without change for any number field K.

REMARK 4.14 Let C be as in the statement of the proposition, and let $P_1, \ldots, P_s$ be a set of representatives for the elements of $E(\mathbb{Q})/2E(\mathbb{Q})$. Let $Q \in E(\mathbb{Q})$ have $\hat{h}(Q) > C$, and write $Q = P_i + 2R$ with $R \in E(\mathbb{Q})$. Then $\hat{h}(R) < \hat{h}(Q)$ because otherwise the argument in the proposition would give a contradiction. Thus C has the property used in the sketch of a proof p. 135.

5 The problem of computing the rank of $E(\mathbb{Q})$

According to André Weil, one of the two oldest outstanding problems in mathematics is that of finding an algorithm for determining the group $E(\mathbb{Q})$.[6] We know that $E(\mathbb{Q})$ is finitely generated, and so

$$E(\mathbb{Q}) \approx E(\mathbb{Q})_{\text{tors}} \oplus \mathbb{Z}^r,$$

[6]In the form of the congruent number problem. The other is the question of whether there is an odd perfect number. I once suggested to Weil that this last was not very interesting, but he responded that we won't know until it has been answered.

for some $r \geq 0$, called the rank. Since we know how to compute $E(\mathbb{Q})_{\text{tors}}$ (see II, 5.1), this amounts to finding an algorithm for determining r, or better, for determining a basis for $E(\mathbb{Q})/E(\mathbb{Q})_{\text{tors}}$. We regard the Selmer group $S^{(n)}(E/\mathbb{Q})$ as giving a computable upper bound for r with $Ш(E/\mathbb{Q})_n$ as the error term (see the fundamental exact sequence (29), p. 146). The problem is to determine the image of $E(\mathbb{Q})$ in $S^{(n)}(E/\mathbb{Q})$.

Fix a prime p. From

$$\begin{array}{ccccccccc} 0 & \longrightarrow & E_{p^n} & \longrightarrow & E & \xrightarrow{p^n} & E & \longrightarrow & 0 \\ & & {\scriptstyle p}\uparrow & & {\scriptstyle p}\uparrow & & {\scriptstyle 1}\uparrow & & \\ 0 & \longrightarrow & E_{p^{n+1}} & \longrightarrow & E & \xrightarrow{p^{n+1}} & E & \longrightarrow & 0 \end{array}$$

we get a commutative diagram,

$$\begin{array}{ccccccccc} 0 & \longrightarrow & E(\mathbb{Q})/p^nE(\mathbb{Q}) & \longrightarrow & H^1(\mathbb{Q}, E_{p^n}) & \longrightarrow & H^1(\mathbb{Q}, E)_{p^n} & \longrightarrow & 0 \\ & & \uparrow & & \uparrow & & {\scriptstyle p}\uparrow & & \\ 0 & \rightarrow & E(\mathbb{Q})/p^{n+1}E(\mathbb{Q}) & \rightarrow & H^1(\mathbb{Q}, E_{p^{n+1}}) & \rightarrow & H^1(\mathbb{Q}, E)_{p^{n+1}} & \rightarrow & 0 \end{array}$$

and hence a commutative diagram,

$$\begin{array}{ccccccccc} 0 & \longrightarrow & E(\mathbb{Q})/pE(\mathbb{Q}) & \longrightarrow & S^{(p)}(E/\mathbb{Q}) & \longrightarrow & Ш(E/\mathbb{Q})_p & \longrightarrow & 0 \\ & & \uparrow & & \uparrow & & {\scriptstyle p}\uparrow & & \\ 0 & \longrightarrow & E(\mathbb{Q})/p^2E(\mathbb{Q}) & \longrightarrow & S^{(p^2)}(E/\mathbb{Q}) & \longrightarrow & Ш(E/\mathbb{Q})_{p^2} & \longrightarrow & 0 \\ & & \uparrow & & \uparrow & & {\scriptstyle p}\uparrow & & \\ & & \vdots & & \vdots & & \vdots & & \\ & & \uparrow & & \uparrow & & {\scriptstyle p}\uparrow & & \\ 0 & \longrightarrow & E(\mathbb{Q})/p^nE(\mathbb{Q}) & \longrightarrow & S^{(p^n)}(E/\mathbb{Q}) & \longrightarrow & Ш(E/\mathbb{Q})_{p^n} & \longrightarrow & 0. \end{array}$$

The vertical maps at left are the natural quotient maps. Let $S^{(p,n)}(E/\mathbb{Q})$ denote the image of $S^{(p^n)}(E/\mathbb{Q})$ in $S^{(p)}(E/\mathbb{Q})$.

PROPOSITION 5.1 *We have*

$$E(\mathbb{Q})/pE(\mathbb{Q}) \subset \bigcap_n S^{(p,n)}(E/\mathbb{Q}),$$

with equality if there is no nonzero element in $Ш(E/\mathbb{Q})$ *divisible by all powers of* p. *In this last case,*

$$E(\mathbb{Q})/pE(\mathbb{Q}) \simeq S^{(p,n_0)}(E/\mathbb{Q})$$

for all sufficiently large n_0.

PROOF. As the left-hand vertical arrows in the above diagram are surjective, the image of $E(\mathbb{Q})/pE(\mathbb{Q})$ in $S^{(p)}$ equals the image of $E(\mathbb{Q})/p^n E(\mathbb{Q})$, which is contained in $S^{(p,n)}$ by the commutativity of the diagram. Conversely, let γ lie in $\bigcap_n S^{(p,n)}$, so that, for each n, there exists an element $\gamma_n \in S^{(p^n)}$ mapping to γ. Let δ_n be the image of γ_n in $Ш_{p^n}$. Then $p^{n-1}\delta_n = \delta_1$ for all n, and so δ_1 is divisible by all powers of p. If the only such element in Ш is zero, then γ is in the image of $E(\mathbb{Q})/pE(\mathbb{Q})$.

The group Ш is torsion and $Ш_{p^n}$ is finite for all n. If there does not exist a nonzero element in Ш divisible by all powers of p, then the p-primary component $Ш(p)$ of Ш is finite.[7] Thus, there exists an n_0 such that $p^{n_0-1}Ш(p) = 0$. Now $S^{(p,n_0)}$ maps to zero in $Ш_p$, and so

$$E(\mathbb{Q})/pE(\mathbb{Q}) \simeq S^{(p,n_0)}(E/\mathbb{Q}). \qquad \square$$

REMARK 5.2 This gives a strategy for computing r. Calculate $S^{(p)}$, and let your computer run overnight to calculate the subgroup $T(1)$ of $E(\mathbb{Q})$ generated by the points with height $h(P) \leq 10$. If $T(1)$ maps onto $S^{(p)}$ we have found r, and even a set of generators for $E(\mathbb{Q})$. If not, calculate $S^{(p^2)}$, and have the computer run overnight again to calculate the subgroup $T(2)$ of $E(\mathbb{Q})$ generated by points with height $h(P) \leq 10^2$. If the image of $T(2)$ in $S^{(p)}$ is $S^{(p,2)}$, then we have found r. If not, we continue ...

Nightmare possibility: The Tate–Shafarevich group contains a nonzero element divisible by all powers of p, in which case "we are doomed to continue computing through all eternity" (Tate 1974, p. 193). This would happen, for example, if $Ш(E/\mathbb{Q})$ contains a copy of $\mathbb{Q}_p/\mathbb{Z}_p$. It is widely conjectured that this does not happen.

CONJECTURE 5.3 *The Tate–Shafarevich group is always finite.*

[7] Let A be a torsion abelian group such that A_{p^n} is finite and nonzero for all n. For each N, let $A(N)$ be the set of infinite sequences $a_1,\ldots,a_n,\ldots$ with $a_n \in A_{p^n}$ for all n and $pa_n = a_{n-1}$ for $n \leq N$. Each $A(N)$ is a nonempty closed subset of the compact space $\prod_n A_{p^n}$ and $A(1) \supset A(2) \supset \cdots$. A standard result in topology shows that $\bigcap_N A(N)$ is nonempty, and so A contains a nonzero element divisible by all powers of p.

Equivalently, $\text{Ш}(E/\mathbb{Q})$ has no nonzero divisible element and $\text{Ш}(E/\mathbb{Q})_p = 0$ for all but finitely many p.

Until the work of Kolyvagin and Rubin about 1987 (Kolyvagin 1988a,b; Rubin 1987), the Tate–Shafarevich group was not known to be finite for a single elliptic curve over $\mathbb{Q}$. Over function fields, examples of elliptic curves with finite Tate–Shafarevich group had been known 20 years earlier (Milne 1968) — see §10.

Selmer's conjecture and the Cassels pairing

The fundamental exact sequence (29), p. 146, shows that Ш_m measures the error when $S^{(m)}$ is used to estimate $E(\mathbb{Q})/mE(\mathbb{Q})$. With the help of an electronic computer, Selmer (1954) found empirically in a large number of cases that the difference between the estimates for the number of generators of $E(\mathbb{Q})$ obtained from $S^{(m)}$ and $S^{(m^2)}$ was even. He conjectured that this is always true.

A possible explanation for Selmer's observation runs as follows. Suppose that $\text{Ш}(E/\mathbb{Q})$ is finite and admits a nondegenerate bilinear form l with values in $\mathbb{Q}/\mathbb{Z}$ that is alternating in the sense that $l(\xi,\xi) = 0$ (all $\xi \in \text{Ш}$). Then Ш has order a square.[8] Moreover, when restricted to Ш_m, the bilinear form l has kernel the subgroup $\text{Ш}_m \cap m\text{Ш}$. This implies that the index of $\text{Ш}_m \cap m\text{Ш}$ in Ш_m is a square, which in turn implies Selmer's conjecture.

This argument motivated Cassels to prove the following theorem.

THEOREM 5.4 (CASSELS 1962b) *Let E be elliptic curve over $\mathbb{Q}$. There is a canonical alternating bilinear form l on $\text{Ш}(E/\mathbb{Q})$ with values in $\mathbb{Q}/\mathbb{Z}$ whose kernel is exactly the group of divisible elements of $\text{Ш}(E/\mathbb{Q})$.*

This is probably the most lasting contribution of Cassels to the subject. Granted the theorem, if $\text{Ш}(E/\mathbb{Q})$ has no element divisible by all powers of p, then Selmer's conjecture is true for p. Put differently, if Selmer's conjecture fails for a single p, then Conjecture 5.3 is false.

NOTES Let C be the group making the sequence

$$0 \to \text{Ш}(E/\mathbb{Q}) \to H^1(\mathbb{Q},E) \to \bigoplus_{p=2,\ldots,\infty} H^1(\mathbb{Q}_p,E) \to C \to 0,$$

[8] If Ш is killed by p, hence is a vector space over $\mathbb{F}_p$, this is a standard result in linear algebra. The general statement for a finite abelian group can be deduced from this by an induction argument.

exact, and endow each group in the sequence with the discrete topology. When $Ш(E/\mathbb{Q})$ is finite, Cassels (1964) showed that the Pontryagin dual of this sequence is an exact sequence

$$0 \leftarrow Ш(E/\mathbb{Q}) \leftarrow \Theta \leftarrow \prod_{p=2,\ldots,\infty} H^1(\mathbb{Q}_p, E) \leftarrow \hat{E}(\mathbb{Q}) \leftarrow 0, \qquad (34)$$

where $\hat{E}(\mathbb{Q})$ is the profinite completion of $E(\mathbb{Q})$, i.e., $\hat{E}(\mathbb{Q}) = \varprojlim_n E(\mathbb{Q})/nE(\mathbb{Q})$.

Tate generalized the work of Cassels by proving a duality between the Tate–Shafarevich groups of an abelian variety and its dual (Cassels–Tate duality), and by constructing the dual exact sequence (34) for abelian varieties (the Cassels–Tate dual exact sequence). See Tate 1963.

Explicit calculations of the rank

Explicitly computing the rank r of $E(\mathbb{Q})$ can be difficult[9] (perhaps impossible), but occasionally it is straightforward. In order to avoid the problem of having to work with a number field L other than $\mathbb{Q}$, we assume that the elliptic curve has all its points of order 2 rational over $\mathbb{Q}$, and so is given by an equation

$$Y^2Z = (X - \alpha Z)(X - \beta Z)(X - \gamma Z), \quad \alpha, \beta, \gamma \text{ distinct integers.}$$

The discriminant of $(X - \alpha)(X - \beta)(X - \gamma)$ is

$$\Delta = (\alpha - \beta)^2(\beta - \gamma)^2(\gamma - \alpha)^2.$$

PROPOSITION 5.5 *The rank r of $E(\mathbb{Q})$ satisfies the inequality*

$$r \leq 2 \times \#\{p \mid p \text{ divides } 2\Delta\}.$$

PROOF. Let $T = E(\mathbb{Q})_{\text{tors}}$. As $E(\mathbb{Q}) \approx T \oplus \mathbb{Z}^r$, we have $E(\mathbb{Q})/2E(\mathbb{Q}) \approx T/2T \oplus (\mathbb{Z}/2\mathbb{Z})^r$. Because T is finite, the kernel and cokernel of $T \xrightarrow{2} T$ have the same order, and so $T/2T \approx (\mathbb{Z}/2\mathbb{Z})^2$. We have an injection

$$E(\mathbb{Q})/2E(\mathbb{Q}) \hookrightarrow (\mathbb{Q}^\times/\mathbb{Q}^{\times 2})^2,$$

and the image is contained in the product of the subgroups of $\mathbb{Q}^\times/\mathbb{Q}^{\times 2}$ generated by -1 and the primes where E has bad reduction, namely, those dividing 2Δ (see 3.7). □

[9] Of course, one can always type `ellgenerators` into Pari, and hope that someone else has done the hard work.

It is possible to improve this estimate. Let T_1 be the set of prime numbers dividing Δ for which the reduction is nodal and T_2 the set for which it is cuspidal. Thus T_1 consists of the prime numbers modulo which two of the roots of $(X-\alpha)(X-\beta)(X-\gamma)$ coincide, and T_2 consists of those modulo which all three coincide. Let t_1 and t_2 respectively be the numbers of elements of T_1 and T_2.

PROPOSITION 5.6 *The rank r of $E(\mathbb{Q})$ satisfies $r \leq t_1 + 2t_2 - 1$.*

PROOF. Define $\varphi_\alpha : E(\mathbb{Q})/2E(\mathbb{Q}) \to \mathbb{Q}^\times/\mathbb{Q}^{\times 2}$ as in (3.7):

$$\varphi_\alpha((x:y:z)) = \begin{cases} (\frac{x}{z}-\alpha)\mathbb{Q}^{\times 2} & z \neq 0, \quad x \neq \alpha z; \\ (\alpha-\beta)(\alpha-\gamma)\mathbb{Q}^\times & z \neq 0, \quad x = \alpha z \\ \mathbb{Q}^\times & (x:y:z) = (0:1:0). \end{cases}$$

Define φ_β similarly — the map

$$P \mapsto (\varphi_\alpha(P), \varphi_\beta(P)) \colon E(\mathbb{Q})/2E(\mathbb{Q}) \to (\mathbb{Q}^\times/\mathbb{Q}^{\times 2})^2$$

is injective. For each prime p, let $\varphi_p(P)$ be the element of $(\mathbb{Z}/2\mathbb{Z})^2$ whose components are

$$\mathrm{ord}_p(\varphi_\alpha(P)) \mod 2, \qquad \text{and } \mathrm{ord}_p(\varphi_\beta(P)) \mod 2$$

and let $\varphi_\infty(P)$ be the element of $\{\pm\}^2$ whose components are

$$\mathrm{sign}(\varphi_\alpha(P)), \qquad \text{and } \mathrm{sign}(\varphi_\beta(P)).$$

To prove the proposition, it suffices to prove the following statements:

(a) if p does not divide Δ, then $\varphi_p(P) = 0$ for all P;

(b) if $p \in T_1$, then $\varphi_p(P)$ is contained in the diagonal of $\mathbb{F}_2^2$ for all P;

(c) when α, β, γ are ordered so that $\alpha < \beta < \gamma$, $\varphi_\infty(P)$ equals $(+,+)$ or $(+,-)$.

Except for $p = 2$, (a) was proved in the paragraph preceding (3.7).

We prove (b) in the case $\alpha \equiv \beta \mod p$ and $P = (x:y:1)$, $x \neq \alpha, \beta, \gamma$. Let

$$a = \mathrm{ord}_p(x-\alpha), \quad b = \mathrm{ord}_p(x-\beta), \quad c = \mathrm{ord}_p(x-\gamma).$$

Because

$$(x-\alpha)(x-\beta)(x-\gamma)$$

is a square, $a + b + c \equiv 0 \mod 2$.

If $a < 0$, then (because $\alpha \in \mathbb{Z}$) p^{-a} occurs as a factor of the denominator of x (in its lowest terms), and it follows that $b = a = c$. Since $a+b+c \equiv 0 \mod 2$, this implies that $a \equiv b \equiv c \equiv 0 \mod 2$, and so $\varphi_p(P) = 0$. The same argument applies if $b < 0$ or $c < 0$.

If $a > 0$, then p divides the numerator of $x-\alpha$. Because p does not divide $(\alpha-\gamma)$, it does not divide $(\alpha-\gamma)+(x-\alpha) = (x-\gamma)$, and so $c = 0$. Now $a+b \equiv 0 \mod 2$ implies that $\varphi_p(P)$ lies in the diagonal of $\mathbb{F}_2^2$. A similar argument applies if $b > 0$ or $c > 0$.

The remaining cases of (b) are proved similarly.

We prove (c). Let $P = (x:y:1)$, $x \neq \alpha,\beta,\gamma$. We may suppose that $\alpha < \beta < \gamma$, so that $(x-\alpha) > (x-\beta) > (x-\gamma)$. Then $\varphi_\infty(P) = (+,+)$, $(+,-)$, or $(-,-)$. However, because

$$(x-\alpha)(x-\beta)(x-\gamma)$$

is a square in $\mathbb{Q}$, the pair $(-,-)$ is impossible. The cases $x = \alpha$, etc. are equally easy. □

EXAMPLE 5.7 The curve

$$E: Y^2Z = X^3 - XZ^2$$

is of the above form with $(\alpha,\beta,\gamma) = (-1,0,1)$. The only bad prime is 2, and here the reduction is nodal. Therefore $r = 0$, and E has no point of infinite order,

$$E(\mathbb{Q}) \approx (\mathbb{Z}/2\mathbb{Z})^2.$$

ASIDE 5.8 It is an old, but still open, question whether the rank of $E(\mathbb{Q})$ is bounded or can be arbitrarily large.[10] At present (2020), the highest known rank is 28 — this has not changed since the first edition of the book (2006). For the current records, see `http://web.math.hr/~duje/tors/tors.html`. For elliptic curves over $k(T)$, k a finite field, it is known that the rank can be arbitrarily large (Tate and Shafarevich 1967).

[10]Cassels (1966, p. 257) wrote: "it has been widely conjectured that there is an upper bound for the rank depending only on the groundfield. This seems to me implausible because the theory makes it clear that an [elliptic curve] can only have high rank if it is defined by equations with very large coefficients." Tate (1974, p. 194) wrote "I would guess that there is no bound on the rank." Since then, opinion has shifted back towards the rank being bounded. Park et al. (2019) present a heuristic that suggests that there are only finitely many elliptic curves over $\mathbb{Q}$ with rank greater than 21. See also Poonen 2018.

EXERCISE 5.9 Do one of the following two problems (those who know the quadratic reciprocity law should do (2)).

(1) Show that $E(\mathbb{Q})$ is finite if E has equation

$$Y^2Z = X^3 - 4XZ^2.$$

Hint: Let P be a point of infinite order in $E(\mathbb{Q})$, and show that, after possibly replacing P with $P+Q$, where $2Q=0$, $\varphi_2(P)$ is zero. Then show that $\varphi_\infty(P) = (+,+)$ — contradiction.

(2) Let E be the elliptic curve

$$Y^2Z = X^3 - p^2XZ^2,$$

where p is an odd prime. Show that the rank r of $E(\mathbb{Q})$ satisfies:

$$\begin{aligned} r &\leq 2 \quad \text{if } p \equiv 1 \mod 8 \\ r &= 0 \quad \text{if } p \equiv 3 \mod 8 \\ r &\leq 1 \quad \text{otherwise.} \end{aligned}$$

Hint: Let P be a point of infinite order in $E(\mathbb{Q})$, and show that, after possibly replacing P with $P+Q$, where $2Q=0$, $\varphi_p(P)$ is zero.

These are standard examples.

6 The Néron–Tate pairing

We saw in §4, that there is a canonical $\mathbb{Z}$-bilinear pairing

$$B\colon E(\mathbb{Q}) \times E(\mathbb{Q}) \to \mathbb{R}, \quad B(x,y) = \hat{h}(x+y) - \hat{h}(x) - \hat{h}(y).$$

This pairing extends uniquely to an $\mathbb{R}$-bilinear pairing

$$B : E(\mathbb{Q}) \otimes \mathbb{R} \times E(\mathbb{Q}) \otimes \mathbb{R} \to \mathbb{R}.$$

If $\{e_1,\ldots,e_r\}$ is a $\mathbb{Z}$-basis for $E(\mathbb{Q})/E(\mathbb{Q})_{\text{tors}}$, then $\{e_1 \otimes 1,\ldots,e_r \otimes 1\}$ is an $\mathbb{R}$-basis for $E(\mathbb{Q}) \otimes_{\mathbb{Z}} \mathbb{R}$, with respect to which B has matrix $\big(B(e_i,e_j)\big)$.

THEOREM 6.1 *The bilinear pairing*

$$B\colon E(\mathbb{Q}) \otimes \mathbb{R} \times E(\mathbb{Q}) \otimes \mathbb{R} \to \mathbb{R}$$

is positive definite; in particular, it is nondegenerate.

This follows from Theorem 4.7 and the next lemma. By a ***lattice*** in a $\mathbb{R}$-vector space V, we mean the $\mathbb{Z}$-submodule generated by a basis for V (sometimes this is called a full lattice).

LEMMA 6.2 *Let $q: V \to \mathbb{R}$ be a quadratic form on a finite-dimensional $\mathbb{R}$-vector space V. If there exists a lattice Λ in V such that*

(a) *for every constant C, the set $\{P \in \Lambda \mid q(P) \le C\}$ is finite,*

(b) *the only $P \in \Lambda$ with $q(P) = 0$ is $P = 0$,*

then q is positive definite on V.

PROOF. There exists a basis for V relative to which q takes the form

$$q(X) = X_1^2 + \cdots + X_s^2 - X_{s+1}^2 - \cdots - X_t^2, \quad t \le \dim V.$$

We assume that $s \ne \dim V$ and derive a contradiction. We use the basis to identify V with $\mathbb{R}^n$. Let λ be the length of the shortest vector in Λ, i.e.,

$$\lambda = \inf\{q(P) \mid P \in \Lambda, P \ne 0\}.$$

The conditions (a,b) imply that $\lambda > 0$. Consider the set

$$B(\delta) = \left\{ (x_i) \in \mathbb{R}^n \;\middle|\; x_1^2 + \cdots + x_s^2 \le \frac{\lambda}{2}, \quad x_{s+1}^2 + \cdots + x_t^2 \le \delta \right\}.$$

The length (using q) of every vector in $B(\delta)$ is $\le \lambda/2$, and so $B(\delta) \cap \Lambda = \{0\}$, but the volume of $B(\delta)$ can be made arbitrarily large by taking δ large, and so this violates the following famous theorem of Minkowski. □

THEOREM 6.3 (MINKOWSKI) *Let Λ be a lattice in $\mathbb{R}^n$ with fundamental parallelepiped D_0, and let B be a subset of $\mathbb{R}^n$ that is compact, convex, and symmetric in the origin. If*

$$\mathrm{Vol}(B) \ge 2^n \,\mathrm{Vol}(D_0)$$

then B contains a point of Λ other than the origin.

PROOF. We first show that a measurable set S in $\mathbb{R}^n$ with $\mathrm{Vol}(S) > \mathrm{Vol}(D_0)$ contains distinct points α, β such that $\alpha - \beta \in \Lambda$. Clearly

$$\mathrm{Vol}(S) = \sum \mathrm{Vol}(S \cap D),$$

where the sum is over all the translates of D_0 by elements of Λ. The fundamental parallelepiped D_0 will contain a unique translate (by an element of

Λ) of each set $S \cap D$. Since $\mathrm{Vol}(S) > \mathrm{Vol}(D_0)$, at least two of these sets will overlap, and so there exist elements $\alpha, \beta \in S$ such that

$$\alpha - \lambda = \beta - \lambda', \quad \text{some distinct } \lambda, \lambda' \in \Lambda.$$

Then $\alpha - \beta = \lambda - \lambda' \in \Lambda \smallsetminus \{0\}$.

We apply this with $S = \frac{1}{2}B \stackrel{\text{def}}{=} \{\frac{x}{2} \mid x \in B\}$. It has volume $\frac{1}{2^n}\mathrm{Vol}(B) > \mathrm{Vol}(D_0)$, and so there exist distinct $\alpha, \beta \in B$ such that $\alpha/2 - \beta/2 \in \Lambda$. Because B is symmetric about the origin, $-\beta \in B$, and because it is convex, $(\alpha + (-\beta))/2 \in B$. □

REMARK 6.4 Systems consisting of a real vector space V, a lattice Λ in V, and a positive definite quadratic form q on V are of great interest in mathematics — they are typically referred to simply as lattices. There exists a basis for V that identifies (V, q) with $(\mathbb{R}^n, X_1^2 + \cdots + X_n^2)$. Finding a dense packing of spheres in $\mathbb{R}^n$ centred on the points of a lattice amounts to finding a lattice Λ such that

$$\frac{\|\text{ shortest vector}\|^n}{\mathrm{Vol(fundamental\ parallelopiped)}}$$

is large. Many lattices, for example, the Leech lattice, have very interesting automorphism groups. See Conway and Sloane 1999.

An elliptic curve E over $\mathbb{Q}$, gives rise to such a system, namely, $V = E(\mathbb{Q}) \otimes \mathbb{R}$, $\Lambda = E(\mathbb{Q})/E(\mathbb{Q})_{\text{tors}}$, $q = \hat{h}$. As far as I know, they are not of interesting — at present no elliptic curve is known with $\mathrm{rank}(E(\mathbb{Q})) > 28$. However, when the number field is replaced by a function field in one variable, one gets infinite families of very interesting lattices. We discuss this in §11 below.

7 Geometric interpretation of the cohomology groups; jacobians

Throughout this section, k is a perfect field and $\Gamma = \mathrm{Gal}(k^{\mathrm{al}}/k)$.

For an elliptic curve E over k and an integer $n > 1$, we have an exact sequence of cohomology groups (see §2):

$$0 \to E(k)/nE(k) \to H^1(k, E_n) \to H^1(k, E)_n \to 0.$$

Here $H^1(k, E_n)$ and $H^1(k, E)$ are defined to be the groups of continuous crossed homomorphisms from Γ to $E(k^{\mathrm{al}})_n$ and $E(k^{\mathrm{al}})$ respectively modulo the principal crossed homomorphisms. In this section, we shall give a geometric interpretation of these groups, and hence also of the Selmer and Tate–Shafarevich groups. We shall attach to a curve W of genus 1 over k, possibly without a rational point, an elliptic curve E, called its jacobian, and we shall see that the Tate–Shafarevich group of an elliptic curve E classifies the curves of genus 1 over k with jacobian E for which the Hasse principle fails, i.e., the curves having a point in each $\mathbb{Q}_p$ and in $\mathbb{R}$ without having a point in $\mathbb{Q}$.

In general, $H^1(k,*)$ classifies objects over k that become isomorphic over k^{al} to a fixed object with automorphism group $*$. We shall see several examples of this.

Principal homogeneous spaces of sets

Let A be a commutative group. A right A-set

$$(w,a) \mapsto w+a\colon W \times A \to W$$

is called a ***principal homogeneous space*** for A if $W \neq \emptyset$ and the map

$$(w,a) \mapsto (w,w+a)\colon W \times A \to W \times W$$

is bijective, i.e., if for each pair $w_1, w_2 \in W$, there is a unique $a \in A$ such that $w_1 + a = w_2$.

EXAMPLE 7.1 (a) Addition $A \times A \to A$ makes A into a principal homogeneous space for A.

(b) A principal homogeneous space for a vector space (for example, the universe according to Newton) is called an ***affine space***. Essentially an affine space is a vector space with no preferred origin.

A ***morphism*** $\varphi\colon W \to W'$ of principal homogeneous spaces is simply a map A-sets. We leave it to the reader to check the following statements.

7.2 Let W and W' be principal homogeneous spaces for A.

(a) For any points $w_0 \in W$, $w_0' \in W'$, there exists a unique morphism $\varphi\colon W \to W'$ sending w_0 to w_0'.

(b) Every morphism $W \to W'$ is an isomorphism (i.e., has an inverse that is also a morphism).

7.3 Let W be a principal homogeneous space for A.

(a) For any point $w_0 \in W$, there is a unique morphism $A \to W$ (of principal homogeneous spaces) sending 0 to w_0.

(b) An $a \in A$ defines an automorphism $w \mapsto w + a$ of W, and every automorphism of W is of this form for a unique $a \in A$.

The last statement shows that $\mathrm{Aut}(W) \simeq A$, and so we have defined a class of objects having A as their groups of automorphisms.

Principal homogeneous spaces of curves

Let E be an elliptic curve over a field k. A ***principal homogeneous space***[11] for E is a curve W over k together with a right action of E given by a regular map

$$(w, P) \mapsto w + P : W \times E \to W$$

such that

$$(w, P) \mapsto (w, w + P) : W \times E \to W \times W$$

is an isomorphism of algebraic varieties. The conditions imply that, for every field $K \supset k$, the set $W(K)$ is either empty or is a principal homogeneous space for the group $E(K)$.

EXAMPLE 7.4 Addition $E \times E \to E$ makes E into a principal homogeneous space for E. A principal homogeneous space isomorphic to this principal homogeneous space is said to be ***trivial.***

A ***morphism*** of principal homogeneous spaces for E is a regular map $\varphi: W \to W'$ such that

$$\begin{array}{ccc} W \times E & \longrightarrow & W \\ \downarrow{\scriptstyle \varphi \times \mathrm{id}_E} & & \downarrow{\scriptstyle \varphi} \\ W' \times E & \longrightarrow & W' \end{array}$$

commutes. The statements in the previous subsection extend mutatis mutandis to principal homogeneous spaces for elliptic curves.

7.5 Let W and W' be principal homogeneous spaces for E.

(a) For any field $K \supset k$ and points $w_0 \in W(K)$, $w_0' \in W'(K)$, there exists a unique morphism $\varphi: W \to W'$ over K sending w_0 to w_0'.

(b) Every morphism $W \to W'$ is an isomorphism.

[11] The word "torsor" is also used. Principal homogeneous spaces are the analogue in arithmetic geometry of principal bundles in topology.

7.6 Let W be a principal homogeneous space for E. For any point $w_0 \in W(k)$, there is a unique homomorphism $E \to W$ (of principal homogeneous spaces) sending 0 to w_0. Thus W is trivial if and only if $W(k) \neq \emptyset$. Since W has a point with coordinates in some finite extension of k (this follows from the Nullstellensatz), it becomes trivial over such an extension.

7.7 Let W be a principal homogeneous space for E. A point $P \in E(K)$ defines an automorphism $w \mapsto w+P$ of W, and every automorphism of W over K is of this form for a unique $P \in E(K)$. In particular, $\mathrm{Aut}(W_{k^{\mathrm{al}}}) \simeq E(k^{\mathrm{al}})$.

The classification of principal homogeneous spaces

Let W be a principal homogeneous space for E over k. Then Γ acts continuously on $W(k^{\mathrm{al}})$. Choose a point $w_0 \in W(k^{\mathrm{al}})$. For $\sigma \in \Gamma$, $\sigma w_0 = w_0 + f(\sigma)$ for a unique $f(\sigma) \in E(k^{\mathrm{al}})$. On comparing the equalities

$$\begin{aligned} (\sigma\tau)w_0 &= w_0 + f(\sigma\tau) \qquad \text{(definition of } f\text{)} \\ (\sigma\tau)w_0 &= \sigma(\tau w_0) = \sigma(w_0 + f(\tau)) = \sigma w_0 + \sigma(f(\tau)) \\ &= w_0 + f(\sigma) + \sigma f(\tau), \end{aligned}$$

we see that f is a crossed homomorphism $\Gamma \to E(k^{\mathrm{al}})$. Because w_0 has coordinates in a finite extension of k, f is continuous. A second point $w_1 \in W(k^{\mathrm{al}})$ defines another crossed homomorphism f_1 by

$$\sigma w_1 = w_1 + f_1(\sigma),$$

but $w_1 = w_0 + P$ for some $P \in E(\mathbb{Q}^{\mathrm{al}})$, and so

$$\begin{aligned} \sigma w_1 &= \sigma(w_0 + P) = \sigma w_0 + \sigma P = w_0 + f(\sigma) + \sigma P \\ &= w_1 + f(\sigma) + \sigma P - P, \end{aligned}$$

which shows that f and f_1 differ by a principal crossed homomorphism. Thus the cohomology class of f depends only on W. If this class is zero, then $f(\sigma) = \sigma P - P$ for some $P \in E(k^{\mathrm{al}})$, and

$$\sigma(w_0 - P) = \sigma w_0 - \sigma P = w_0 + \sigma P - P - \sigma P = w_0 - P.$$

This implies that $w_0 - P \in W(k)$, and so W is a trivial principal homogeneous space (see 7.6).

THEOREM 7.8 *The map $W \mapsto [f]$ defines a bijection*

$$\frac{\{\text{principal homogeneous spaces for } E\}}{\approx} \to H^1(k,E) \qquad (35)$$

sending the trivial principal homogeneous space to the zero element.

PROOF. Let $\varphi\colon W \to W'$ be an isomorphism of principal homogeneous spaces for E (over k), and let $w_0 \in W(k^{\mathrm{al}})$. Then (W,w_0) and $(W',\varphi(w_0))$ define the same crossed homomorphism, and so the map

$$\{\text{principal homogeneous spaces for } E\} \to H^1(k,E)$$

is constant on isomorphism classes.

If W and W' define the same cohomology class, then we can choose $w_0 \in W(k^{\mathrm{al}})$ and $w_0' \in W'(k^{\mathrm{al}})$ so that (W,w_0) and (W',w_0') define the same crossed homomorphism. According to 7.5, there is a unique isomorphism $\varphi\colon W \to W'$ of principal homogeneous spaces over k^{al} sending w_0 to w_0'. Let $w \in W(k^{\mathrm{al}})$, and let $w = w_0 + P$ with $P \in E(k^{\mathrm{al}})$. Then

$$\begin{aligned}\varphi(\sigma w) &= \varphi(\sigma(w_0+P)) = \varphi(\sigma w_0 + \sigma P) = \varphi(w_0 + f(\sigma) + \sigma P)\\ &= w_0' + f(\sigma) + \sigma P = \sigma w_0' + \sigma P\\ &= \sigma\varphi(w),\end{aligned}$$

which implies that the map φ is defined over k (see I, 5.5). Hence the map (35) is one-to-one.

Finally let $f\colon \Gamma \to E(k^{\mathrm{al}})$ be a continuous crossed homomorphism. Let W denote the algebraic curve $E_{k^{\mathrm{al}}}$ over k^{al}, but with Γ acting according to the rule

$${}^{\sigma}P = \sigma P + f(\sigma), \quad P \in W(k^{\mathrm{al}}) = E(k^{\mathrm{al}}).$$

Then

$$\begin{cases} {}^{\sigma\tau}P = \sigma\tau P + f(\sigma\tau) = \sigma\tau P + f(\sigma) + \sigma f(\tau)\\ {}^{\sigma}({}^{\tau}P) = {}^{\sigma}(\tau P + f(\tau)) = \sigma\tau P + \sigma f(\tau) + f(\sigma),\end{cases}$$

and so this is an action of Γ on $W(k^{\mathrm{al}})$. The action is continuous (because f is) and regular, and so arises from a model W_0 of W over k (I, 5.5). The curve W_0 has a natural structure of a principal homogeneous space for E over k. □

ASIDE 7.9 Let $\mathrm{WC}(E/k)$ denote the set of isomorphism classes of principal homogeneous spaces for E over k. The bijection in the theorem defines a commutative group structure on $\mathrm{WC}(E/k)$, which can be described as follows: for principal homogeneous spaces W and W', define $W \wedge W'$ to be the quotient of $W \times W'$ under the diagonal action of E, so that

$$(W \wedge W')(k^{\mathrm{al}}) = \left(W(k^{\mathrm{al}}) \times W'(k^{\mathrm{al}})\right)/\sim,$$
$$(w, w') \sim (w + P, w' + P),\ P \in E(k^{\mathrm{al}}).$$

Then $W \wedge W'$ has a natural structure of a principal homogeneous space, and represents the sum of W and W' in $\mathrm{WC}(E/k)$.

Geometric interpretation of $H^1(\mathbb{Q}, E_n)$

We now give a geometric interpretation of $H^1(k, E_n)$. An ***n-covering*** is a pair (W, α) consisting of a principal homogeneous space W for E and a regular map $\alpha: W \to E$ (defined over k) with the following property:

> there exists a $w_1 \in W(k^{\mathrm{al}})$ such that $\alpha(w_1 + P) = nP$ for all $P \in E(k^{\mathrm{al}})$.

A ***morphism*** $(W, \alpha) \to (W', \alpha')$ of n-coverings is a morphism $\varphi: W \to W'$ of principal homogeneous spaces such that $\alpha = \alpha' \circ \varphi$.

For $\sigma \in \Gamma$, write

$$\sigma w_1 = w_1 + f(\sigma), \quad f(\sigma) \in E(k^{\mathrm{al}}). \tag{36}$$

As before, f is a crossed homomorphism with values in $E(k^{\mathrm{al}})$. On applying α to both sides of (36) and using that

$$\alpha(\sigma w_1) = \sigma(\alpha w_1) = \sigma(\alpha(w_1 + O)) = O,$$

we find that $nf = 0$, and so f takes values in $E_n(k^{\mathrm{al}})$. The element w_1 is uniquely determined up to a replacement by $w_1 + Q$, $Q \in E_n(k^{\mathrm{al}})$, from which it follows that the class of f in $H^1(k, E_n)$ is independent of the choice of w_1.

THEOREM 7.10 *The map $(W, \alpha) \mapsto [f]$ defines a bijection*

$$\frac{\{n\text{-coverings}\}}{\approx} \to H^1(k, E_n).$$

PROOF. Write $\mathrm{WC}(E_n/k)$ for the set of n-coverings modulo isomorphism. The map "forget α" $(W, \alpha) \mapsto W$ defines a surjection

$$\mathrm{WC}(E_n/k) \to \mathrm{WC}(E/k)_n$$

whose fibres are, in a natural way principal homogeneous spaces for the group $E(k)/nE(k)$. For example, if W is trivial, so that there exists a $w_0 \in W(k)$, then $\alpha(w_0) \in E(k)$; if $w_0' \in W(k)$ also, then $w_0' = w_0 + P$ for some $P \in E(k)$, and $\alpha(w_0') = \alpha(w_0) + nP$, and so $\alpha(w_0)$ is well-defined as an element of $E(k)/nE(k)$. Now consider

$$\begin{array}{ccccccccc} & & & & \mathrm{WC}(E_n/k) & \xrightarrow{b} & \mathrm{WC}(E/k)_n & & \\ & & & & \downarrow a & & \downarrow \simeq & & \\ 0 & \longrightarrow & E(k)/nE(k) & \longrightarrow & H^1(k,E_n) & \xrightarrow{c} & H^1(k,E)_n & \longrightarrow & 0. \end{array}$$

The diagram commutes, and so a maps the fibres of b into the fibres of c. As these fibres are principal homogeneous spaces for $E(k)/nE(k)$, a is bijective on each fibre (by 7.2), and hence is bijective on the entire sets. □

The diagram in the proof gives a geometric interpretation of the exact sequence (28), p. 145.

Traditionally, when carrying out the strategy in Remark 5.2, one works with p^n-coverings rather than cohomology classes, and the construction of p^n-coverings is usually called the ***nth descent*** in the literature. See Cassels 1966, §26.

ASIDE 7.11 Let $\varphi: E \to E'$ be a homomorphism of elliptic curves. A principal homogeneous space W for E gives a well-defined principal homogeneous space $W' = \varphi_* W$ for E' and a φ-equivariant map $W \to W'$. To give an n-covering amounts to giving a principal homogeneous space W for E and a trivialization of $n_* E$.

Twists of elliptic curves

In this subsection we study the following problem:

> given an elliptic curve E_0 over k, find all elliptic curves E over k that become isomorphic to E_0 over k^{al}.

Such a curve E is called a ***twist*** of E_0. Remember that an elliptic curve E over k has a distinguished point $O \in E(k)$. For simplicity, we assume that the characteristic of k is $\neq 2, 3$, so that we can write our elliptic curve as

$$E(a,b): Y^2 Z = X^3 + aXZ^2 + bZ^3, \quad a,b \in k, \quad \Delta \neq 0.$$

Recall (I, 2.1) that $E(a,b)$ and $E(a',b')$ are isomorphic if and only if there exists a $c \in k^\times$ such that $a' = c^4 a$, $b' = c^6 b$. Every such c defines an isomorphism

$$(x{:}\,y{:}\,z) \mapsto (c^2x{:}\,c^3y{:}\,z){:}\; E(a,b) \to E(a',b'),$$

and all isomorphisms are of this form.

EXAMPLE 7.12 For all $d \in k^\times$,

$$E_d : dY^2Z = X^3 + aXZ^2 + bZ^3,$$

is a twist of $E(a,b)$. Indeed, after making a change of variables $dZ \leftrightarrow Z$, the equation becomes

$$Y^2Z = X^3 + \frac{a}{d^2}XZ^2 + \frac{b}{d^3}Z^3,$$

and so E_d becomes isomorphic to $E(a,b)$ over any field in which d is a square.

In order to be able to apply cohomology, we need to compute the group $\mathrm{Aut}(E,0)$ of automorphisms of E fixing the zero element. Such an automorphism commutes with the group structure (I, 1.5), and so we can apply II, 2.1b, with $(a',b') = (a,b)$.

CASE $ab \neq 0$: Here we seek $c \in k^\times$ such that $c^4 = 1 = c^6$. These equations imply that $c = \pm 1$, and so the only automorphisms of (E,O) are the identity map and

$$(x : y : z) \mapsto (x : -y : z).$$

CASE $a = 0$: Here c can be any 6th root ζ of 1 in k, and the automorphisms of (E,O) are the maps

$$(x : y : z) \mapsto (\zeta^{2i}x : \zeta^{3i}y : z), \quad i = 0,1,2,3,4,5.$$

CASE $b = 0$: Here c can be any 4th root ζ of 1 in k, and the automorphisms of (E,O) are the maps

$$(x : y : z) \mapsto (\zeta^{2i}x : \zeta^{3i}y : z), \quad i = 0,1,2,3.$$

PROPOSITION 7.13 *Let (E,O) be an elliptic curve over a field k of characteristic $\neq 2,3$. The automorphism group of (E,O) is $\{\pm 1\}$ unless $j(E)$ is 0 or 1728, in which cases it is isomorphic to $\mu_6(k)$ or $\mu_4(k)$ respectively.*

PROOF. As $j(E) = \frac{1728(4a^3)}{4a^3+27b^2}$, this follows from the above discussion. □

For an elliptic curve E_0 over k, we write $\mathrm{Aut}_{k^{\mathrm{al}}}(E)$ for $\mathrm{Aut}(E_{k^{\mathrm{al}}})$.

REMARK 7.14 (a) Note that the proposition is consistent with III, 3.17, which says that (over $\mathbb{C}$), $\mathrm{End}(E)$ is isomorphic to $\mathbb{Z}$ or to a subring of the ring of integers in a field $\mathbb{Q}[\sqrt{-d}]$, $d > 0$. The only units in such rings are roots of 1, and only $\mathbb{Q}[\sqrt{-1}]$ and $\mathbb{Q}[\sqrt{-3}]$ contain roots of 1 other than ± 1.

(b) If we allow k to have characteristic 2 or 3, then it is still true that $\mathrm{Aut}(E, O) = \{\pm 1\}$ when $j(E) \neq 0, 1728$.

However, if k is algebraically closed of characteristic 2 and $j = 0 = 1728$, then E is isomorphic to the curve $Y^2 - Y = X^3$. In this case, $\mathrm{Aut}(E, O)$ is a noncommutative group of order 24. Its action on $E_3(k) \approx \mathbb{F}_3^2$ defines an isomorphism $\mathrm{Aut}(E, O) \approx \mathrm{SL}_2(\mathbb{F}_3)$.

If k is algebraically closed of characteristic 2 and $j = 0 = 1728$, then E is isomorphic to the curve $Y^2 = X^3 - X$. In this case, $\mathrm{Aut}(E, O)$ is a noncommutative group of order 12. Its action on $E_2(k) \approx \mathbb{F}_2^2$ defines an isomorphism $\mathrm{Aut}(E, O)/\{\pm 1\} \approx \mathrm{SL}_2(\mathbb{F}_2) \simeq S_3$.

Fix an elliptic curve E_0 over k, and let E be an elliptic curve over k that becomes isomorphic to E_0 over k^{al}. Choose an isomorphism $\varphi\colon E_0 \to E$ over k^{al}. For every $\sigma \in \Gamma$, we obtain a second isomorphism $\sigma\varphi \overset{\mathrm{def}}{=} \sigma \circ \varphi \circ \sigma^{-1}\colon E_0 \to E$ over k^{al}. For example, if φ is $(x : y : z) \mapsto (c^2 x : c^3 y : z)$, then $\sigma\varphi$ is $(x : y : z) \mapsto ((\sigma c)^2 x : (\sigma c)^3 y : z)$. The two isomorphisms $\varphi, \sigma\varphi\colon E_0 \to E$ (over k^{al}) differ by an automorphism of E_0 over k^{al}:

$$\sigma\varphi = \varphi \circ \alpha(\sigma), \quad \alpha(\sigma) \in \mathrm{Aut}_{k^{\mathrm{al}}}(E_0, O).$$

On comparing the equalities

$$\begin{aligned}
(\sigma\tau)\varphi &= \varphi \circ \alpha(\sigma\tau) \quad \text{(by definition)} \\
(\sigma\tau)\varphi &= \sigma(\tau\varphi) = \sigma(\varphi \circ \alpha(\tau)) \\
&= \varphi \circ \alpha(\sigma) \circ \sigma\alpha(\tau),
\end{aligned}$$

we see that α is a crossed homomorphism $\Gamma \to \mathrm{Aut}_{k^{\mathrm{al}}}(E_0, O)$. Choosing a different isomorphism φ replaces $\alpha(\sigma)$ by its composite with a principal crossed homomorphism.

THEOREM 7.15 *The map $E \mapsto [\alpha]$ defines a one-to-one correspondence*

$$\frac{\{\text{twists of } E_0\}}{\approx} \overset{1:1}{\longleftrightarrow} H^1(\Gamma, \mathrm{Aut}_{k^{\mathrm{al}}}(E_0)).$$

PROOF. The proof is similar to that of Theorem 7.8. □

COROLLARY 7.16 *If $j(E_0) \neq 0, 1728$, then the twists of E_0 are exactly the curves E_d in Example 7.12.*

PROOF. In this case, $\mathrm{Aut}_{k^{\mathrm{al}}}(E, O) = \mu_2$, and so $H^1(\Gamma, \mu_2) \simeq k^\times / k^{\times 2}$ (see 1.8). Under the correspondence in the theorem, $E_d \leftrightarrow d \mod k^{\times 2}$. □

REMARK 7.17 The same arguments can be used to obtain the description of the twisted multiplicative groups on p. 71. The endomorphisms of the algebraic group $\mathbb{G}_m = \mathbb{A}^1 \smallsetminus \{0\}$ are the maps $t \mapsto t^m$, $m \in \mathbb{Z}$. Hence $\mathrm{End}(\mathbb{G}_m) \simeq \mathbb{Z}$ and $\mathrm{Aut}(\mathbb{G}_m) = (\mathrm{End}(\mathbb{G}_m))^\times = \{\pm 1\}$. The twisted forms of $\mathbb{G}_m$ are classified by

$$H^1(k, \{\pm 1\}) \simeq H^1(k, \mu_2) \overset{1.8\mathrm{b}}{\simeq} k^\times / k^{\times 2}.$$

The twisted multiplicative group corresponding to $a \in k^\times / k^{\times 2}$ is $\mathbb{G}_m[a]$.

REMARK 7.18 Let $\mathrm{Aut}(E)$ be the group of all automorphisms of E, not necessarily preserving O. The map $Q \mapsto t_Q$, where t_Q is the translation $P \mapsto P + Q$, identifies $E(k)$ with a subgroup of $\mathrm{Aut}(E)$. I claim that $\mathrm{Aut}(E)$ is a semi-direct product,

$$\mathrm{Aut}(E) = E(k) \rtimes \mathrm{Aut}(E, O),$$

i.e., that

(a) $E(k)$ is a normal subgroup of $\mathrm{Aut}(E)$;

(b) $E(k) \cap \mathrm{Aut}(E, O) = \{0\}$;

(c) $\mathrm{Aut}(E) = E(k) \cdot \mathrm{Aut}(E, O)$.

Let $Q \in E(k)$ and let $\alpha \in \mathrm{Aut}(E, O)$. Then α is a homomorphism, and so, for every $P \in E$,

$$(\alpha \circ t_Q \circ \alpha^{-1})(P) = \alpha(\alpha^{-1}(P) + Q) = P + \alpha(Q) = t_{\alpha(Q)}(P),$$

which proves (a). Assertion (b) is obvious. For (c), let $\gamma \in \mathrm{Aut}(E)$, and let $\gamma(0) = Q$; then $\gamma = t_Q \circ (t_{-Q} \circ \gamma)$, and $t_{-Q} \circ \gamma \in \mathrm{Aut}(E, O)$.

Curves of genus 1

Let W be a principal homogeneous space for an elliptic curve E over k. Then W becomes isomorphic to E over k^{al}, and so W is projective, nonsingular, and of genus 1 (at least over k^{al}, which implies that it is also over k). The next theorem shows that, conversely, every projective nonsingular curve W of genus 1 over k occurs as a principal homogeneous space for some elliptic curve over k.

THEOREM 7.19 *Let W be a nonsingular projective curve over k of genus 1. There exists an elliptic curve E_0 over k such that W is a principal homogeneous space for E_0. Moreover, E_0 is unique up to an isomorphism (over k).*

PROOF. Choose a point in $W(k^{\mathrm{al}})$, and use it to define an isomorphism

$$W_{k^{\mathrm{al}}} \xrightarrow{\varphi} E \subset \mathbb{P}^2$$

from $W_{k^{\mathrm{al}}}$ onto an elliptic curve

$$Y^2Z + a_1XYZ + a_3YZ^2 = X^3 + a_2X^2Z + a_4XZ^2 + a_6Z^3, \quad a_i \in k^{\mathrm{al}}$$

(see the proof of (d)→(c) of II, 1.1). Let $\sigma \in \Gamma$. Then $\sigma\varphi$ is an isomorphism $\sigma(W_{k^{\mathrm{al}}}) \to \sigma E$. Here $\sigma(W_{k^{\mathrm{al}}})$ and σE are obtained from $W_{k^{\mathrm{al}}}$ and E by applying σ to the coefficients of the polynomials defining them. For example, σE is defined by the same equation as E but with σa_i for a_i. As W is defined by polynomials with coefficients in k, $\sigma(W_{k^{\mathrm{al}}}) = W_{k^{\mathrm{al}}}$. Therefore $E \approx W_{k^{\mathrm{al}}} \approx \sigma E$, and so $j(E) = j(\sigma E) = \sigma j(E)$. As this is true for all $\sigma \in \Gamma$, $j(E)$ lies in k. According to II, 2.3, there exists a curve E_0 over k with $j(E_0) = j(E)$. In fact, there will be many such curves over k, and so we have to make sure we have the correct one. We choose one, E_0, and then twist it to get the correct one.

Choose an isomorphism $\varphi\colon E_0 \to W$ over k^{al}, and for $\sigma \in \Gamma$, let $\sigma\varphi = \varphi \circ \alpha(\sigma)$, where $\alpha(\sigma) \in \mathrm{Aut}_{k^{\mathrm{al}}}(E_0)$. Then $\sigma \mapsto \alpha(\sigma)$ is a crossed homomorphism into $\mathrm{Aut}_{k^{\mathrm{al}}}(E_0)$, and hence defines a class $[\alpha]$ in $H^1(k, \mathrm{Aut}_{k^{\mathrm{al}}}(E_0))$. According to 7.18, there is an exact sequence

$$1 \to E_0(k^{\mathrm{al}}) \to \mathrm{Aut}_{k^{\mathrm{al}}}(E_0) \to \mathrm{Aut}_{k^{\mathrm{al}}}(E_0, O) \to 1.$$

If $[\alpha]$ lies in the subgroup $H^1(k, E_0)$ of $H^1(k, \mathrm{Aut}_{k^{\mathrm{al}}}(E_0))$, then W is a principal homogeneous space for E_0. If not, we use the image of $[\alpha]$

in $H^1(k, \mathrm{Aut}_{k^{\mathrm{al}}}(E_0, O))$ to twist E_0 to obtain a second curve E_1 over k with the same j-invariant. Now one can check that the class of the crossed homomorphism $[\alpha]$ lies in $H^1(k, E_1)$, and so W is a principal homogeneous space for E_1. □

The curve E_0 given by the theorem is called the ***jacobian*** of W. It is characterized by the following property: there exists an isomorphism $\varphi: E_0 \to W$ over k^{al} such that, for every $\sigma \in \Gamma$, there exists a point $Q_\sigma \in E_0(k^{\mathrm{al}})$ such that

$$(\sigma\varphi)(P) = \varphi(P + Q_\sigma), \qquad \text{all } P \in E(k^{\mathrm{al}}).$$

To construct the jacobian of a curve W over $\mathbb{Q}$, use the following steps:

(a) by a change of variables over $\mathbb{Q}^{\mathrm{al}}$, obtain an isomorphism $W \approx E$, where E is an elliptic curve over $\mathbb{Q}^{\mathrm{al}}$ in standard form;

(b) write down an elliptic curve E_0 over $\mathbb{Q}$ in standard form that becomes isomorphic to E over $\mathbb{Q}^{\mathrm{al}}$;

(c) modify E_0 if necessary so that it has the property characterizing the jacobian.

REMARK 7.20 In the above proof we used crossed homomorphisms with values in a group $\mathrm{Aut}_{k^{\mathrm{al}}}(E_0)$ that need not be commutative. However, we can still define $H^1(G, M)$ when M is not commutative. Write M multiplicatively. As in the commutative case, a crossed homomorphism is a map $f: G \to M$ such that $f(\sigma\tau) = f(\sigma) \cdot \sigma f(\tau)$. Two crossed homomorphisms f and g are said to be equivalent if there exists an $m \in M$ such that $g(\sigma) = m^{-1} \cdot f(\sigma) \cdot \sigma m$ for all $\sigma \in G$, and $H^1(G, M)$ is defined to be the set of equivalence classes of crossed homomorphisms. It is a set with a distinguished element, namely, the class of the constant map $\sigma \mapsto 1$.

NOTES Let C be a curve of genus 1 over k and D a divisor on C of degree $n+1 \geq 2$. According to the Riemann–Roch theorem, $L(D)$ has dimension $n+1$. A basis $f_0, f_1, \ldots f_n$ for $L(D)$ defines a morphism $P \mapsto (f_0(P): \cdots : f_n(P)): C \to \mathbb{P}^n$. This raises two questions: how to find equations describing the image of C in $\mathbb{P}^n$; and from these equations, how to find the Weierstrass equation for the jacobian of C. For a recent article on these questions, see Fisher 2018.

The classification of curves of genus 1 over $\mathbb{Q}$

We summarize the above results.

7.21 Let (E, O) be an elliptic curve over $\mathbb{Q}$. We attach to it the invariant $j(E) \in \mathbb{Q}$. Every element of $\mathbb{Q}$ occurs as the j-invariant of an elliptic curve over $\mathbb{Q}$, and two elliptic curves over $\mathbb{Q}$ have the same j-invariant if and only if they become isomorphic over $\mathbb{Q}^{\mathrm{al}}$ (II, 2.1).

7.22 Fix a $j \in \mathbb{Q}$, and consider the elliptic curves (E, O) over $\mathbb{Q}$ with $j(E) = j$. Choose one, for example, that in II, 2.3. The isomorphism classes of such curves are in natural one-to-one correspondence with the elements of $H^1(\mathbb{Q}, \mathrm{Aut}(E, O))$. For example, in the case that $j \neq 0, 1728$, $\mathrm{Aut}(E, O) = \mu_2$, $H^1(\mathbb{Q}, \mathrm{Aut}(E, O)) \simeq \mathbb{Q}^\times/\mathbb{Q}^{\times 2}$, and the curve corresponding to $d \in \mathbb{Q}^\times$ is the curve E_d of Example 7.12.

7.23 Fix an elliptic curve (E, O) over $\mathbb{Q}$, and consider the curves of genus 1 over $\mathbb{Q}$ having E as their jacobian. Such a curve has the structure of a principal homogeneous space for E, and every principal homogeneous space for E has E as its jacobian. The principal homogeneous spaces for E are classified by the group $H^1(\mathbb{Q}, E)$ (which is a very large group).

7.24 Every curve of genus 1 over $\mathbb{Q}$ has a jacobian, which is an elliptic curve E over $\mathbb{Q}$, and the curve is a principal homogeneous space for E.

EXERCISE 7.25 Find the jacobian of the curve

$$W : aX^3 + bY^3 + cZ^3 = 0, \quad a, b, c \in \mathbb{Q}^\times.$$

Hint: The curve $E : X^3 + Y^3 + dZ^3 = 0$, $d \in \mathbb{Q}^\times$, has the point $O : (1 : -1 : 0)$ — the pair (E, O) is an elliptic curve over $\mathbb{Q}$. It can be put in standard form by the change of variables $X = X' + Y'$, $Y = X' - Y'$.

8 Failure of the Hasse (local-global) principle

Let E be an elliptic curve over $\mathbb{Q}$. Recall that its Tate–Shafarevich group is the kernel of $H^1(\mathbb{Q}, E) \to \prod_p H^1(\mathbb{Q}_p, E)$. Thus, the cohomology class of a principal homogeneous space W for E lies in the Tate–Shafarevich group if and only if $W(\mathbb{Q}_p)$ is nonempty for all p (including $p = \infty$). Therefore the nonzero elements of $\text{Ш}(E/\mathbb{Q})$ are represented by curves over $\mathbb{Q}$ with jacobian E that fail the Hasse principle.

In this section, we describe a family of curves whose Tate–Shafarevich groups are nonzero, and which therefore give examples of curves of genus 1 for which the Hasse principle fails.

PROPOSITION 8.1 *If $p \equiv 1 \mod 8$, then the 2-Selmer group $S^{(2)}(E/\mathbb{Q})$ of the elliptic curve*

$$E: Y^2Z = X^3 + pXZ^2$$

is isomorphic to $(\mathbb{Z}/2\mathbb{Z})^3$.

The family of curves in the statement is similar to that in Exercise 5.9(2), but as only one of the points of order 2 on E has coordinates in $\mathbb{Q}$, we do not have a simple description of $H^1(\mathbb{Q}, E_2)$. Of course, we could pass to $\mathbb{Q}[\sqrt{p}]$, but it is easier to proceed as follows. Let E' be the quotient of E by the subgroup generated by its point $P = (0{:}0{:}1)$ of order 2. There are homomorphisms

$$E \xrightarrow{\phi} E' \xrightarrow{\psi} E$$

whose composite is multiplication by 2. From a study of the cohomology sequences of

$$0 \to \langle P \rangle \to E(\mathbb{Q}^{\mathrm{al}}) \xrightarrow{\phi} E'(\mathbb{Q}^{\mathrm{al}}) \to 0$$

$$0 \to \operatorname{Ker}\psi \to E'(\mathbb{Q}^{\mathrm{al}}) \xrightarrow{\psi} \mathbb{Q}^{\mathrm{al}}) \to 0,$$

it is possible to draw information about $E(\mathbb{Q})/2E(\mathbb{Q})$, $S^{(2)}(E/\mathbb{Q})$, and $Ш(E/\mathbb{Q})_2$. For example, the kernel-cokernel sequence of the maps

$$E(\mathbb{Q}) \xrightarrow{\phi} E'(\mathbb{Q}) \xrightarrow{\psi} E(\mathbb{Q})$$

is an exact sequence

$$E(\mathbb{Q})/\phi(E(\mathbb{Q})) \to E(\mathbb{Q})/2E(\mathbb{Q}) \to E(\mathbb{Q})/\psi(E(\mathbb{Q})) \to 0.$$

See Silverman 2009, X, 6.2.

Assume the proposition. As $E(\mathbb{Q})_2 \simeq \mathbb{Z}/2\mathbb{Z}$, it follows from the fundamental exact sequence (29), p. 146, that

$$\operatorname{rank}(E(\mathbb{Q})) + 1 + \dim_{\mathbb{F}_2} Ш(E/\mathbb{Q})_2 = \dim_{\mathbb{F}_2} S^{(2)}(E/\mathbb{Q}).$$

Thus the rank r of E is $= 0, 1$, or 2. If $Ш(E/\mathbb{Q})(2)$ is finite, its order is known to be a square (see 5.4), and so, conjecturally, the only possibilities are $r = 0, 2$.

PROPOSITION 8.2 *Let E be as in Proposition 8.1. If 2 is not a fourth power modulo p, then* $\operatorname{rank}(E(\mathbb{Q})) = 0$ *and* $Ш(E/\mathbb{Q})_2 \approx (\mathbb{Z}/2\mathbb{Z})^2$.

We discuss the proof below.

REMARK 8.3 In order to apply the proposition, we have to find the primes p such that 2 is a fourth power modulo p. Gauss found an efficient test, which we now explain. From basic algebra, we know that the ring of Gaussian integers, $\mathbb{Z}[i]$, is a principal ideal domain. An odd prime p either remains prime in $\mathbb{Z}[i]$ or it factors as $p = (A+iB)(A-iB)$. The first case occurs exactly when $\mathbb{Z}[i]/p\mathbb{Z}[i]$ is a field of degree 2 over $\mathbb{F}_p$. Therefore p remains prime if and only if $\mathbb{F}_p$ does not contain a primitive 4th root of 1. Because $\mathbb{F}_p^\times$ is cyclic, it contains an element of order 4 if and only if 4 divides its order $p-1$. We conclude that the odd primes p that can be expressed as $p = A^2 + B^2$ with $A, B \in \mathbb{Z}$ are exactly those such that $p \equiv 1 \mod 4$. For a prime $p \equiv 1$ modulo 8, Gauss shows that 2 is a 4th power modulo p if and only if $8|AB$. Therefore, the proposition applies to the primes,

$$17 = 1^2 + 4^2, \quad 41 = 5^2 + 4^2, \quad 97 = 9^2 + 4^2, \quad 193 = 7^2 + 12^2, \ldots$$

The proof of Gauss's criterion, which is quite elementary, can be found in Silverman 2009, X, 6.6. Number theorists will wish to prove that there are infinitely many such primes p (and find their density).

To show that the rank of an elliptic curve is smaller than the bound given by the Selmer group, we must exhibit nontrivial elements of $\text{Ш}(E/\mathbb{Q})_2$. For the curve $E: Y^2Z = X^3 + pZ^3$ in the proposition, there are the following curves:

$$Y^2 = 4pX^4 - 1, \quad Y^2 = 2pX^4 - 2, \quad -Y^2 = 2pX^4 - 2.$$

It is possible to check directly that these are principal homogeneous spaces for $E: Y^2Z = X^3 + pZ^3$, but it is more easily deduced from the proof of Proposition 8.1 (Silverman 2009, X, 6.2(b)).

REMARK 8.4 We should explain what we mean by the curves listed above. Consider, more generally, the curve

$$C: Y^2 = aX^4 + bX^3 + cX^2 + dX + e.$$

We assume that the polynomial on the right has no repeated roots and that the characteristic is $\neq 2, 3$. Then this is a nonsingular affine curve, but its projective closure

$$\bar{C}: Y^2Z^2 = aX^4 + bX^3Z + cX^2Z^2 + dXZ^3 + eZ^4$$

is singular because, on setting $Y = 1$, we obtain the equation

$$Z^2 = aX^4 + bX^3Z + cX^2Z^2 + dXZ^3 + eZ^4$$

which is visibly singular at $(0,0)$. The genus of a plane projective curve of degree d is

$$g = \frac{(d-1)(d-2)}{2} - \sum_{P \text{ singular}} \delta_P.$$

For $P = (0,0)$, $\delta_P = 2$, and so the genus of $\bar{C}$ is $3-2=1$. One can resolve the singularity to obtain a nonsingular curve C' and a regular map $C' \to \bar{C}$ which is an isomorphism except over the singular point. It is really C' that one means when one writes C.

We shall prove that the curve

$$C : Y^2 = 2 - 2pX^4$$

has no points in $\mathbb{Q}$, but has points in $\mathbb{R}$ and $\mathbb{Q}_p$ for all p. For this we need to use the quadratic reciprocity law. For an integer a not divisible by the prime p, the ***Legendre symbol*** $\left(\frac{a}{p}\right)$ is $+1$ if a is a square modulo p (i.e., a quadratic residue) and is -1 otherwise.

THEOREM 8.5 (QUADRATIC RECIPROCITY LAW) *For odd primes p, q,*

$$\left(\frac{p}{q}\right)\left(\frac{q}{p}\right) = (-1)^{\frac{p-1}{2}\frac{q-1}{2}}.$$

Moreover,

$$\left(\frac{2}{p}\right) = (-1)^{\frac{p^2-1}{8}}.$$

PROOF. The theorem may have more published proofs than any other in mathematics.[12] The first six complete proofs were found by Gauss. For two elementary proofs, see Serre 1973, Chapter I. For a short proof using cyclotomic fields, see ANT, Chap. 8, Examples, or Samuel 2008, §6.5. □

---

[12]Rivals include the Pythagorean theorem and the parallel postulate (all of whose proofs are false).

We now prove that

$$C : Y^2 = 2 - 2pX^4$$

has no points with coordinates in $\mathbb{Q}$. Suppose that (x, y) is a point on the curve. Let $x = r/t$ with r and t integers having no common factor. Then

$$y^2 = \frac{2t^4 - 2pr^4}{t^4}.$$

The numerator and denominator on the right are integers with no common factor, and so $2t^4 - 2pr^4$ is the square of an integer, which must be even. Therefore, there exists an integer s such that

$$\boxed{2s^2 = t^4 - pr^4.}$$

Let q be an odd prime dividing s. Then $t^4 \equiv pr^4 \mod q$, and so $\left(\frac{p}{q}\right) = 1$. According to the quadratic reciprocity law, this implies that $\left(\frac{q}{p}\right) = 1$. From the quadratic reciprocity law, $\left(\frac{2}{p}\right) = 1$, and so all prime factors of s are squares modulo p. Hence s^2 is a 4th power modulo p. The equation

$$2s^2 \equiv t^4 \mod p$$

now shows that 2 is a 4th power modulo p, which contradicts our hypothesis.

We should also make sure that there is no point lurking at infinity. The projective closure of C is

$$\bar{C} : Y^2 Z^2 = 2Z^4 - 2pX^4,$$

and we have just shown that $\bar{C}$ has no rational point with $Z = 1$. For $Z = 0$, there is a rational solution, namely, $(0:1:0)$, but this is the singular point $(0,0)$ on the curve

$$Z^2 = 2Z^4 - 2pX^4.$$

On the desingularization $C' \to \bar{C}$ of $\bar{C}$, no $\mathbb{Q}$-point lies over $(0{:}1{:}0)$.

The curve C obviously has points in $\mathbb{R}$. Hensel's lemma (I, 2.12) shows that C has a point in $\mathbb{Q}_q$ if its reduction modulo the prime q has a nonsingular point with coordinates in $\mathbb{F}_q$. For $q \neq 2, p$, the curve C has good reduction at q, and Corollary 9.3 below shows that it has a point with coordinates in $\mathbb{F}_p$. Therefore, C automatically has a point with coordinates in $\mathbb{Q}_q$ except possibly for q equal to 2 or p. These two primes require a more elaborate application of Hensel's lemma, which we leave as an exercise to the reader.

NOTES The first example of a curve of genus 1 that fails the Hasse principle is

$$X^4 - 17 = 2Y^2$$

(Lind 1940). Here, following Cassels 1966, is Lind's argument that it has no $\mathbb{Q}$-rational point. Suppose that (u, v, w) is a solution of

$$u^4 - 17v^2 = 2w^2$$

in coprime integers. If p is an odd prime dividing w, then $u^4 - 17v^2 \equiv 0 \bmod p$, and so 17 is a quadratic residue of p. According to the quadratic reciprocity law, this implies that p is a quadratic residue of 17. As 2 is a quadratic residue of 17 (quadratic reciprocity law again), we see that all prime factors of w are quadratic residues of 17, and so w is a square modulo 17. Hence w^2 is a 4th power modulo 17, and the equation implies that 2 is a 4th power modulo 17, which is not true.

9 Elliptic curves over finite fields

We fix an algebraic closure $\mathbb{F}$ of $\mathbb{F}_p$, and let $\Gamma = \mathrm{Gal}(\mathbb{F}/\mathbb{F}_p)$. The map $a \mapsto a^p$ is a topological generator of Γ, called the ***Frobenius automorphism*** (FT, 7.16).

We let $\mathbb{F}_{p^m}$ denote the subfield of $\mathbb{F}$ whose elements are the roots of $X^{p^m} - X$. Thus $\mathbb{F}_{p^m}$ has p^m elements, and $\mathbb{F}_{p^m} \subset \mathbb{F}_{p^n}$ if and only if $m|n$. The irreducible factors of $X^{p^m} - X$ in $\mathbb{F}_p[X]$ are exactly the monic irreducible polynomials in $\mathbb{F}_p[X]$ of degree dividing m (FT, 4.22, 4.23).

We shall make use of the fact that many of the power series identities in calculus are valid over $\mathbb{Q}$. For example, if we define

$$\begin{aligned} \log(1+T) &= T - \tfrac{1}{2}T^2 + \tfrac{1}{3}T^3 - \tfrac{1}{4}T^4 + \tfrac{1}{5}T^5 - \cdots \\ \exp(T) &= 1 + T + \tfrac{1}{2!}T^2 + \cdots + \tfrac{1}{n!}T^n + \cdots \end{aligned}$$

in the power series ring $\mathbb{Q}[[T]]$, then

$$\begin{aligned} \log\frac{1}{1-T} &= -\log(1-T) = T + \tfrac{1}{2}T^2 + \tfrac{1}{3}T^3 + \tfrac{1}{4}T^4 + \tfrac{1}{5}T^5 + \cdots \\ \exp(\log(1+T)) &= 1 + T. \end{aligned}$$

The numbers a_p

Let $E(a,b) : Y^2Z = X^3 + aXZ^2 + bZ^3$ be an elliptic curve over $\mathbb{F}_p$, $p \neq 2$, and let N_p be the number of points on E with coordinates in $\mathbb{F}_p$.

There is one such point (0:1:0) "at infinity", and the remainder are the solutions of

$$Y^2 = X^3 + aX + b$$

in $\mathbb{F}_p$. One way of counting them is to make a list of the squares in $\mathbb{F}_p$, and then check the values of $x^3 + ax + b$, $x \in \mathbb{F}_p$, against the list. For $z \in \mathbb{F}_p^\times$, let $\chi(z) = 1$ if z is a square in $\mathbb{F}_p^\times$ and -1 otherwise; extend χ to $\mathbb{F}_p$ by setting $\chi(0) = 0$. Then

$$\begin{aligned} N_p &= 1 + \sum_{x \in \mathbb{F}_p} (\chi(x^3 + ax + b) + 1) \\ &= p + 1 - a_p, \end{aligned}$$

where

$$-a_p = \sum_{x \in \mathbb{F}_p} \chi(x^3 + ax + b). \tag{37}$$

As $\mathbb{F}_p^\times$ is cyclic of even order, exactly half of its elements are squares. Since there is no reason to expect that $x^3 + ax + b$ is more likely to be a square than not, we might expect that $-a_p$ is a sum of p terms randomly distributed between $+1$ and -1. The expected value of $|a_p|$ would then be of the order $\sqrt{p}$. Calculations support this when we fix $a, b \in \mathbb{Z}$ and compute a_p for the curve $E(a,b)$ modulo the different primes p. However, for good reduction we always find that

$$|a_p| < 2\sqrt{p} \tag{38}$$

whereas, if the terms in the sum (37) were truly random, then every value between 0 and $p-1$ would be possible. We shall prove the inequality (38) presently, and, in the remainder of the book, we shall see the innocuous-looking numbers a_p turn into key players.

The Frobenius map

Let C be a projective plane curve of degree d over $\mathbb{F}_p$, so that C is defined by a polynomial

$$F(X,Y,Z) = \sum_{i+j+k=d} a_{ijk} X^i Y^j Z^k, \quad a_{ijk} \in \mathbb{F}_p.$$

If $P = (x:y:z) \in C(\mathbb{F})$, then

$$\sum_{i+j+k=d} a_{ijk} x^i y^j z^k = 0.$$

On raising this equation to the pth power, and using that we are in characteristic p and that $a^p = a$ for $a \in \mathbb{F}_p$, we obtain the equation

$$\sum_{i+j+k=d} a_{ijk} x^{ip} y^{jp} z^{kp} = 0,$$

which says that $(x^p : y^p : z^p)$ also lies on C. We therefore have a map

$$(x:y:z) \mapsto (x^p : y^p : z^p) : C(\mathbb{F}) \to C(\mathbb{F}),$$

which, being defined by polynomials, is regular. It is called the ***Frobenius map.***

PROPOSITION 9.1 *(a) The degree of the Frobenius map is p.*
(b) The Frobenius map is zero on the tangent line at every point of E.

PROOF. (a) From the diagram

$$\begin{array}{ccc} k(x,y) & \xrightarrow{(x,y)\mapsto(x^p,y^p)} & k(x,y) \\ \Big|{\scriptstyle 2} & & \Big|{\scriptstyle 2} \\ k(X) & \xrightarrow{X \mapsto X^p} & k(X) \end{array}$$

we see that

$$\deg(\mathrm{Frob}) \stackrel{\mathrm{def}}{=} [k(x,y) : k(x^p, y^p)] = [k(X) : k(X^p)] = p.$$

(b) Consider a regular map

$$\alpha : \mathbb{A}^2 \to \mathbb{A}^2, \quad (a,b) \mapsto (f(a,b), g(a,b))$$

defined by polynomials $f(X,Y)$ and $g(X,Y)$. The map $(d\alpha)_P$ on the tangent space at a point P of $\mathbb{A}^2$ is given by the matrix $\begin{pmatrix} \left(\frac{\partial f}{\partial X}\right)_P & \left(\frac{\partial f}{\partial Y}\right)_P \\ \left(\frac{\partial g}{\partial X}\right)_P & \left(\frac{\partial g}{\partial Y}\right)_P \end{pmatrix}$. If α maps a curve C into a curve C', then $(d\alpha)_P$ maps the tangent space to C into that of C'. For the Frobenius map, $(f(X,Y), g(X,Y)) = (X^p, Y^p)$, and so $\frac{\partial f}{\partial X}, \frac{\partial f}{\partial Y}, \frac{\partial g}{\partial X}, \frac{\partial g}{\partial Y}$ are all identically zero. Hence the maps on tangent spaces are all zero. □

Curves of genus 1 over $\mathbb{F}_p$

PROPOSITION 9.2 *Let E be an elliptic curve over $\mathbb{F}_p$. Then $H^1(\mathbb{F}_p, E) = 0$, and so all principal homogeneous spaces for E over $\mathbb{F}_p$ are trivial*

PROOF. To prove the proposition, we must show that every continuous crossed homomorphism $f: \Gamma \to E(\mathbb{F})$ is principal. Let $\varphi: E \to E$ be the Frobenius map. It acts on $E(\mathbb{F})$ as the Frobenius automorphism $\sigma \in \Gamma$. The regular map $\varphi - \mathrm{id}_E: E \to E$ is nonconstant, and so

$$P \mapsto \varphi(P) - P: E(\mathbb{F}) \to E(\mathbb{F})$$

is surjective (I, 4.23). In particular, there exists a $P \in E(\mathbb{F})$ such that $\varphi(P) - P = f(\sigma)$, i.e., such that $f(\sigma) = \sigma P - P$. Now

$$f(\sigma^2) = f(\sigma) + \sigma f(\sigma) = \sigma P - P + \sigma^2 P - \sigma P = \sigma^2 P - P,$$

and, inductively,

$$f(\sigma^n) = f(\sigma) + \sigma f(\sigma^{n-1}) = \sigma P - P + \sigma(\sigma^{n-1} P - P) = \sigma^n P - P.$$

Therefore f and the principal crossed homomorphism $\tau \mapsto \tau P - P$ agree on σ^n for all n. Because both are continuous, this implies that they agree on the whole of Γ.

The second part of the statement follows from the first because of Theorem 7.8. □

COROLLARY 9.3 *Every nonsingular projective curve C of genus 1 over $\mathbb{F}_p$ has a point with coordinates in $\mathbb{F}_p$.*

PROOF. According to Theorem 7.19 there exists an elliptic curve E over $\mathbb{F}_p$ such that C has the structure of a principal homogeneous space for E. According to the proposition, C is trivial, i.e., $C(\mathbb{F}_p) \neq \emptyset$. □

Proof of the Riemann hypothesis for elliptic curves

THEOREM 9.4 (RIEMANN HYPOTHESIS) *For an elliptic curve E over $\mathbb{F}_p$,*

$$|\#E(\mathbb{F}_p) - (p+1)| \leq 2\sqrt{p}.$$

PROOF. The kernel of

$$\mathrm{id}_E - \varphi: E(\mathbb{F}) \to E(\mathbb{F})$$

is the set of points $(x:y:z)$ on E such that $(x^p:y^p:z^p) = (x:y:z)$. These are precisely the points having a representative $(x:y:z)$ with $x,y,z \in \mathbb{F}_p$ (see 1.4). Because φ acts as zero on the tangent space at O, $\mathrm{id}_E - \varphi$ acts as the identity map, and so $\mathrm{id}_E - \varphi$ is separable (I, 4.24b). It follows (I, 4.24a) that

$$\#E(\mathbb{F}_p) = \deg(\mathrm{id}_E - \varphi).$$

Let $f(T) = T^2 - sT + t$ be the characteristic polynomial of φ as an endomorphism of E. Then $s,t \in \mathbb{Z}$, $4t \geq s^2$, and $f(n) = \deg(n-\varphi)$ for all $n \in \mathbb{Z}$ (II, 6.2). In particular,

$$t = f(0) = \deg(\varphi) = p,$$
$$1 - s + p = f(1) = \deg(1-\varphi) = \#E(\mathbb{F}_p),$$

and so

$$\left|\#E(\mathbb{F}_p) - (p+1)\right| = |s| \leq 2\sqrt{p}. \qquad \square$$

After this short proof, we spend the rest of the section explaining why the inequality is called the Riemann hypothesis.

Everything in the last three sections holds with p replaced by q.

ASIDE 9.5 Let E be an elliptic curve over $\mathbb{Q}$. According to Theorem 9.4, for a prime p where E as good reduction, the number of points on E modulo p is $N_p = p + 1 - a_p$ with $|a_p| \leq 2\sqrt{p}$. We can write

$$a_p = 2\sqrt{p}\cos\theta_p, \quad 0 \leq \theta_p \leq \pi.$$

When E does not have complex multiplication over $\mathbb{C}$, i.e., when $\mathrm{End}_{\mathbb{C}}(E) = \mathbb{Z}$, Mikio Sato[13] found computationally (in March 1963) that the θ_p appeared to have a density distribution $\frac{2}{\pi}\sin^2\theta$, i.e., for $0 \leq c < d \leq \pi$,

$$\lim_{N\to\infty} \frac{\#\{p \mid p \text{ prime}, p \leq N, c \leq \theta_p \leq d\}}{\#\{p \mid p \text{ prime}, p \leq N\}} = \frac{2}{\pi}\int_c^d \sin^2\theta \, d\theta.$$

In December 1963, Tate announced his famous conjectures relating the algebraic cycles on an algebraic variety to the poles of its zeta function. In examining the powers of an elliptic curve over $\mathbb{Q}$, Tate found (heuristically) that Sato's distribution is the only symmetric density distribution for which the poles of the zeta functions are as predicted by his conjecture. Serre (1968, Appendix to Chap. I) gave a rigorous treatment of Tate's argument, and referred to the statement that the θ_p are distributed as $\frac{2}{\pi}\sin^2\theta$ as the Sato–Tate conjecture. Sato did not continue working

[13] According to Sarnak (lecture 2018), Sato did much more than compute, and may have been the first to understand the use of symmetric tensor powers in this context.

in number theory, but the names of Sato and Tate were linked again forty years later when they were both awarded the Wolf prize for 2002/3.

The Sato–Tate conjecture has now been proved by Michael Harris, Richard Taylor, and their collaborators. See Barnet-Lamb et al. 2011, which states: "In passing we also note that the Sato–Tate conjecture can now be proved for any elliptic curve over a totally real field."

Zeta functions of number fields

First recall that the original Riemann zeta function is

$$\zeta(s) = \prod_{p \text{ prime}} \frac{1}{1-p^{-s}} = \sum_{n \geq 1} n^{-s}, \quad s \in \mathbb{C}, \quad \Re(s) > 1. \tag{39}$$

The second expression is obtained from the first by expanding it out

$$\prod_p \frac{1}{1-p^{-s}} = \prod_p (1 + p^{-s} + (p^{-s})^2 + (p^{-s})^3 + \cdots)$$

to get a sum of terms

$$(p_1^{-s})^{r_1}(p_2^{-s})^{r_2} \cdots (p_t^{-s})^{r_t} = (p_1^{r_1} \cdots p_t^{r_t})^{-s}$$

and applying unique factorization.

Both the sum and the product in (39) converge for $\Re(s) > 1$, and so $\zeta(s)$ is holomorphic and nonzero for $\Re(s) > 1$. In fact, $\zeta(s)$ extends to a meromorphic function on the whole complex plane with a simple pole at $s = 0$. Moreover, the function $\xi(s) = \pi^{-\frac{s}{2}}\Gamma(\frac{s}{2})\zeta(s)$ satisfies the functional equation $\xi(s) = \xi(1-s)$, has simple poles at $s = 0, 1$, and is otherwise holomorphic. Here $\Gamma(s)$ is the gamma function. Since $\Gamma(s)$ has poles at $s = 0, -1, -2, -3, \ldots$, this forces ζ to be zero at $s = -2n$, $n > 0$, $n \in \mathbb{Z}$. These are called the trivial zeros of the zeta function.

CONJECTURE 9.6 (RIEMANN HYPOTHESIS) *The nontrivial zeros of $\zeta(s)$ lie on the line $\Re(s) = \frac{1}{2}$.*

This is probably the most famous open problem in mathematics.

Dedekind extended Riemann's definition by attaching a zeta function $\zeta_K(s)$ to every number field K. He defined

$$\zeta_K(s) = \prod_{\mathfrak{p}} \frac{1}{1-\mathbb{N}(\mathfrak{p})^{-s}} = \sum_{\mathfrak{a}} \mathbb{N}(\mathfrak{a})^{-s}, \quad s \in \mathbb{C}, \quad \Re(s) > 1,$$

where $\mathfrak{p}$ runs over the nonzero prime ideals in $\mathcal{O}_K$ and $\mathfrak{a}$ runs over the nonzero ideals in $\mathcal{O}_K$. Recall that the numerical norm $\mathbb{N}\mathfrak{a}$ of an ideal $\mathfrak{a}$ is the order of the quotient ring $\mathcal{O}_K/\mathfrak{a}$. The second expression is obtained from the first by using that every nonzero ideal $\mathfrak{a}$ has a unique factorization $\mathfrak{a} = \prod \mathfrak{p}_i^{r_i}$ into a product of powers of prime ideals (see 3.9) and that $\mathbb{N}(\mathfrak{a}) = \prod \mathbb{N}(\mathfrak{p}_i)^{r_i}$. Note that for $K = \mathbb{Q}$, this definition gives back $\zeta(s)$. The function $\zeta_K(s)$ extends to a meromorphic function on the whole complex plane with a simple pole at $s = 1$, and a certain multiple $\xi_K(s)$ of it satisfies a functional equation $\xi_K(s) = \xi_K(1-s)$. It is conjectured that the nontrivial zeros of $\zeta_K(s)$ lie on the line $\Re(s) = \frac{1}{2}$ (generalized Riemann hypothesis).

Zeta functions of affine plane curves over finite fields

Consider a nonsingular affine plane curve

$$C : f(X,Y) = 0$$

over the field $\mathbb{F}_p$. In analogy with Dedekind's definition, we define

$$\zeta(C,s) = \prod\nolimits_{\mathfrak{p}} \frac{1}{1-\mathbb{N}\mathfrak{p}^{-s}}, \qquad \Re(s) > 1,$$

where $\mathfrak{p}$ runs over the nonzero prime ideals in

$$\mathbb{F}_p[x,y] \stackrel{\text{def}}{=} \mathbb{F}_p[X,Y]/(f(X,Y))$$

and $\mathbb{N}\mathfrak{p}$ is the order of the quotient ring $\mathbb{F}_p[x,y]/\mathfrak{p}$.

Because $\mathbb{F}_p[x,y]/\mathfrak{p}$ is finite and an integral domain, it is a field, and we let $\deg(\mathfrak{p})$ denote its degree over $\mathbb{F}_p$. Then $\mathbb{N}\mathfrak{p} = p^{\deg \mathfrak{p}}$, which allows us to make a change of variables in the zeta function. When we define[14]

$$Z(C,T) = \prod\nolimits_{\mathfrak{p}} \frac{1}{1-T^{\deg \mathfrak{p}}},$$

then

$$\zeta(C,s) = Z(C,p^{-s}).$$

The product $\prod \frac{1}{1-T^{\deg \mathfrak{p}}}$ converges for all small T, but usually we just regard it as a formal power series

$$Z(C,T) = \prod\nolimits_{\mathfrak{p}} (1 + T^{\deg \mathfrak{p}} + T^{2\deg \mathfrak{p}} + T^{3\deg \mathfrak{p}} + \cdots) \in \mathbb{Z}[[T]].$$

[14]The Z is an upper case zeta.

EXAMPLE 9.7 We can realize $\mathbb{A}^1$ as the affine plane curve $Y = 0$, so that

$$\mathbb{F}_p[x, y] = \mathbb{F}_p[X, Y]/(Y) = \mathbb{F}_p[X].$$

The nonzero ideals on $\mathbb{F}_p[X]$ are in one-to-one correspondence with the monic polynomials $f \in \mathbb{F}_p[X]$, and the ideal (f) is prime if and only if f is irreducible. Thus

$$Z(\mathbb{A}^1, T) = \prod_{f \text{ monic irreducible}} \frac{1}{1 - T^{\deg f}}.$$

Take logs of both sides,

$$\log Z(\mathbb{A}^1, T) = -\sum_f \log(1 - T^{\deg f}),$$

and then differentiate

$$\begin{aligned} \frac{Z'(\mathbb{A}^1, T)}{Z(\mathbb{A}^1, T)} &= \sum_f \frac{\deg f \cdot T^{\deg f - 1}}{1 - T^{\deg f}} \\ &= \sum_f \sum_{n \geq 0} \deg f \cdot T^{(n+1)\deg f - 1}. \end{aligned}$$

In this power series, the coefficient of T^{m-1} is $\sum \deg f$, where f runs over all monic irreducible polynomials $f \in \mathbb{F}_p[T]$ of degree dividing m. As we observed, these polynomials are exactly the irreducible factors of $X^{p^m} - X$, and so the coefficient of T^{m-1} is p^m, and we have shown that

$$\frac{Z'(\mathbb{A}^1, T)}{Z(\mathbb{A}^1, T)} = \sum p^m T^{m-1}.$$

On integrating, we find that

$$\log Z(\mathbb{A}^1, T) = \sum \frac{p^m T^m}{m} = \log \frac{1}{1 - pT}, \tag{40}$$

and so

$$Z(\mathbb{A}^1, T) = \frac{1}{1 - pT}.$$

Expressing $Z(C,T)$ in terms of the points of C

Let $N_m = \#\mathbb{A}^1(\mathbb{F}_{p^m}) = p^m$. Then (40) can be rewritten as

$$\log Z(\mathbb{A}^1,T) = \sum \frac{N_m T^m}{m}.$$

We show that a similar formula holds for every nonsingular affine curve over $\mathbb{F}_p$.

Let $C\colon f(X,Y) = 0$ be a nonsingular affine curve over $\mathbb{F}_p$, and let $\mathbb{F}_p[x,y] = \mathbb{F}_p[X,Y]/(f)$. As for $\mathbb{A}^1$,

$$\begin{aligned}\frac{Z'(C,T)}{Z(C,T)} &= \sum_{\mathfrak{p}} \frac{\deg\mathfrak{p}\cdot T^{\deg\mathfrak{p}-1}}{1-T^{\deg\mathfrak{p}}} \\ &= \sum_{\mathfrak{p}}\sum_{n\geq 0} \deg\mathfrak{p}\cdot T^{(n+1)\deg\mathfrak{p}}/T.\end{aligned}$$

In this power series, the coefficient of T^{m-1} is $\sum \deg\mathfrak{p}$, where $\mathfrak{p}$ runs over the nonzero prime ideals of $k[x,y]$ such that $\deg\mathfrak{p}$ divides m. But

$$\deg\mathfrak{p} \stackrel{\text{def}}{=} [\mathbb{F}_p[x,y]/\mathfrak{p}\colon\mathbb{F}_p],$$

and so the condition $\deg\mathfrak{p}|m$ means that there exists a homomorphism $\mathbb{F}_p[C]/\mathfrak{p} \hookrightarrow \mathbb{F}_{p^m}$. There will in fact be exactly $\deg\mathfrak{p}$ such homomorphisms because $\mathbb{F}_p[C]/\mathfrak{p}$ is separable over $\mathbb{F}_p$. Conversely, every homomorphism $\mathbb{F}_p[x,y] \to \mathbb{F}_{p^m}$ factors through $\mathbb{F}_p[x,y]/\mathfrak{p}$ for some prime ideal with $\deg\mathfrak{p}|m$. Therefore, the coefficient of T^{m-1} in the above power series is the number of homomorphisms (of $\mathbb{F}_p$-algebras)

$$\mathbb{F}_p[x,y] \to \mathbb{F}_{p^m}.$$

But a homomorphism $\mathbb{F}_p[x,y] \to \mathbb{F}_{p^m}$ is determined by the images a,b of x,y, and conversely the homomorphism

$$P(X,Y) \mapsto P(a,b)\colon \mathbb{F}_p[X,Y] \to \mathbb{F}_{p^m}$$

defined by a pair $(a,b) \in \left(\mathbb{F}_{p^m}\right)^2$ factors through $\mathbb{F}_p[x,y]$ if and only if $f(a,b) = 0$. Therefore there is a natural one-to-one correspondence

$$\{\text{homomorphisms } \mathbb{F}_p[C] \to \mathbb{F}_{p^m}\} \stackrel{1:1}{\longleftrightarrow} C(\mathbb{F}_{p^m}),$$

and we have proved that

$$\log Z(C,T) = \sum_{m\geq 1} N_m \frac{T^m}{m}, \quad N_m = \#C(\mathbb{F}_{p^m}).$$

PROPOSITION 9.8 *Let $Z(C,T)$ be the zeta function of a nonsingular affine plane curve C over $\mathbb{F}_p$; then*

$$Z(C,T) = \exp\left(\sum_{m\geq 1} N_m \frac{T^m}{m}\right).$$

PROOF. Apply exp to the preceding equality. □

Zeta functions of function fields

Let C be a nonsingular affine plane curve over $\mathbb{F}_p$. If $\mathfrak{p}$ is a prime ideal in $k[C]$, then $k[C]_{\mathfrak{p}}$ is a discrete valuation ring in $k(C)$ with maximal ideal $\mathfrak{p}k[C]_{\mathfrak{p}}$ and residue field $k[C]_{\mathfrak{p}}/\mathfrak{p}k[C]_{\mathfrak{p}} \simeq k[C]/\mathfrak{p}$. Thus,

$$\zeta(C,s) = \prod_R \frac{1}{1-(R{:}\mathfrak{m}_R)^{-s}},$$

where R runs over the discrete valuation rings corresponding to prime ideals in $k[C]$ and $\mathfrak{m}_R$ is the maximal ideal in R.

Let K be a function field in one variable over $\mathbb{F}_p$. The above discussion suggests defining

$$\zeta(K,s) = \prod_R \frac{1}{1-(R{:}\mathfrak{m}_R)^{-s}}$$

where R runs over the discrete valuation rings in K with field of fractions K. Define $Z(K,T)$ similarly. Then, for example,

$$\zeta(\mathbb{F}_p(T),s) = \frac{1}{(1-p^{-s})}\zeta(\mathbb{A}^1,s) = \frac{1}{(1-p^{-s})(1-p^{1-s})}$$

$$Z(\mathbb{F}_p(T),s) = \frac{1}{1-T}Z(\mathbb{A}^1,T) = \frac{1}{(1-T)(1-pT)}.$$

Zeta functions of projective curves

Let C be a nonsingular projective curve over $\mathbb{F}_p$. We define the zeta function of C to be the zeta function of its function field. For example,

$$Z(\mathbb{P}^1,T) = \frac{1}{(1-T)(1-pT)}$$

$$\zeta(\mathbb{P}^1,s) = \frac{1}{(1-p^{-s})(1-p^{1-s})}.$$

As in the affine case, we have expressions

$$\begin{cases} Z(C,T) = \exp\left(\sum_{m\geq 1} N_m \frac{T^m}{m}\right), & N_m = \#C(\mathbb{F}_{p^m}), \\ \zeta(C,s) = Z(C,p^{-s}). \end{cases}$$

For example,

$$N_m(\mathbb{P}^1) = N_m(\mathbb{A}^1) + 1, \text{ for all } m,$$

and so

$$\log Z(\mathbb{P}^1,T) = \log Z(\mathbb{A}^1,T) + \log \frac{1}{1-T}.$$

Therefore

$$Z(\mathbb{P}^1,T) = \frac{1}{1-T} Z(\mathbb{A}^1,T) = \frac{1}{(1-T)(1-pT)}.$$

Similarly, for E an elliptic curve over $\mathbb{F}_p$,

$$Z(E,T) = \frac{1}{1-T} Z(E^{\text{aff}},T),$$

where E^{aff} is the affine curve $E \cap \{Z \neq 0\}$,

Counting positive divisors of fixed degree

We need the next result in the proof that $Z(E,T)$ is a rational function of T.

PROPOSITION 9.9 *Let E be an elliptic curve over $\mathbb{F}_p$. For all $m \geq 1$, there are exactly $\#E(\mathbb{F}_p)(p^m - 1)/(p-1)$ positive divisors of degree m.*

PROOF. Fix a divisor D_0, and let $P(D_0)$ be the set of all positive divisors D on E such that $D \sim D_0$, i.e., $D = D_0 + (f)$ for some $f \in \mathbb{F}_p(E)^\times$. Then

$$f \mapsto D_0 + (f) \colon L(D_0) \smallsetminus \{0\} \to P(D_0)$$

is surjective, and two functions have the same image if and only if one is a constant multiple of the other. We therefore have a bijection

$$(L(D_0) \smallsetminus \{0\})/\mathbb{F}_p^\times \to P(D_0).$$

When $\deg D_0 = m \geq 1$, the Riemann–Roch theorem (I, 4.13) shows that $L(D_0)$ has dimension m, and so $P(D_0)$ has $(p^m - 1)/(p-1)$ elements.

Let $\mathrm{Pic}(E) = \mathrm{Div}(E)/\{(f) \mid f \in k(E)^{\times}\}$ — according to 1.10, this agrees with the definition in I, §4. Because the degree of a principal divisor is zero, the degree map factors through $\mathrm{Pic}(E)$,

$$\deg\colon \mathrm{Pic}\, E \longrightarrow \mathbb{Z}.$$

Note that the map deg is surjective because $\mathfrak{p}_{\infty} \mapsto 1$. We define

$$\mathrm{Pic}^{m}\, E = \{\mathfrak{d} \in \mathrm{Pic}(E) \mid \deg \mathfrak{d} = m\}.$$

Then,

(a) the map $\mathrm{Pic}^{0}(E) \to \mathrm{Pic}^{m}(E)$, $\mathfrak{d} \mapsto \mathfrak{d} + m\mathfrak{p}_{\infty}$, is a bijection (obvious);

(b) there exists a bijection $E(k) \to \mathrm{Pic}^{0}(E)$ ((8), p. 46).

We are now able to count the positive divisor classes of degree m on E. We saw above that each class in $\mathrm{Pic}^{m}(E)$ has $\frac{p^m-1}{p-1}$ elements, and that there are

$$\#\, \mathrm{Pic}^{m}(E) \overset{(a)}{=} \#\, \mathrm{Pic}^{0}(E) \overset{(b)}{=} \#E(\mathbb{F}_p)$$

such classes. Therefore, altogether, there are

$$\#E(\mathbb{F}_p)\frac{p^m-1}{p-1}$$

positive divisors of degree m, as claimed. □

The rationality of the zeta function of an elliptic curve

THEOREM 9.10 *Let E be an elliptic curve over $\mathbb{F}_p$. Then*

$$Z(E,T) = \frac{1+(N_1-p-1)T+pT^2}{(1-T)(1-pT)}, \quad N_1 = \#E(\mathbb{F}_p).$$

PROOF. Here

$$Z(E,T) = \frac{1}{1-T} Z(E^{\mathrm{aff}},T) = \frac{1}{1-T}\prod_{\mathfrak{p}} \frac{1}{1-T^{\deg \mathfrak{p}}},$$

where the $\mathfrak{p}$ run through the prime ideals of $\mathbb{F}_p[C]$. On expanding the product we find that

$$Z(E,T) = \sum d_m T^m,$$

where $d_0 = 1$ and d_m is the number of positive divisors of degree m. According to Proposition 9.9, $d_m = N_1 \frac{p^m - 1}{p-1}$. Therefore,

$$\begin{aligned} Z(E,T) &= 1 + \sum_{m>0} N_1 \frac{p^m - 1}{p-1} T^m \\ &= 1 + \frac{N_1}{p-1} \left(\frac{1}{1-pT} - \frac{1}{1-T} \right) \\ &= \frac{1 + (N_1 - p - 1)T + pT^2}{(1-T)(1-pT)}. \end{aligned}$$

□

REMARK 9.11 (a) Write

$$1 + (N_1 - p - 1)T + pT^2 = (1 - \alpha T)(1 - \beta T),$$

so that α, β are algebraic integers such that

$$N_1 - p - 1 = -\alpha - \beta, \quad \alpha\beta = p.$$

Then

$$\log Z(E,T) = \log \frac{(1-\alpha T)(1-\beta T)}{(1-T)(1-pT)} = \sum_m (1 + p^m - \alpha^m - \beta^m) \frac{T^m}{m},$$

and so

$$N_m(E) = 1 + p^m - \alpha^m - \beta^m.$$

Thus, if one knows N_1, one can find α and β, and the whole of the sequence

$$N_1(E), N_2(E), N_3(E), \ldots$$

(b) With the notation in (a),

$$\zeta(E,s) = \frac{(1-\alpha p^{-s})(1-\beta p^{-s})}{(1-p^{-s})(1-p^{1-s})}.$$

It has simple poles at $s = 0$ and $s = 1$, and zeros where $p^s = \alpha$ and $p^s = \beta$. Write $s = \sigma + it$. Then $|p^s| = p^\sigma$, and the zeros of $\zeta(E,s)$ have real part $\frac{1}{2}$ if and only if α and β have absolute value $p^{\frac{1}{2}}$.

By definition α and β are the inverse roots of a polynomial

$$1 + bT + pT^2, \quad b = N_1 - p - 1.$$

If $b^2 - 4p \le 0$, then α and β are complex conjugates. Since their product is p, this implies that they each have absolute value $p^{\frac{1}{2}}$. Conversely, if $|\alpha| = p^{\frac{1}{2}} = |\beta|$, then

$$|N_1 - p - 1| = |\alpha + \beta| \le 2\sqrt{p}.$$

Thus, granted Theorem 9.10, the Riemann hypothesis for E is equivalent to the statement

$$|N_1 - p - 1| \le 2\sqrt{p}.$$

EXERCISE 9.12 Prove that the zeta function of an elliptic curve E over $\mathbb{F}_p$ satisfies the functional equation

$$\zeta(E,s) = \zeta(E,1-s).$$

EXERCISE 9.13 Compute the zeta functions for the curve

$$E : Y^2Z + YZ^2 = X^3 - X^2Z$$

over the fields $\mathbb{F}_2$, $\mathbb{F}_3$, $\mathbb{F}_5$, $\mathbb{F}_7$, and verify the Riemann hypothesis in each case. How many points does the curve have over the field with 625 elements?

EXERCISE 9.14 (a) Let E be the elliptic curve

$$E : Y^2Z = X^3 - 4X^2Z + 16Z^3.$$

Compute $N_p \stackrel{\text{def}}{=} \#E(\mathbb{F}_p)$ for all primes $3 \le p \le 13$ (more if you use a computer).

(b) Let $F(q)$ be the (formal) power series given by the infinite product

$$F(q) = q\prod_{n=1}^{\infty}(1-q^n)^2(1-q^{11n})^2 = q - 2q^2 - q^3 + 2q^4 + \cdots.$$

Calculate the coefficient M_n of q^n in $F(q)$ for $n \le 13$ (more if you use a computer).

(c) For each prime p, compute the sum $M_p + N_p$. Formulate a conjecture as to what $M_p + N_p$ should be in general.

(d) Prove your conjecture. [This is probably very difficult, perhaps even impossible, using only the information covered in the course.]

(Your conjecture is, or, at least, should be a special case of a theorem of Eichler and Shimura. The big result of Wiles (and others) is that the theorem of Eichler and Shimura applies to all elliptic curves over $\mathbb{Q}$.)

A brief history of zeta

Since the first edition of this book was published, Roquette's extensive account of the early history of Riemann hypothesis in characteristic p has become available (Roquette 2018). I have drawn on it, and Oort and Schappacher 2016, in revising these brief notes, and recommend both to readers. For the story from "Weil to the present day", see Milne 2016.

The story begins, as do most stories in number theory, with Gauss.

GAUSS 1814

From 1796 to 1814, Gauss kept a mathematical notebook, whose last entry reads,

> If $a+bi$ is a prime number with $a-1+bi$ divisible by $2+2i$, then the number of solutions of the congruence
>
> $$1 = xx + yy + xxyy \mod a+bi,$$
>
> including $x=\infty$, $y=\pm i$, $x=\pm i$, $y=\infty$, equals $(a-1)^2 + bb$.

Gauss's curve $X^2+Y^2+X^2Y^2=1$ is affine and singular, but its associated nonsingular curve is the elliptic curve $E: Y^2Z = X^3+4XZ^2$, and Gauss's statement implies that

$$|N-p-1| < 2p^{1/2}, \quad N = \#E(\mathbb{F}_p),$$

when p is a prime congruent to 1 mod 4. For the remaining odd primes, it is easy to see that $N = p+1$, and so Gauss is asserting that the Riemann hypothesis holds for E. A proof of Gauss's statement was provided in 1921 by Herglotz.[15] See Roquette 2018, 3.4

EMIL ARTIN 1921

In his thesis, Artin studied the algebraic number theory of function fields of the form $\mathbb{F}_p(X)[\sqrt{f(X)}]$, $f(X)\in\mathbb{F}_p[X]$, and defined their zeta functions. He proved that the zeta function is a rational function of p^{-s} and satisfies a functional equation, and he checked that its zeros are on the line $R(s)=1/2$ for about 40 polynomials f.

[15]The curve E has complex multiplication by i, and Gauss's more precise statement can be regarded as the begining of the theory of complex multiplication for elliptic curves.

Artin completed his thesis in 1921. In the same year he obtained further results, which he did not publish, but included in letters to his advisor Herglotz. For example, he proved the following criterion:

> Let C be an affine curve over $\mathbb{F}_p$, and let $N_m = \#C(\mathbb{F}_{p^m})$; the Riemann hypothesis holds for C if and only if
>
> $$|N_m - p^m| < Ap^{m/2}, \quad m = 1,2,3,\ldots,$$
>
> for some constant A, independent of m.

F.K. SCHMIDT 1925–1931

Schmidt proved the Riemann–Roch theorem for a nonsingular projective curve C over an arbitrary field k. When $k = \mathbb{F}_q$, he used it to show, as we did for elliptic curves, that

$$Z(C,T) = \frac{P(T)}{(1-T)(1-qT)}, \quad P(T) = 1 + \cdots \in \mathbb{Z}[T], \quad \deg P(T) = 2g.$$

Moreover,

$$Z(1/qT) = q^{1-g} T^{2-2g} Z(T),$$

and so

$$\zeta(1-s) = q^{1-g}(q^{2-2g})^{-s}\zeta(s).$$

Write

$$P(T) = \prod(1-\alpha_j T).$$

Then the formulas show that

$$N_m = 1 + q^m - \sum \alpha_j^m.$$

Thus, once one knows $\alpha_1, \ldots, \alpha_{2g}$, then one knows N_m for all m. However, unlike the elliptic curve case, N_1 does not determine the α_j — one needs to know several of the N_m.

DAVENPORT, MORDELL, HASSE 1932.

Davenport and Mordell were interested in the following problem: let $f(X,Y) \in \mathbb{Z}[X,Y]$, and let p be a prime number such that f is geometrically irreducible modulo p; find an estimate of the form

$$|N - p| \leq Ap^{\gamma} \tag{41}$$

for the number of solutions N of $f(X,Y) = 0$ in $\mathbb{F}_p$, where A is a positive constant and the exponent $\gamma < 1$ is as small as possible; neither γ nor A should depend on p.

They obtained a number of results by ad hoc methods. For example, for the polynomial

$$aX^m + bY^n + c,$$

Davenport proved that the estimate (41) holds with the best possible γ, namely, $1/2$. Hasse improved Davenport's result by showing that the estimate continues to hold, with the same constant A, when p is replaced by p^r. When told of this statement, Artin explained to Hasse that it is equivalent to the Riemann hypothesis for the curve $aX^m + bY^n + c = 0$, and, indeed, that the Davenport-Mordell problem, when extended to arbitrary finite base fields $\mathbb{F}_{p^r}$, is equivalent to the characteristic p Riemann hypothesis. It was this that prompted Hasse's interest in the Riemann hypothesis for function fields.[16]

HASSE, DEURING, ROQUETTE 1932–1951

Hasse proved the Riemann hypothesis for elliptic curves over finite fields. His proof, as in the proof we gave, is based on the endomorphism ring of the elliptic curve. Deuring correctly saw that the key to extending Hasse's proof to curves of arbitrary genus was to replace the endomorphism ring of the elliptic curve with the ring of correspondences of the curve with itself. However, the function field methods used by the German school were badly adapted to the study of correspondences, which required working with "double fields", and it was not until 1951 that Hasse's student Roquette completed a proof of the Riemann hypothesis for curves over finite fields along these lines.

WEIL 1940–1948

In 1940, Weil announced a proof of the Riemann hypothesis for all curves, i.e., that $|\alpha_i| = p^{1/2}$ for $1 \leq i \leq 2g$, where α_i is above. His proof assumed the existence of a theory of algebraic geometry over arbitrary fields, including of jacobian and abelian varieties, which at the time was known only over $\mathbb{C}$. He spent most of the 1940s developing the algebraic geometry he

[16]Given this, it is surprising that the formula $Z(C,T) = \exp(\sum N_m T^m/m)$ seems not to have appeared in the literature before 1940 (Oort and Schappacher 2016, 2.3).

needed, and gave a detailed proof of the Riemann hypothesis for curves in a book published in 1948.

WEIL 1949

Weil's knowledge of the zeta functions of curves, abelian varieties, and some other special algebraic varieties led him to state his famous "Weil conjectures". For a nonsingular projective variety V over $\mathbb{F}_p$, one can define (as for curves)

$$\begin{cases} Z(V,T) = \exp\left(\sum_{m\geq 1} N_m \frac{T^m}{m}\right), & N_m = \#V(\mathbb{F}_{p^m}), \\ \zeta(V,s) = Z(V,p^{-s}). \end{cases}$$

Weil conjectured that $Z(V,T)$ is a rational function of T, that $\zeta(V,s)$ satisfies a functional equation, and that a "Riemann hypothesis" holds for $\zeta(V,s)$. Weil knew that the rationality and functional equation would follow from the existence of a cohomology theory for varieties over arbitrary fields giving the "correct" Betti numbers and admitting a Lefschetz fixed point formula and a Poincaré duality theorem.[17]

DWORK 1960

After Weil stated his conjectures, the search began for a cohomology theory with the properties needed to prove the rationality and functional equation. Dwork, unexpectedly, found an "elementary" proof that $Z(V,T)$ is a rational function of T.

GROTHENDIECK ET AL. 1963/64

Grothendieck defined étale cohomology and, with the help of M. Artin and Verdier, developed it sufficiently to prove that $Z(V,T)$ is rational and satisfies a functional equation. The results of Grothendieck, Artin, and Verdier show that, for a nonsingular projective variety V over $\mathbb{F}_p$,

$$Z(V,T) = \frac{P_1(T)P_3(T)\cdots P_{2d-1}(T)}{(1-T)P_2(T)P_4(T)\cdots P_{2d-2}(T)(1-p^dT)},$$

[17] It is not clear whether Weil believed that such a cohomology should exist. The results of Hasse (II, 6.9) show that no such cohomology can exist for varieties in characteristic p with coefficients in $\mathbb{Q}$ (or even in $\mathbb{Q}_p$ or in $\mathbb{R}$).

where $d = \dim V$ and $P_i(T) \in \mathbb{Z}[T]$. Moreover

$$Z(V, 1/p^d T) = \pm T^\chi p^{d\chi/2} Z(V,T),$$

where χ is the self-intersection number of the diagonal in $V \times V$.

DELIGNE 1973

Deligne used étale cohomology to prove the remaining Weil conjecture, namely, the Riemann hypothesis,

$$P_i(T) = \prod\nolimits_j (1 - \alpha_{ij} T), \quad |\alpha_{ij}| = p^{i/2}.$$

This last statement says that $\zeta(V,s)$ has its zeros on the lines

$$\Re(s) = \frac{1}{2}, \frac{3}{2}, \ldots, \frac{2d-1}{2}$$

and its poles on the lines

$$\Re(s) = 0, 1, 2, \ldots, d.$$

EXERCISE 9.15 Let C be a nonsingular projective curve over $\mathbb{F}_q$ of genus g.
(a) Show that

$$Z(C,T) = \frac{P(T)}{(1-T)(1-qT)}, \quad P(T) = 1 + \cdots \in \mathbb{Z}[T], \quad \deg P(T) = 2g.$$

(b) Show that

$$Z\left(\frac{1}{qT}\right) = q^{1-g} T^{2-2g} Z(T),$$

Deduce that if α is an inverse root of P, then so also is q/α.

(c) (Artin's criterion) Show that the Riemann hypothesis holds for C if and only if

$$|N_m - q^m - 1| \le 2g\sqrt{q^m},$$

all $m \ge 1$, where $N_m = \#C(\mathbb{F}_{q^m})$.

Hints. For (a), see the proof of Theorem 9.10. For (b), let

$$Z(C,T) = \prod\nolimits_{\mathfrak{p}} \frac{1}{1 - T^{\deg(\mathfrak{p})}} = \sum\nolimits_{n \ge 0} a_n T^n, \quad a_n = \sum\nolimits_{\mathfrak{d}} \frac{q^n - 1}{q-1},$$

where $\mathfrak{d}$ runs over the divisor classes of degree n, and apply the Riemann–Roch theorem. For (c), use the inequality to show that

$$\log P(T^{-1}) = \log \prod\nolimits_{1 \le i \le 2g} (1 - \alpha_i T^{-1}) = \sum_{m \ge 1} (N_m - q^m - 1) \frac{T^{-m}}{m},$$

converges for $q^{\frac{1}{2}} |T|^{-1} < 1$, and deduce that $P(T^{-1})$ has no zero in this region.

NOTES The work of Hasse and Deuring in the 1930s laid the foundations for the modern theory of elliptic curves in arbitrary characteristic. Beyond their calculation of the possible endomorphism rings of elliptic curves (p. 97) and Hasse's proof of the Riemann hypothesis (p. 207), Deuring obtained a number of other important results.

10 The conjecture of Birch and Swinnerton-Dyer

Introduction

We return to the problem of computing the rank of $E(\mathbb{Q})$. Our purely algebraic approach provides only an upper bound for the rank, in terms of the Selmer group, and the difference between the upper bound and the actual rank is measured by the mysterious Tate–Shafarevich group. It is difficult to decide whether an element of $S^{(2)}(E/\mathbb{Q})$ comes from an element of infinite order or survives to give a nontrivial element of $Ш(E/\mathbb{Q})$ — in fact, there is no proven algorithm for doing this. Clearly, it would be helpful to have another approach.

Let E be an elliptic curve over $\mathbb{Q}$. For each prime p of good reduction, we let

$$N_p = \#E_p(\mathbb{F}_p),$$

where E_p denotes the reduction of E over $\mathbb{F}_p$.

One idea is that perhaps the rank $E(\mathbb{Q})$ should be related to the orders of the groups $E_p(\mathbb{F}_p)$. For a good prime p, there is a reduction map

$$E(\mathbb{Q}) \to E_p(\mathbb{F}_p)$$

but, in general, this will be far from injective or surjective. For example, if $E(\mathbb{Q})$ is infinite, then so also is the kernel, and if $E(\mathbb{Q})$ is finite (and hence has order ≤ 16) then it will fail to be surjective for all large p (because $\#E_p(\mathbb{F}_p) \ge p + 1 - 2\sqrt{p}$).

In the late fifties, Birch and Swinnerton-Dyer had the idea[18] that if $E(\mathbb{Q})$ is large then this should force the N_p to be larger than usual. Since they had access to one of the few computers then in existence, they were able to test this computationally. For P a large number (large, depending on the speed of your computer), let

$$f(P) = \prod_{p \leq P} \frac{N_p}{p}.$$

Recall that N_p is approximately p. Their calculations led them to the following conjecture.

CONJECTURE 10.1 *For each elliptic curve E over $\mathbb{Q}$, there exists a constant $c(E)$ such that*

$$\lim_{P \to \infty} f(P) = c(E) \log(P)^r,$$

where $r = \operatorname{rank}(E(\mathbb{Q}))$.

We can write the conjecture more succinctly as

$$f(P) \sim c(E) \log(P)^r \text{ as } P \to \infty.$$

Note the remarkable nature of this conjecture: it predicts that one can determine the rank of $E(\mathbb{Q})$ from the sequence of numbers N_p. Moreover, together with an estimate for the error term, it would provide an algorithm for finding r.

Birch and Swinnerton-Dyer were, in practice, able to predict r from this conjecture with fairly consistent success, but they found that $f(P)$ oscillates vigorously as P increases, and that there seemed to be little hope of finding $c(E)$ with an error of less than, say, 10%. Instead, prompted by Davenport and Cassels, they re-expressed their conjecture in terms of the zeta function of E.

---

[18] In more detail, Birch had learnt, while in Princeton during the academic year 1956–57, of the "beautiful reformulation of Siegel's work on quadratic forms in terms of a natural 'Tamagawa measure' for linear algebraic groups", and they wondered whether there was a similar phenomenon for elliptic curves E over $\mathbb{Q}$. As a substitute for the Tamagawa number of E, they took the product of the p-adic measures of $E(\mathbb{Q}_p)$ over the primes p less than some bound P. For a p of good reduction, the measure is N_p/p, and so this is essentially $f(P)$.

The zeta function of a variety over $\mathbb{Q}$

Let V be a nonsingular projective variety over $\mathbb{Q}$. Such a variety is the zero set of a collection of homogeneous polynomials

$$F(X_0,\ldots,X_n) \in \mathbb{Q}[X_0,\ldots,X_n].$$

Scale each such polynomial so that its coefficients lie in $\mathbb{Z}$ but have no common factor, and let $\bar{F}(X_0,\ldots,X_n) \in \mathbb{F}_p[X_0,\ldots,X_n]$ be the reduction of the scaled polynomial modulo p. If the polynomials $\bar{F}$ define a nonsingular variety V_p over $\mathbb{F}_p$, then we say V has ***good reduction*** at p or that p is ***good*** for V. All but finitely many primes will be good for a given variety.

For each good prime p we have a zeta function

$$\begin{cases} Z(V_p,T) = \exp\left(\sum_{m\geq 1} N_m \dfrac{T^m}{m}\right), & N_m = \#V_p(\mathbb{F}_{p^m}), \\ \zeta(V_p,s) = Z(V_p,p^{-s}), \end{cases}$$

and we define

$$\zeta(V,s) = \prod\nolimits_p \zeta(V_p,s).$$

Because the Riemann hypothesis holds for V_p, the product converges for $\Re(s) > \dim V + 1$ by comparison with $\prod_p \frac{1}{1-p^{-s}}$.

Let $*$ be the point over $\mathbb{Q}$, i.e., $* = \mathbb{A}^0 = \mathbb{P}^0$. For this variety, all primes are good, and

$$\log Z(*_p,T) = \sum 1\frac{T^m}{m} = \log\frac{1}{1-T}.$$

Therefore

$$\zeta(*,s) = \prod_p \frac{1}{1-p^{-s}}$$

which is just the Riemann zeta function, already an interesting function.

Let $V = \mathbb{P}^n$. Again all primes are good, and

$$\#\mathbb{P}^n(\mathbb{F}_q) = \frac{q^{n+1}-1}{q-1} = 1+q+\cdots+q^n, \quad q = p^m,$$

from which it follows that

$$\zeta(\mathbb{P}^n,s) = \zeta(s)\zeta(s-1)\cdots\zeta(s-n)$$

with $\zeta(s)$ the Riemann zeta function.

CONJECTURE 10.2 (HASSE–WEIL) *For any nonsingular projective variety V over $\mathbb{Q}$, $\zeta(V,s)$ can be analytically continued to a meromorphic function on the whole complex plane, and satisfies a functional equation relating $\zeta(V,s)$ with $\zeta(V,d+1-s)$, where $d = \dim V$.*

The conjecture is widely believed to be true, but it is known in only a few cases. Note that the above calculations show that, for $V = \mathbb{P}^n$, it follows from the similar statement for $\zeta(s)$.

Of course, $\zeta(V,s)$ is expected to satisfy a Riemann hypothesis, but this is not known for any variety, even a point.

NOTES According to Weil's recollections (Collected Papers, II, p. 529), Hasse defined the Hasse–Weil zeta function for an elliptic curve over $\mathbb{Q}$, and set the Hasse–Weil conjecture in this case as a thesis problem! Initially, Weil was sceptical of the conjecture, but he proved it for curves of the form $Y^m = aX^n + b$ over number fields by expressing their zeta functions in terms of Hecke L-functions. In particular, Weil showed that the zeta functions of the elliptic curves $Y^2 = aX^3 + b$ and $Y^2 = aX^4 + b$ can be expressed in terms of Hecke L-functions, and he suggested that the same should be true for all elliptic curves with complex multiplication. This was proved by Deuring in a "beautiful series" of papers. Deuring's result was extended to all abelian varieties with complex multiplication by Shimura and Taniyama and Weil. In a different direction, Eichler and Shimura proved the Hasse–Weil conjecture for elliptic modular curves by identifying their zeta functions with the Mellin transforms of modular forms. See Chapter V.

The zeta function of an elliptic curve over $\mathbb{Q}$

Let E be an elliptic curve over $\mathbb{Q}$, and let S be the set of primes where E has bad reduction. According to the above definition

$$\begin{aligned}\zeta(E,s) &= \prod_{p \notin S} \frac{1+(N_p - p - 1)p^{-s} + p^{1-2s}}{(1-p^{-s})(1-p^{1-s})} \\ &= \frac{\zeta_S(s)\zeta_S(s-1)}{L_S(s)},\end{aligned}$$

where $\zeta_S(s)$ is Riemann's zeta function except that the factors corresponding to the primes in S have been omitted, and

$$L_S(E,s) = \prod_{p \notin S} \frac{1}{1+(N_p - p - 1)p^{-s} + p^{1-2s}}.$$

Write

$$1+(N_p-p-1)T+pT^2=(1-\alpha_pT)(1-\beta_pT),$$

so that

$$L_S(E,s)=\prod_p \frac{1}{1-\alpha_p p^{-s}}\frac{1}{1-\beta_p p^{-s}}.$$

The product $\prod_p \frac{1}{1-p^{-s}}$ converges for $\Re(s)>1$, and $|\alpha_p|=p^{\frac{1}{2}}=|\beta_p|$, and so the product $L_S(E,s)$ converges for $\Re(s)>\frac{3}{2}$ by comparison with

$$\prod_p \frac{1}{1-p^{\frac{1}{2}}p^{-s}}.$$

We add factors to $L_S(E,s)$ for the bad primes. Define

$$L_p(T)=\begin{cases} 1-a_pT+pT^2, & \text{if } p \text{ is good,} \quad a_p=p+1-N_p,\\ 1-T, & \text{if } E \text{ has split multiplicative reduction,}\\ 1+T, & \text{if } E \text{ has non-split multiplicative reduction,}\\ 1, & \text{if } E \text{ has additive reduction.}\end{cases}$$

In the four cases

$$L_p(p^{-1})=\frac{N_p}{p},\frac{p-1}{p},\frac{p+1}{p},\frac{p}{p},$$

which, in each case, equals $\#E_p^{\mathrm{ns}}(\mathbb{F}_p)/p$, where E_p^{ns} is the nonsingular part of the elliptic curve modulo p (see the table p. 78). Define

$$L(E,s)=\prod_p \frac{1}{L_p(p^{-s})},$$

where the product is now over all prime numbers. The ***conductor*** $N_{E/\mathbb{Q}}$ of E is defined to be $\prod_{p \text{ bad}} p^{f_p}$, where

$$f_p=\begin{cases} f_p=1 \text{ if } E \text{ has multiplicative reduction at } p\\ f_p\geq 2 \text{ if } E \text{ has additive reduction at } p, \text{ and equals 2 if } p\neq 2,3.\end{cases}$$

The precise definition of f_p when E has additive reduction is a little complicated. However, there is a formula of Ogg that can be used to compute it in all cases,

$$f_p=\mathrm{ord}_p(\Delta)+1-m_p,$$

where m_p is the number of irreducible components of the Néron model (not counting multiplicities) and Δ is the discriminant of the minimal Weierstrass equation (11), p. 67, of E.

Define

$$\Lambda(E,s) = N_{E/\mathbb{Q}}^{s/2}(2\pi)^{-s}\Gamma(s)L(E,s).$$

The completed L-function $\Lambda(E,s)$ can be defined similarly for an elliptic curve over a number field. The following is the precise version of the Hasse–Weil conjecture for the case of an elliptic curve.

CONJECTURE 10.3 *The function $\Lambda(E,s)$ can be analytically continued to a meromorphic function on the whole of complex plane, and it satisfies a functional equation*

$$\Lambda(E,s) = w_E\Lambda(E,2-s), \quad w_E = \pm 1.$$

There is even a recipe for what w_E should be.

As noted earlier, for curves with complex multiplication, the conjecture was proved by Deuring in the early 1950s. The key point is that there is always a "formula" for N_p similar to that proved by Gauss for the curve $Y^2Z = X^3 + 4XZ^2$ (see p. 205) which allows one to identify the L-function of E with an L-function of a type already defined and studied by Hecke (a "Hecke L-function") and for which one knows analytic continuation and a functional equation.

A much more important result, which we shall spend most of the rest of the book discussing, is the following. Let

$$\Gamma_0(N) = \left\{ \begin{pmatrix} a & b \\ c & d \end{pmatrix} \in \mathrm{SL}_2(\mathbb{Z}) \middle| c \equiv 0 \mod N \right\}.$$

Then $\Gamma_0(N)$ acts on the complex upper half-plane, and the quotient $\Gamma_0(N)\backslash\mathbb{H}$ has the structure of a Riemann surface. An elliptic curve E over $\mathbb{Q}$ is said to be ***modular*** if there is a nonconstant map of Riemann surfaces

$$\Gamma_0(N)\backslash\mathbb{H} \to E(\mathbb{C})$$

for some N. Eichler and Shimura (in the fifties and sixties) proved a slightly weaker form of Conjecture 10.3 for modular elliptic curves (see Chap. V, §7).

Recall that a curve is said to have semistable reduction at p if it has good or multiplicative reduction at p; equivalently, if p^2 does not divide the conductor. Wiles (with Richard Taylor) proved that an elliptic curve with square free conductor is modular. This work was extended to *all* elliptic curves over $\mathbb{Q}$ by various authors (see Chapter V). Thus, Conjecture 10.3 is known for all elliptic curves over $\mathbb{Q}$.

REMARK 10.4 (a) Deuring's result is valid for elliptic curves with complex multiplication over any number field. The results of Eichler, Shimura, Wiles, et al. are valid only for elliptic curves over $\mathbb{Q}$. Even today, little is known about the L-functions of elliptic curves over number fields apart from some totally real fields.

(b) Both results prove much more than the simple statement of Conjecture 10.3 — they succeed in identifying the function $\Lambda(E,s)$.

Statement of the conjecture of Birch and Swinnerton-Dyer

Let E be an elliptic curve over $\mathbb{Q}$. Let

$$Y^2Z + a_1XYZ + a_3YZ^2 = X^3 + a_2X^2Z + a_4XZ^2 + a_6Z^3$$

be a minimal Weierstrass equation for E over $\mathbb{Q}$, and let

$$\omega = \frac{dx}{2y + a_1x + a_3}.$$

Recall (§6) that there is a canonical Néron–Tate height pairing

$$\langle\,,\rangle\colon E(\mathbb{Q}) \times E(\mathbb{Q}) \to \mathbb{R}, \quad \langle P, Q\rangle = \hat{h}(P+Q) - \hat{h}(P) - \hat{h}(Q).$$

CONJECTURE 10.5 (BIRCH AND SWINNERTON-DYER) *Let r be the rank of $E(\mathbb{Q})$, and let $P_1, \ldots, P_r$ be $\mathbb{Z}$-linearly independent elements of $E(\mathbb{Q})$; then*

$$L(E,s) \sim \left(\Omega \prod_{p \text{ bad}} c_p\right) \frac{[\text{Ш}(E/\mathbb{Q})]\det(\langle P_i, P_j\rangle)}{(E(\mathbb{Q}) : \sum \mathbb{Z}P_i)^2}(s-1)^r \text{ as } s \to 1,$$

where

$[*]$ = order of $*$ (elsewhere written $\#*$);

$\Omega = \int_{E(\mathbb{R})} |\omega|$;

$c_p = (E(\mathbb{Q}_p) : E^0(\mathbb{Q}_p))$.

In particular, the conjecture predicts that the rank of $E(\mathbb{Q})$ is equal to the order of the zero of $L(E,s)$ at $s=1$. In the literature, this statement is called the ***weak Birch–Swinnerton-Dyer (weak BSD) conjecture***, while 10.5 is called the ***full Birch–Swinnerton-Dyer (full BSD) conjecture***.[19]

REMARK 10.6 (a) For a modular elliptic curve, all terms in the conjecture are computable except for the Tate–Shafarevich group, and, in fact, can be computed by Pari (see II, §7).

(b) The quotient

$$\frac{\det(\langle P_i, P_j\rangle)}{(E(\mathbb{Q}):\sum \mathbb{Z}P_i)^2}$$

is independent of the choice of $P_1,\ldots,P_r$. When they are chosen to form a basis for $E(\mathbb{Q})$ modulo torsion, the quotient equals

$$\frac{D}{[E(\mathbb{Q})_{\text{tors}}]^2},$$

where D is the discriminant of the Néron–Tate pairing.

(c) The integral $\int_{E(\mathbb{Q}_p)}|\omega|$ makes sense, and, in fact it is equal to $(E(\mathbb{Q}_p):E^1(\mathbb{Q}_p))/p$. Indeed (see II, 2.7) there is a bijection $E^1(\mathbb{Q}_p)\leftrightarrow p\mathbb{Z}_p$ under which ω corresponds to the Haar measure on $\mathbb{Z}_p$ for which $\mathbb{Z}_p$ has measure 1 and (therefore) $p\mathbb{Z}_p$ has measure $1/(\mathbb{Z}_p:p\mathbb{Z}_p)=1/p$. Hence,

$$\int_{E(\mathbb{Q}_p)}|\omega| = (E(\mathbb{Q}_p):E^1(\mathbb{Q}_p))\int_{E^1(\mathbb{Q}_p)}|\omega| = \frac{(E(\mathbb{Q}_p):E^1(\mathbb{Q}_p))}{p}$$
$$= \frac{c_pN_p}{p}.$$

For any finite set S of prime numbers including all those for which E has bad reduction, define

$$L_S^*(s) = \left(\prod_{p\in S\cup\{\infty\}}\int_{E(\mathbb{Q}_p)}|\omega|\right)^{-1}\prod_{p\notin S}\frac{1}{L_p(p^{-s})},\quad \text{where}$$
$$L_p(T) = 1+(N_p-p-1)T+pT^2.$$

[19] A proof of the "weak" conjecture already suffices to win the million-dollar Millenium Prize.

Recall that $\mathbb{Q}_p = \mathbb{R}$ when $p = \infty$. When p is good,

$$L_p(p^{-1}) = \frac{N_p}{p} = \left(\int_{E(\mathbb{Q}_p)} |\omega|\right),$$

and so the behaviour of $L_S^*(s)$ near $s = 1$ is independent[20] of S satisfying the condition, and the conjecture of Birch and Swinnerton-Dyer can be stated as

$$L_S^*(E,s) \sim \frac{[\text{Ш}(E/\mathbb{Q})]\cdot D}{[E(\mathbb{Q})_{\text{tors}}]^2}(s-1)^r \text{ as } s \to 1.$$

ASIDE 10.7 Formally, $L_p(1) = \prod \frac{p}{N_p}$, and so Conjecture 10.5 has an air of compatibility with Conjecture 10.1. Assuming E to be modular, Goldfeld (1982) showed that, if $\prod_{p\leq x} N_p/p \sim C(\log x)^{-r}$ for some $C > 0$ and r, then

- ◇ $L(E,s) \sim C'\cdot(s-1)^r$ as $s \to 1$ *and*
- ◇ the generalized Riemann hypothesis holds for $L(E,s)$, i.e., $L(E,s) \neq 0$ for $\Re(s) > 1$.

Needless to say, Conjecture 10.1 has not been proved for a single curve. In fact, it is *stronger* than the generalized Riemann hypothesis: for a fixed modular elliptic curve E over $\mathbb{Q}$, the following statements are equivalent,

- ◇ $\prod_{p\leq x} N_p/p \sim C(\log x)^r$ for some nonzero C and some r;
- ◇ $\psi_E(x) = o(x\log x)$, where $\psi_E(x) = \sum_{p\leq x}(\alpha_p^k + \bar{\alpha}_p^k)\log p$.

The generalized Riemann hypothesis for the L-function $L(E,s)$ is equivalent to $\psi_E(x) = O(x(\log x)^2)$, and so $\psi_E(x) = o(x\log x)$ can be considered a deeper (but still plausible) form of it.

ASIDE 10.8 One difference between elliptic curves and higher dimensional abelian varieties is that abelian varieties come in pairs: to each abelian variety A there is a dual abelian variety $A^\vee$ (the Picard variety of A). For elliptic curves, A and $A^\vee$ are canonically isomorphic. The Néron–Tate height gives a pairing $A^\vee(\mathbb{Q}) \times A(\mathbb{Q}) \to \mathbb{R}$, and the Cassels–Tate pairing gives a duality between $\text{Ш}(A^\vee/\mathbb{Q})$ and $\text{Ш}(A/\mathbb{Q})$. For an abelian variety A over $\mathbb{Q}$, the conjecture of Birch and Swinnerton-Dyer generalizes to a statement

$$L_S^*(A,s) \sim \frac{[\text{Ш}(A/\mathbb{Q})]\cdot D}{[A^\vee(\mathbb{Q})_{\text{tors}}][A(\mathbb{Q})_{\text{tors}}]}(s-1)^r \text{ as } s \to 1$$

where D is the discriminant of the height pairing $A^\vee(\mathbb{Q}) \times A(\mathbb{Q}) \to \mathbb{R}$. We have $L_S^*(A^\vee,s) = L_S^*(A,s)$ because $A^\vee$ is isogenous to A, and $[\text{Ш}(A^\vee/\mathbb{Q})] = [\text{Ш}(A/\mathbb{Q})]$ because of the Cassels–Tate duality, and so the statement is symmetric between A and $A^\vee$. See Tate 1966b.

[20]More precisely, $\lim_{s\to 1} L_S^*(s)/L_{S'}^*(s) = 1$ for any two such sets S, S'.

What is known about the conjecture of Birch and Swinnerton-Dyer

Beginning with the work of Birch and Swinnerton-Dyer (1963, 1965), a massive amount of computational evidence has accumulated in support of the conjectures: all the terms in the conjecture except Ш have been computed for thousands of curves; the value of [Ш] predicted by the computations is always a square; whenever the order of some p-primary component of [Ш] has been computed, it has agreed with the conjecture.

For a pair of isogenous elliptic curves over $\mathbb{Q}$, most of the terms in the full BSD conjecture will differ for the two curves, but Cassels (1965) showed that if the conjecture is true for one curve, then it is true for both, i.e., the validity of the full BSD conjecture for an elliptic curve depends only on the isogeny class of the curve.

For certain elliptic curves over function fields, the conjecture is known (see the next section).

By the mid-1970s, little progress had been made toward proving the full BSD conjecture over $\mathbb{Q}$. As Tate (1974) put it, "This remarkable conjecture relates the behaviour of a function L at a point where it is not at present known to be defined to the order of a group Ш which is not known to be finite."

Coates and Wiles (1977) proved that if E has complex multiplication by an imaginary quadratic field of class number 1 and $E(\mathbb{Q})$ is infinite, then $L(E,1)=0$.

From now on, we assume that $E/\mathbb{Q}$ is modular. Thus $L(E/\mathbb{Q},s)$ extends to the whole complex plane, and satisfies a functional equation (Conjecture 10.3).

Heegner (1952) was the first to find a systematic way of constructing points of infinite order on elliptic curves over number fields. Modular curves contain certain points, called "special", which correspond to elliptic curves with complex multiplication, and these can be pushed forward to give rational points on the modular elliptic curves. Heegner's paper was written in an unpersuasive style, and assumed a knowledge of Weber 1908 that few possessed, and so was largely ignored until Birch took it up in the late 1960s.

For a modular elliptic curve $E/\mathbb{Q}$ and a complex quadratic extension K of $\mathbb{Q}$, Birch (1969a, 1969b, 1970, 1975) defined a "Heegner point" $P_K \in E(K)$, and suggested that it should often be of infinite order. Based on numerical evidence, he observed that the heights of the Heegner points seemed to be related to the value at 1 of the L-functions of certain elliptic curves and their first derivatives (Harvard seminar, 1973). Confirming Birch's observation, Gross and Zagier (1986) proved that if E is a modular elliptic curve over $\mathbb{Q}$, then

$$L'(E/K,1) = C \cdot \hat{h}(P_K), \quad C \neq 0.$$

Thus P_K has infinite order if and only if $L'(E/K,1) \neq 0$. For a historical account of this work, see Birch 2004.

Kolyvagin (1988a, 1988b) showed that the system of Heegner points on an elliptic curve controls the size and structure of the Selmer group. By about 1990, the following was known (thanks to the work of Gross, Zagier, Kolyvagin, and others):

> (*) if the order of the zero of $L(E/\mathbb{Q},s)$ at $s = 1$ is at most 1, then the weak BSD conjecture holds for E and $Ш(E/\mathbb{Q})$ is finite.

In fact, Kolyvagin shows more.

Under the hypothesis of (*), in order to complete the proof of the full BSD conjecture for E, it remains to show that the p-primary component of $Ш(E/\mathbb{Q})$ has the predicted order for all p. Using Iwasawa theory, Kato, Rubin, Skinner, Urban, and others have done this for many E and p.

Without the hypothesis on the order of the zero of $L(E/\mathbb{Q},s)$, little is known. For example, it is not known how to prove that $\text{rank}(E(\mathbb{Q})) \leq 1$ implies the weak BSD conjecture.

The preprint Bhargava et al. 2019 shows that, when ordered by height, a majority of isomorphism classes of elliptic curves E over $\mathbb{Q}$ satisfy the hypothesis of (*).

For recent reviews of work on the BSD conjecture for elliptic curves over $\mathbb{Q}$, see Coates 2013, 2016.

Congruent numbers

Recall that a natural number n is congruent[21] if it is the area of a right-angle triangle with rational sides. There are two basic questions.

(a) Is there an algorithm for deciding whether a given n is congruent?

(b) Are all integers n congruent to 5, 6, or 7 modulo 8 congruent numbers (as the numerical evidence suggests)?

[21] The name comes from Fibonacci, *Liber Quadratorum* (Book of Squares), 1225.

We first explain the relation of congruent numbers to elliptic curves. Suppose that n is congruent, so that there are positive rational numbers a, b, c such that $a^2 + b^2 = c^2$ and $ab/2 = n$. Then $x = n(a+c)/b$, $y = 2n^2(a+c)/b^2$ is a point on the elliptic curve

$$E^{(n)} : Y^2 = X^3 - n^2 X$$

with $y \neq 0$. Conversely, suppose that there is a point $(x, y) \in E^{(n)}(\mathbb{Q})$ with $y \neq 0$. Then

$$a = |(n^2 - x^2)/y|, \quad b = |2nx/y|, \quad c = |(n^2 + x^2)/y|$$

are the sides of a right-angle triangle whose area is n.

The only rational points of finite order on the projective closure of $E^{(n)}$ are $(0:1:0)$, $(0:0:1)$, $(n:0:1)$, and $(-n:0:1)$, which all have order 2. We conclude that

$$n \text{ is congruent} \iff E^{(n)} \text{ has a rational point of infinite order.}$$

In his 1952 paper, Heegner was in fact interested in congruent numbers. He proved (using Heegner points) that if p is a prime congruent to 5 or 7 modulo 8, then p is a congruent number, and if p is congruent to 3 or 7 modulo 8, then $2p$ is a congruent number.

Let n be congruent to $5, 6$, or 7 modulo 8. Then the constant w_E in the functional equation of $L(E^{(n)}, s)$ is -1, and so $L(E^{(n)}, s)$ has a zero at $s = 1$. Therefore, the weak BSD conjecture implies that n is congruent. Alternatively, n can be shown to be congruent directly from Selmer's conjecture (Stephens 1975). It is known that, among the square free integers congruent to $5, 6$, or 7 modulo 8, those that are congruent numbers have density 1 (Kriz 2020).

If the Tate–Shafarevich group of $E^{(n)}$ is finite, then there is an algorithm for finding the rational points on $E^{(n)}$, and hence for deciding if n is congruent. Tunnell has given a more explicit algorithm. Let n be odd and square free; if n is congruent, then the number of triples satisfying $2x^2 + y^2 + 8z^2 = n$ is twice the number satisfying $2x^2 + y^2 + 32z^2 = n$, and the converse is true if the weak BSD conjecture holds for $E^{(n)}$. There is a similar statement for even n.

For a recent review of work on congruent numbers, see Coates 2017.

11 Elliptic curves and sphere packings

The conjecture of Birch and Swinnerton-Dyer is expected to hold, not just for elliptic curves over $\mathbb{Q}$, but for all elliptic curves over global fields (finite extensions of $\mathbb{Q}$ or $\mathbb{F}_p(T)$). For functions fields, the conjecture has been proved in some important cases, and Elkies, Shioda, and others used this to show that the lattices $(E(K), \langle\ ,\ \rangle)$ arising in this way give very dense sphere packings.[22]

The conjecture of Birch and Swinnerton-Dyer over function fields

Let K be a finite extension of $\mathbb{F}_p(T)$, and let $\mathbb{F}_q$ be the algebraic closure of $\mathbb{F}_p$ in K. There exists a nonsingular projective curve C over $\mathbb{F}_q$ such that $K = \mathbb{F}_q(C)$ (I, 6.3). As we discussed in §9, the zeta function of C over $\mathbb{F}_q$, is of the form

$$Z(C,T) = \frac{\prod_{i=1}^{2g}(1-\omega_i T)}{(1-T)(1-qT)}, \quad |\omega_i| = q^{\frac{1}{2}},$$

where g is the genus of C. Let E be a ***constant*** elliptic curve over K, i.e., the curve defined by an equation

$$Y^2Z + a_1XYZ + a_3YZ^2 = X^3 + a_2X^2Z + a_4XZ^2 + a_6Z^3$$

with the $a_i \in \mathbb{F}_q \subset K$. Write the zeta function of E over $\mathbb{F}_q$ as

$$Z(E,T) = \frac{(1-\alpha_1 T)(1-\alpha_2 T)}{(1-T)(1-qT)}.$$

THEOREM 11.1 *The weak conjecture of Birch and Swinnerton-Dyer holds for E/K.*

PROOF. In this case, the weak conjecture says that the rank r of $E(K)$ is equal to the number of pairs (i, j), $1 \le i \le 2$, $1 \le j \le 2g$, such that $\alpha_i = \omega_j$. The rank of $E(K)$ is equal to the rank of $\mathrm{Hom}(J, E)$, where J is the jacobian of C. Tate (1966a) proved that the rank of $\mathrm{Hom}(J, E)$ equals the number of pairs (i, j) with $\alpha_i = \omega_j$. □

[22] "One of the most exciting developments has been the construction by Elkies and Shioda of lattice packings from the Mordell–Weil groups of elliptic curves over function fields. Such lattices have a greater density than any previously known in dimensions from about 80 to 4096." Conway and Sloane 1999, p. xviii.

THEOREM 11.2 *The full conjecture of Birch and Swinnerton-Dyer holds for E/K.*

PROOF. In this case, the full conjecture says that

$$q^g \prod_{\alpha_i \neq \omega_j} \left(1 - \frac{\omega_j}{\alpha_i}\right) = [Ш(E/K)] \cdot D$$

where D is the discriminant of the pairing

$$\mathrm{Hom}(J,E) \times \mathrm{Hom}(E,J) \xrightarrow{a,b \mapsto b \circ a} \mathrm{End}(J) \xrightarrow{\text{trace}} \mathbb{Z}.$$

This is proved in Milne 1968. □

More generally, for a *constant* abelian variety over a function field, the weak BSD conjecture is a theorem of Tate (1966a) and the full BSD conjecture is a theorem of Milne (1968). Moreover, for an elliptic curve over a function field, it is known that the weak conjecture implies the full conjecture (Milne 1975, 1981). The weak BSD conjecture is known for an elliptic curve defined over a field $\mathbb{F}_q(T)$ by a Weierstrass equation in which the coefficients $a_i = a_i(T)$ are polynomials with $\deg a_i(T) \leq 2i$ (Artin and Swinnerton-Dyer 1973).

Let E_1 and E_2 be elliptic curves over $\mathbb{F}_q$, and regard E_2 as an elliptic curve over the field $K = \mathbb{F}_q(E_1)$. If E_1 and E_2 are nonisogenous, then the Tate–Shafarevich group $Ш(E/K)$ has order $(n_1 - n_2)^2$ where $n_1 = \#E_1(k))$ and $n_2 = \#E_2(k)$. If $E_1 = E_2 = E$ is nonsupersingular, then the Tate–Shafarevich group $Ш(E/K)$ has order $(\mathrm{End}(E){:}\mathbb{Z}[F])^2$ where F is the Frobenius map. (Exercise from the above formulas.)

Sphere packings

As we noted in Remark 6.4, pairs consisting of a free $\mathbb{Z}$-module of finite rank L and a positive definite quadratic form q on $V \stackrel{\text{def}}{=} L \otimes \mathbb{R}$ are of great interest. We can choose a basis for V that identifies (V,q) with $(\mathbb{R}^r, X_1^2 + \cdots + X_r^2)$. The bilinear form associated with q is

$$\langle x,y \rangle = q(x+y) - q(x) - q(y).$$

Given such a pair (L,q), the numbers one needs to compute are

(a) the rank r of L;

(b) the square of the length of the shortest vector

$$m(L) = \inf_{v \in L, v \neq 0} \langle v, v \rangle;$$

(c) the discriminant of L,

$$\operatorname{disc} L = \det(\langle e_i, e_j \rangle),$$

where $e_1, \ldots, e_r$ is a basis for L.

The discriminant is independent of the choice of a basis for L. Let

$$\gamma(L) = m(L)/\operatorname{disc}(L)^{\frac{1}{r}}.$$

The volume of a fundamental parallelepiped for L is $\sqrt{\operatorname{disc} L}$. The sphere packing associated with L is formed of spheres of radius $\frac{1}{2}\sqrt{m(L)}$, and therefore its density is

$$d(L) = 2^{-r} b_r \gamma(L)^{\frac{r}{2}},$$

where $b_r = \pi^{r/2}/\Gamma(\frac{r+2}{2})$ is the volume of the r-dimensional unit ball. To maximize $d(L)$, we need to maximise $\gamma(L)$.

Let E be a constant elliptic curve over a field $K = \mathbb{F}_q(C)$ as before, and let $L = E(K)/E(K)_{\text{tors}}$ with the quadratic form $q = 2\hat{h}$. If we know the numbers ω_i and α_j, then Theorem 11.1 gives us r, and Theorem 11.2 gives us an upper bound for disc L,

$$\operatorname{disc} L = q^g \cdot \prod_{\alpha_i \neq \omega_j} \left(1 - \frac{\omega_j}{\alpha_i}\right) \cdot [\text{Ш}]^{-1} \leq q^g \cdot \prod_{\alpha_i \neq \omega_j} \left(1 - \frac{\omega_j}{\alpha_i}\right).$$

Finally, an easy, but nonelementary argument,[23] shows that

$$m(L) \geq 2[C(k)]/[E(k)]$$

for all finite $k \supset \mathbb{F}_q$.

---

[23]An element P of $E(K)$ defines a map $u: C \to E$, and $\hat{h}(P)$ is related to the degree of u. Thus, we get a lower bound for $m(L)$ in terms of the ω_i and α_j.

Example

Consider the curve

$$C : X^{q+1} + Y^{q+1} + Z^{q+1} = 0$$

over $\mathbb{F}_{q^2}$ (note, not over $\mathbb{F}_q$).

LEMMA 11.3

(a) *The curve* C *is nonsingular, of genus* $g = \frac{q(q-1)}{2}$.

(b) $\#C(\mathbb{F}_{q^2}) = q^3 + 1$.

(c) $Z(C,T) = \frac{(1+qT)^{q(q-1)}}{(1-T)(1-q^2T)}$.

PROOF. (a) The partial derivatives of the defining equation are X^q, Y^q, Z^q, and these have no common zero in $\mathbb{P}^2$. Therefore, the curve is nonsingular, and so the formula (6), p. 41, shows that it has genus $q(q-1)/2$.

(b) The group $\mathbb{F}_{q^2}^\times$ is cyclic of order $q^2 - 1 = (q+1)(q-1)$, and $\mathbb{F}_q^\times$ is its subgroup of order $q-1$. Thus, $x^{q+1} \in \mathbb{F}_q^\times$ and $x \mapsto x^{q+1} : \mathbb{F}_{q^2}^\times \to \mathbb{F}_q^\times$ is a surjective homomorphism with kernel a cyclic group of order $q+1$. As x runs through $\mathbb{F}_{q^2}$, x^{q+1} takes the value 0 once and each nonzero value in $\mathbb{F}_q^\times$ exactly $q+1$ times. A similar remark applies to y^{q+1} and z^{q+1}. We can scale each solution of $X^{q+1} + Y^{q+1} + Z^{q+1} = 0$ so that $x = 0$ or 1.

Case 1: $x = 1$, $1 + y^{q+1} \neq 0$. There are $q^2 - q - 1$ possibilities for y, and then $q+1$ possibilities for z. Hence $(q^2 - q - 1)(q+1) = q^3 - 2q - 1$ solutions.

Case 2: $x = 1$, $1 + y^{q+1} = 0$. There are $q+1$ possibilities for y, and then one for z. Hence $q+1$ solutions.

Case 3: $x = 0$. We can take $y = 1$, and then there are $q+1$ possibilities for z.

In sum, there are $q^3 + 1$ solutions.

(c) We know that

$$\#C(\mathbb{F}_q) = 1 + q^2 - \sum_{i=1}^{2g} \omega_i.$$

Therefore $\sum_{i=1}^{2g} \omega_i = q^2 - q^3 = -2gq$. Because $|\omega_i| = q$, this forces $\omega_i = -q$. □

For all q, it is known that there is an elliptic curve E over $\mathbb{F}_{q^2}$, such that $E(\mathbb{F}_{q^2})$ has q^2+2q+1 elements (the maximum allowed by the Riemann hypothesis). For such a curve

$$Z(E,T)=\frac{(1+qT)^2}{(1-T)(1-q^2T)}.$$

PROPOSITION 11.4 *Let $L=E(K)/E(K)_{\text{tors}}$ with E and $K=\mathbb{F}_q(C)$ as above. Then:*

(a) *The rank r of L is $2q(q-1)$;*

(b) $m(L)\geq 2(q-1)$;

(c) $[\text{Ш}(E/K)]\,\text{disc}(L)=q^{q(q-1)}$;

(d) $\gamma(L)\geq 2(q-1)/\sqrt{q}$.

PROOF. (a) Since all α_i and ω_j equal $-q$, if follows from (11.2a) that the rank is $2\times 2g=2q(q-1)$.

(b) We have

$$m(L)\geq\frac{[C(\mathbb{F}_{q^2})]}{[E(\mathbb{F}_{q^2})]}=\frac{q^3+1}{(q^2+1)^2}>q-2.$$

(c) This is a special case of (11.2), taking count that our field is $\mathbb{F}_{q^2}$ (not $\mathbb{F}_q$) and $g=q(q-1)/2$.

(d) Follows immediately from the preceding. □

REMARK 11.5 (a) Dummigan (1995) has obtained information on the Tate–Shafarevich group in the above, and a closely related, situation. For example, $\text{Ш}(E/K)$ is zero if $q=p$ or p^2, and has cardinality at least $p^{p^3(p-1)^3/2}$ if $q=p^3$.

(b) For $q=2$, L is isomorphic to the lattice denoted D_4, for $q=3$, to the Coxeter-Todd lattice K_{12}, and for $q=3$ it is similar to the Leech lattice.

For a more detailed account of the applications of elliptic curves to lattices, see Oesterlé 1990.

EXERCISE 11.6 Consider the curve $E: Y^2Z+YZ^2=X^3$.

(a) Show that E is a nonsingular curve over $\mathbb{F}_2$.

(b) Compute $\#E(\mathbb{F}_4)$, $\mathbb{F}_4$ being the field with 4 elements.

(c) Let K be the field of fractions of the integral domain $\mathbb{F}_4[X,Y]/(X^3+Y^3+1)$. Let $L=E(K)/E(K)_{\text{tors}}$, and regard it as a lattice in $V=L\otimes\mathbb{R}$ endowed with the height pairing. Compute the rank of L, $m(L)$, and $\gamma(L)$.

Chapter V

Elliptic Curves and Modular Forms

Fermat is one of the most fascinating mathematical personalities of all time, the creator (with Descartes) of analytic geometry, one of the founders of the calculus, and undisputed founder of modern number theory.

Weil 1973.

We wish to understand the L-function $L(E/\mathbb{Q},s)$ of an elliptic curve E over $\mathbb{Q}$, i.e., we wish to understand the sequence of numbers

$$N_2, N_3, N_5, \dots, N_p, \dots \qquad N_p = \#E_p(\mathbb{F}_p),$$

or, equivalently, the sequence of numbers

$$a_2, a_3, a_5, \dots, a_p, \dots \qquad a_p = p+1-N_p.$$

There is no direct way of doing this. Instead, we shall see how the study of modular curves and modular forms leads to functions that are candidates for being the L-function of an elliptic curve over $\mathbb{Q}$, and then we shall see how Wiles and others showed that all the L-functions of elliptic curves over $\mathbb{Q}$ do in fact arise from modular forms.

1 The Riemann surfaces $X_0(N)$

Quotients of Riemann surfaces by group actions

We shall need to define Riemann surfaces as the quotients of other simpler Riemann surfaces by group actions. This can be quite complicated. The following examples will help.

EXAMPLE 1.1 Let $n \in \mathbb{Z}$ act on $\mathbb{C}$ by $z \mapsto z+n$. Topologically $\mathbb{C}/\mathbb{Z}$ is a cylinder. Write $\pi:\mathbb{C} \to \mathbb{C}/\mathbb{Z}$ for the quotient map. Then $\mathbb{C}/\mathbb{Z}$ has a unique complex structure for which π is a local isomorphism of Riemann surfaces. A function $f:U \to \mathbb{C}$ on an open subset of $\mathbb{C}/\mathbb{Z}$ is holomorphic (resp. meromorphic) for this structure if and only if $f \circ \pi$ is holomorphic (resp. meromorphic). Thus the holomorphic functions f on $U \subset \mathbb{C}/\mathbb{Z}$ can be identified with the holomorphic functions g on $\pi^{-1}(U)$ invariant under $\mathbb{Z}$, i.e., such that $g(z+1) = g(z)$. For example, $q(z) = e^{2\pi i z}$ defines a holomorphic function on $\mathbb{C}/\mathbb{Z}$. In fact, it gives an isomorphism $\mathbb{C}/\mathbb{Z} \to \mathbb{C}^{\times}$ whose in inverse $\mathbb{C}^{\times} \to \mathbb{C}/\mathbb{Z}$ is (by definition) $(2\pi i)^{-1} \cdot \log$.

EXAMPLE 1.2 Let D be the open unit disk $\{z \mid |z| < 1\}$, and let Δ be a finite group acting on D and fixing 0. The Schwarz lemma (Cartan 1963, III, 3) shows that the automorphisms of D fixing 0 are the maps $z \mapsto \lambda z$, $|\lambda| = 1$, and so

$$\mathrm{Aut}(D) = \{z \in \mathbb{C} \mid |z| = 1\} \simeq \mathbb{R}/\mathbb{Z}.$$

It follows that Δ is a finite cyclic group. Let $z \mapsto \zeta z$ generate it and suppose that ζ has order m. Then z^m is invariant under Δ, and so defines a function on $\Delta \backslash D$, which in fact is a homeomorphism $\Delta \backslash D \to D$, and therefore defines a complex structure on $\Delta \backslash D$.

Let $\pi: D \to \Delta \backslash D$ be the quotient map. The space of holomorphic functions on an open $U \subset \Delta \backslash D$ can be identified with the space of holomorphic functions on $\pi^{-1}(U)$ such that $g(\zeta z) = g(z)$ for all z, and so are of the form $g(z) = h(z^m)$ with h holomorphic. Note that if $\pi(Q) = P = 0$, then $\mathrm{ord}_P(f) = \frac{1}{m} \mathrm{ord}_Q(f \circ \pi)$.

Let Γ be a group acting on a Riemann surface X. A ***fundamental domain*** for Γ is a connected open subset D of X such that

(a) no two points of D lie in the same orbit of Γ;

(b) the closure $\bar{D}$ of D contains at least one element from each orbit.

For example,

$$D = \{z \in \mathbb{C} \mid 0 < \Re(z) < 1\}$$

is a fundamental domain for $\mathbb{Z}$ acting on $\mathbb{C}$ as in 1.1, and

$$D_0 = \{z \in D \mid 0 < \arg z < 2\pi/m\}$$

is a fundamental domain for $\mathbb{Z}/m\mathbb{Z}$ acting on the unit disk as in 1.2.

The Riemann surfaces $X(\Gamma)$

Let Γ be a subgroup of finite index in $\mathrm{SL}_2(\mathbb{Z})$. We want to define the structure of a Riemann surface on the quotient $\Gamma\backslash\mathbb{H}$. This we can do, but the resulting surface will not be compact. Instead, we need to form a quotient $\Gamma\backslash\mathbb{H}^*$, where $\mathbb{H}^*$ properly contains $\mathbb{H}$.

THE ACTION OF $\mathrm{SL}_2(\mathbb{Z})$ ON THE UPPER HALF-PLANE

Recall (III, §1) that $\mathrm{SL}_2(\mathbb{Z})$ acts on $\mathbb{H} = \{z \mid \Im(z) > 0\}$ according to the rule

$$\begin{pmatrix} a & b \\ c & d \end{pmatrix} z = \frac{az+b}{cz+d}.$$

Note that $-I = \left(\begin{smallmatrix} -1 & 0 \\ 0 & -1 \end{smallmatrix}\right)$ acts trivially on $\mathbb{H}$, and so the action factors through the ***modular group*** $\mathrm{PSL}_2(\mathbb{Z}) \stackrel{\text{def}}{=} \mathrm{SL}_2(\mathbb{Z})/\{\pm I\}$.

Let

$S = \begin{pmatrix} 0 & -1 \\ 1 & 0 \end{pmatrix}$, so that $z \mapsto Sz = \dfrac{-1}{z}$ is the "inversion-symmetry", and

$T = \begin{pmatrix} 1 & 1 \\ 0 & 1 \end{pmatrix}$, so that $z \mapsto Tz = z+1$ is translation.

Then, in $\mathrm{PSL}_2(\mathbb{Z})$,

$$S^2 = 1, \quad (ST)^3 = 1.$$

PROPOSITION 1.3 *Let*

$$D = \left\{z \in \mathbb{H} \quad \middle| \quad |z| > 1, \quad -\tfrac{1}{2} < \Re(z) < \tfrac{1}{2}\right\}.$$

(a) *D is a fundamental domain for* $\mathrm{PSL}_2(\mathbb{Z})$*; moreover, two elements $z \neq z'$ of the closure $\bar{D}$ of D are in the same orbit if and only if*

i) $\Re(z) = \pm\frac{1}{2}$ *and* $z' = z \pm 1$ *(so* $z' = Tz$ *or* $z = Tz'$*); or*

ii) $|z| = 1$ *and* $z' = -1/z$ $(= Sz)$.

In particular, z and z' lie in the boundary of $\bar{D}$.

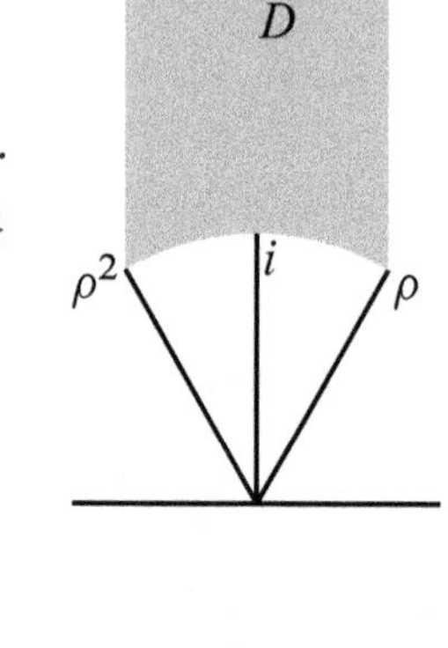

(b) *Let $z \in \bar{D}$. The stabilizer of z in* $\mathrm{PSL}_2(\mathbb{Z})$ *is* $\{1\}$ *except in the following three cases:*

i) $z = i$, *stabilizer* $\langle S \rangle$;

ii) $z = \rho = e^{2\pi i/6}$, *stabilizer* $\langle TS \rangle$;

iii) $z = \rho^2 = e^{2\pi i/3}$, *stabilizer* $\langle ST \rangle$.

The group $\mathrm{PSL}_2(\mathbb{Z})$ *is generated by S and T.*

PROOF. Define G to be the subgroup $\langle S, T\rangle$ of $\mathrm{PSL}_2(\mathbb{Z})$ generated by S and T, and let $z \in \mathbb{H}$. We shall show that there exists a $g \in G$ such that $gz \in \bar{D}$. If $g = \left(\begin{smallmatrix} a & b \\ c & d \end{smallmatrix}\right) \in G$, then

$$\Im(gz) = \frac{\Im(z)}{|cz+d|^2}.$$

For any N, $|cz+d| < N$ for only finitely many pairs $c, d \in \mathbb{Z}$, and so can choose g so that $|cz+d|$ is minimum; then $\Im(gz)$ is maximum. We next choose n so that

$$-\tfrac{1}{2} \leq \Re(T^n gz) \leq \tfrac{1}{2}.$$

Then $T^n gz \in \bar{D}$, because otherwise $|T^n gz| < 1$, and

$$\mathrm{Im}(ST^n gz) = \frac{\mathrm{Im}(T^n gz)}{|T^n gz|^2} > \mathrm{Im}(T^n gz) = \mathrm{Im}(gz),$$

contradicting the choice of g. We have shown that $G \cdot \bar{D} = \mathbb{H}$; in particular, $\mathrm{PSL}_2(\mathbb{Z}) \cdot \bar{D} = \mathbb{H}$.

The proof of (b) and of the rest of (a) is routine (Serre 1973, VII, 1.2).

For (c), let $h \in \mathrm{PSL}_2(\mathbb{Z})$, and choose a point z_0 in D. There exists a g in G such that $ghz_0 \in \bar{D}$. As ghz_0 and z_0 are in the same $\mathrm{PSL}_2(\mathbb{Z})$-orbit, it follows from (a) that they are equal, and now (b) shows that $gh = 1$. Hence $h \in G$. □

REMARK 1.4 (a) It can be shown that the modular group $\mathrm{PSL}_2(\mathbb{Z})$ has a presentation $\langle S,T|S^2,(ST)^3\rangle$. In other words, it is a free product of the group $\langle S\rangle$ of order 2 and the group $\langle ST\rangle$ of order 3. Many interesting finite groups, for example, all but four of the sporadic finite simple groups, are generated by an element of order 2 and an element of order 3, and hence are quotients of $\mathrm{PSL}_2(\mathbb{Z})$.

(b) Let Γ be a subgroup of finite index in $\mathrm{SL}_2(\mathbb{Z})$, and write

$$\mathrm{SL}_2(\mathbb{Z}) = \Gamma\gamma_1 \cup \ldots \cup \Gamma\gamma_m \quad \text{(disjoint union)}.$$

Then $D' = \bigcup \gamma_i D$ satisfies the conditions to be a fundamental domain for Γ, except that it need not be connected. However, it is possible to choose the γ_i so that the closure of D' is connected, in which case the interior of the closure will be a fundamental domain for Γ.

THE EXTENDED UPPER HALF-PLANE

The elements of $\mathrm{SL}_2(\mathbb{Z})$ act on $\mathbb{P}^1(\mathbb{C})$ by projective linear transformations,

$$\begin{pmatrix} a & b \\ c & d \end{pmatrix}(z_1 \colon z_2) = (az_1+bz_2 \colon cz_1+dz_2).$$

Identify $\mathbb{H}$, $\mathbb{Q}$, and $\{\infty\}$ with subsets of $\mathbb{P}^1(\mathbb{C})$ according to

$$\begin{array}{ccll} z & \leftrightarrow & (z\colon 1) & z \in \mathbb{H} \\ r & \leftrightarrow & (r\colon 1) & r \in \mathbb{Q} \\ \infty & \leftrightarrow & (1\colon 0). & \end{array}$$

The action of $\mathrm{SL}_2(\mathbb{Z})$ on $\mathbb{P}^1(\mathbb{C})$ stabilizes $\mathbb{H}^* \stackrel{\text{def}}{=} \mathbb{H}\cup\mathbb{Q}\cup\{\infty\}$. In fact,

$$\text{if } z \in \mathbb{H}, \quad \begin{pmatrix} a & b \\ c & d \end{pmatrix}(z\colon 1) = (az+b\colon cz+d) = \left(\frac{az+b}{cz+d}\colon 1\right),$$

$$\text{if } r \in \mathbb{Q}, \quad \begin{pmatrix} a & b \\ c & d \end{pmatrix}(r\colon 1) = (ar+b\colon cr+d) = \begin{cases} (\frac{ar+b}{cr+d} : 1) & r \neq -\frac{d}{c} \\ \infty & r = -\frac{d}{c}, \end{cases}$$

$$\begin{pmatrix} a & b \\ c & d \end{pmatrix}\infty = (a:c) = \begin{cases} (\frac{a}{c} : 1) & c \neq 0 \\ \infty & c = 0. \end{cases}$$

In passing from $\mathbb{H}$ to $\mathbb{H}^*$, we have added one additional $\mathrm{SL}_2(\mathbb{Z})$ orbit. The points in $\mathbb{H}^*$ not in $\mathbb{H}$ are called the ***cusps***.

We make $\mathbb{H}^*$ into a topological space as follows: the topology on $\mathbb{H}$ is that inherited from $\mathbb{C}$; the sets

$$\{z \mid \Im(z) > M\}, \quad M > 0$$

form a fundamental system of neighbourhoods of ∞; the sets

$$\{z \mid |z-(a+ir)| < r\} \cup \{a\}$$

form a fundamental system of neighbourhoods of $a \in \mathbb{Q}$. One shows that $\mathbb{H}^*$ is Hausdorff, and that the action of $\mathrm{SL}_2(\mathbb{Z})$ is continuous.

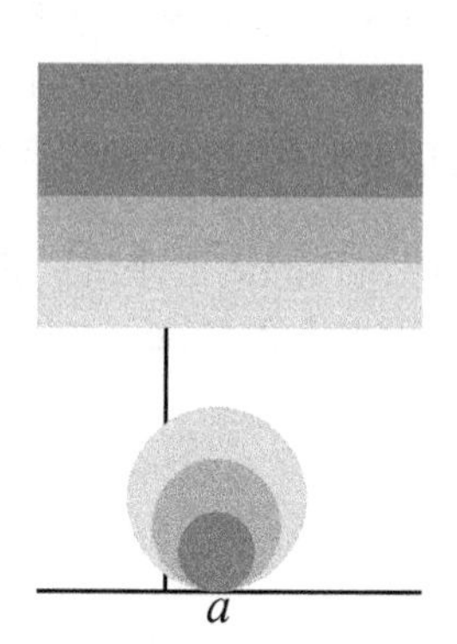

The topology on $\Gamma\backslash\mathbb{H}^*$

Recall that if $\pi: X \to Y$ is a surjective map and X is a topological space, then the quotient topology on Y is that for which a set U is open if and only if $\pi^{-1}(U)$ is open. In general the quotient of a Hausdorff space by a group action will not be Hausdorff even if the orbits are closed — one needs that distinct orbits have disjoint open neighbourhoods.

Let Γ be a subgroup of finite index in $\mathrm{SL}_2(\mathbb{Z})$. The discrete group Γ acts continuously on $\mathbb{H}$, and one can show that the action is proper,[1] i.e., for any pair of points $x, y \in \mathbb{H}$, there exist neighbourhoods U of x and V of y such that

$$\{\gamma \in \Gamma \mid \gamma U \cap V \neq \emptyset\}$$

is finite. In particular, this implies that the stabilizer of every point of $\mathbb{H}$ is finite (which we knew anyway).

PROPOSITION 1.5 (a) *For all compact sets A and B of $\mathbb{H}$, $\{\gamma \in \Gamma \mid \gamma A \cap B \neq \emptyset\}$ is finite.*

(b) *Every $z \in \mathbb{H}$ has a neighbourhood U such that γU and U are disjoint for $\gamma \in \Gamma$ unless $\gamma z = z$.*

(c) *For any points x, y of $\mathbb{H}$ not in the same Γ-orbit, there exist neighbourhoods U of x and V of y such that $\gamma U \cap V = \emptyset$ for all $\gamma \in \Gamma$.*

---

[1] Thus, Γ may be said to act properly continuously on $\mathbb{H}$. In the literature however, Γ is usually said to act "properly discontinuously"! See Lee 2003, p. 225, for a discussion of this terminology.

PROOF. (a) This follows easily from the fact that Γ acts continuously and properly.

(b) Let V be compact neighbourhood of z, and let S be the set of $\gamma \in \Gamma$ such that $V \cap \gamma V \neq \emptyset$ but γ does not fix z. From (a), we know that S is finite. For each $\gamma \in S$, choose disjoint neighbourhoods V_γ of z and W_γ of γz, and set

$$U = V \cap \bigcap_{\gamma \in S} (V_\gamma \cap \gamma^{-1} W_\gamma).$$

For $\gamma \in S$, $\gamma U \subset W_\gamma$. As W_γ is disjoint from V_γ and V_γ contains U, this implies that γU is disjoint from U.

(c) Choose compact neighbourhoods A of x and B of y, and let S be the (finite) set of $\gamma \in \Gamma$ such that $\gamma A \cap B \neq \emptyset$. Because $\gamma x \neq y$, for each $\gamma \in S$, there exist disjoint neighbourhoods U_γ and V_γ of γx and y. Now $U = A \cap \bigcap_{\gamma \in S} \gamma^{-1} U_\gamma$ and $V = B \cap \bigcap_{\gamma \in S} V_\gamma$ are neighbourhoods of x and y respectively such that γU and V are disjoint for all γ in Γ. □

COROLLARY 1.6 *The space $\Gamma \backslash \mathbb{H}$ is Hausdorff.*

PROOF. Let x and y be points of $\mathbb{H}$ not in the same Γ-orbit, and choose neighbourhoods U and V of x and y as in (c) of the last proposition. Then ΓU and ΓV are disjoint neighbourhoods of Γx and Γy. □

PROPOSITION 1.7 *The space $\Gamma \backslash \mathbb{H}^*$ is Hausdorff and compact.*

PROOF. After 1.6, to show that $\Gamma \backslash \mathbb{H}^*$ is Hausdorff, it only remains to examine it near the cusps, which we leave to the reader (cf. the next subsection). The space $\Gamma \backslash \mathbb{H}^*$ is a quotient of $\mathrm{SL}_2(\mathbb{Z}) \backslash \mathbb{H}^*$, which is compact because $\bar{D} \cup \{\infty\}$ is compact. □

The complex structure on $\Gamma_0(N) \backslash \mathbb{H}^*$

The subgroups of $\mathrm{SL}_2(\mathbb{Z})$ that we shall be especially interested in are the ***Hecke subgroups***

$$\Gamma_0(N) = \left\{ \begin{pmatrix} a & b \\ c & d \end{pmatrix} \middle| \, c \equiv 0 \mod N \right\}.$$

We let $\Gamma_0(1) = \mathrm{SL}_2(\mathbb{Z})$. The integer N is the ***level*** of $\Gamma_0(N)$.

We now define a complex structure on $\Gamma_0(N) \backslash \mathbb{H}^*$. For $z_0 \in \mathbb{H}$, choose a neighbourhood V of z_0 such that

$$\gamma V \cap V \neq \emptyset \Rightarrow \gamma z_0 = z_0,$$

and let $U = \pi(V)$ — it is open because $\pi^{-1} U = \bigcup \gamma V$ is open.

If the stabilizer of z_0 in $\Gamma_0(N)$ is $\pm I$, then $\pi: V \to U$ is a homeomorphism, with inverse φ, say, and we require (U, φ) to be a coordinate neighbourhood.

If the stabilizer of z_0 in $\Gamma_0(N)$ is $\neq \{\pm I\}$, then it is a cyclic group of order $2m$ with $m = 2$ or 3 (and its stabilizer in $\Gamma_0(N)/\{\pm I\}$ has order 2 or 3) — see (1.3b). The fractional linear transformation

$$\lambda : \mathbb{H} \to D, \quad z \mapsto \frac{z - z_0}{z - \bar{z}_0},$$

carries z_0 to 0 in the unit disk D. There is a well-defined map $\varphi: U \to \mathbb{C}$ such that $\varphi(\pi(z)) = \lambda(z)^m$, and we require (U, φ) to be a coordinate neighbourhood (cf. Example 1.2).

Next consider $z_0 = \infty$. Choose V to be the neighbourhood $\{z \mid \Im(z) > 2\}$ of ∞, and let $U = \pi(V)$. If

$$z \in V \cap \gamma V, \quad \gamma = \begin{pmatrix} a & b \\ c & d \end{pmatrix} \in \Gamma_0(N),$$

then

$$2 \leq \Im(\gamma z) = \frac{\Im(z)}{|cz + d|^2} \leq \frac{1}{|c|^2 \Im(z)} \leq \frac{1}{2|c|^2}$$

and so $c = 0$. Therefore

$$\gamma = \pm \begin{pmatrix} 1 & m \\ 0 & 1 \end{pmatrix},$$

and so there is a well-defined map $\varphi: U \to \mathbb{C}$ such that $\varphi(\pi(z)) = e^{2\pi i z}$, and we require (U, φ) to be a coordinate neighbourhood (cf. Example 1.1).

For $z_0 \in \mathbb{Q}$, we choose a $\beta \in \mathrm{SL}_2(\mathbb{Z})$ such that $\beta(z_0) = \infty$, and proceed similarly.

PROPOSITION 1.8 *The coordinate neighbourhoods defined above are compatible, and therefore define on $\Gamma_0(N)\backslash\mathbb{H}^*$ the structure of a Riemann surface.*

PROOF. Routine exercise. □

Write $X_0(N)$ for the Riemann surface $\Gamma_0(N)\backslash\mathbb{H}^*$, and $Y_0(N)$ for its open subsurface $\Gamma_0(N)\backslash\mathbb{H}$.

The genus of $X_0(N)$

The genus of a Riemann surface can be computed by "triangulating" it, and using the formula

$$2-2g = V - E + F,$$

where V is the number of vertices, E is the number of edges, and F is the number of faces. For example, the sphere can be triangulated by projecting out from a regular tetrahedron. Then $V = 4$, $E = 6$, and $F = 4$, so that $g = 0$ as expected.

PROPOSITION 1.9 *The Riemann surface $X_0(1)$ has genus zero.*

PROOF. One gets a fake triangulation of the sphere by taking as vertices three points on the equator, and the upper and lower hemispheres as the faces. This gives the correct genus

$$2 = 3-3+2$$

but it violates the usual definition of a triangulation, which requires that any two triangles intersect in a single side, a single vertex, or not at all. It can be made into a valid triangulation by adding the north pole as a vertex, and joining it to the three vertices on the equator.

One gets a fake triangulation of $X_0(1)$ by taking the three vertices ρ, i, and ∞ and the obvious curves joining them (two on the boundary of D and one the imaginary axis from i to ∞). It can be turned into a valid triangulation by adding a fourth point in D with real part > 0, and joining it to ρ, i, and ∞. □

For a finite mapping $\pi : Y \to X$ of compact Riemann surfaces, the Riemann–Hurwitz formula relates the genus g_Y of Y to the genus g_X of X,

$$2g_Y - 2 = (2g_X - 2)m + \sum_{Q \in Y} (e_Q - 1).$$

Here m is the degree of the mapping, so that $\pi^{-1}(P)$ has m elements except for finitely many P, and e_Q is the ramification index, so that $e_Q = 1$ unless at least two sheets come together at Q above $\pi(Q)$ in which case it is the number of such sheets.

For example, if E is the elliptic curve

$$E : Y^2 Z = X^3 + aXZ^2 + bZ^3, \quad a, b \in \mathbb{C}, \quad \Delta \neq 0,$$

and π is the map

$$\infty \mapsto \infty\,,\ (x\!:\!y\!:\!z) \mapsto (x\!:\!z) : E(\mathbb{C}) \to \mathbb{P}^1(\mathbb{C}),$$

then $m = 2$ and $e_Q = 1$ except for $Q = \infty$ or one of the three points of order 2 on E, in which case $e_Q = 2$. This is consistent with $E(\mathbb{C})$ having genus 1 and $\mathbb{P}^1(\mathbb{C})$ (the Riemann sphere) having genus 0.

The Riemann–Hurwitz formula is proved by triangulating Y in such a way that the ramification points are vertices and the triangulation of Y lies over a triangulation of X.

Now we can compute the genus of $X_0(N)$ by studying the quotient map $X_0(N) \to X_0(1)$. The only (possible) ramification points are those $\Gamma_0(1)$-equivalent to one of i, ρ, or ∞. Explicit formulas can be found in Shimura 1971, pp. 23–25. For example, one finds that, for p a prime > 3,

$$\text{genus}(X_0(p)) = \begin{cases} n-1 & \text{if } p = 12n+1 \\ n & \text{if } p = 12n+5, 12n+7 \\ n+1 & \text{if } p = 12n+11. \end{cases}$$

Moreover,

$$g = 0 \text{ if } N = 1,2,3,\ldots,10,12,13,16,18,25;$$
$$g = 1 \text{ if } N = 11,14,15,17,19,20,21,24,27,32,36,49;$$
$$g = 2 \text{ if } N = 22,23,26,28,29,31,37,50.$$

EXERCISE 1.10 For a prime p, show that the natural action of $\Gamma_0(p)$ on $\mathbb{P}^1(\mathbb{Q})$ has only two orbits, represented by 0 and $\infty = (1:0)$. Deduce that $X_0(p) \smallsetminus Y_0(p)$ has exactly two elements.

2 $X_0(N)$ as an algebraic curve over $\mathbb{Q}$

In the last section, we defined compact Riemann surfaces $X_0(N)$. A general theorem states that every compact Riemann surface X can be identified with the set of complex points of a unique nonsingular projective algebraic curve C over $\mathbb{C}$. However, in general C cannot be defined over $\mathbb{Q}$ (or even $\mathbb{Q}^{\text{al}}$) — consider for example a Riemann surface $\mathbb{C}/\Lambda$ as in Chapter III whose j-invariant is transcendental — and when C can be defined over $\mathbb{Q}$, in general, it cannot be defined in any canonical way — consider an elliptic curve E over $\mathbb{C}$ with $j(E) \in \mathbb{Q}$.

In this section, we shall see that $X_0(N)$ has the remarkable property that it *is* the set of complex points of a *canonical* curve over $\mathbb{Q}$.

Modular functions

For a connected compact Riemann surface X, the meromorphic functions on X form a field of transcendence degree 1 over $\mathbb{C}$. We shall determine this field for $X = X_0(N)$.

For a subgroup Γ of finite index in $\mathrm{SL}_2(\mathbb{Z})$, the meromorphic functions on $\Gamma\backslash\mathbb{H}^*$ are called the ***modular functions*** for Γ. If $\pi\colon\mathbb{H}\to\Gamma\backslash\mathbb{H}^*$ is the quotient map, then $g\mapsto\pi\circ g$ identifies the modular functions for Γ with the functions f on $\mathbb{H}$ such that

(a) f is meromorphic on $\mathbb{H}$;

(b) for all $\gamma\in\Gamma$, $f(\gamma z)=f(z)$;

(c) f is meromorphic at the cusps (i.e., at the points of $\mathbb{H}^*\smallsetminus\mathbb{H}$).

The meromorphic functions on $X_0(1)$

Let S be the Riemann sphere $S=\mathbb{C}\cup\{\infty\}$ (III, 3.2). The meromorphic functions on S are the rational functions of z, and the automorphisms of S are the fractional-linear transformations,

$$z\mapsto\frac{az+b}{cz+d},\quad a,b,c,d\in\mathbb{C},\quad ad-bc\neq 0.$$

In fact, $\mathrm{Aut}(S)=\mathrm{PGL}_2(\mathbb{C})\overset{\text{def}}{=}\mathrm{GL}_2(\mathbb{C})/\mathbb{C}^\times$. Moreover, given two sets of distinct points on S, $\{P_1,P_2,P_3\}$ and $\{Q_1,Q_2,Q_3\}$, there is a unique fractional-linear transformation sending each P_i to Q_i. (The proof of the last statement is an easy exercise in linear algebra: given two sets $\{L_1,L_2,L_3\}$ and $\{M_1,M_2,M_3\}$ of distinct lines through the origin in $\mathbb{C}^2$, there is a linear transformation carrying each L_i to M_i, and the linear transformation is unique up to multiplication by a nonzero constant.)

Recall that ∞, i, and ρ are points in $\mathbb{H}^*$. We use the same symbols to denote their images in $X_0(1)$.

PROPOSITION 2.1 *There exists a unique meromorphic function J on $X_0(1)$ that is holomorphic except at ∞, where it has a simple pole, and takes the values*

$$J(i)=1,\quad J(\rho)=0.$$

The meromorphic functions on $X_0(1)$ are the rational functions of J.

PROOF. We saw in the last section that $X_0(1)$ is isomorphic (as a Riemann surface) to the Riemann sphere S. Let $f\colon X_0(1)\to S$ be an isomorphism, and let P, Q, R be the images of ρ, i, ∞. There is a unique fractional-linear transformation L sending P, Q, R to $0, 1, \infty$, and the composite $J = L \circ f$ has the required properties. If J' is a second function with the required properties, then the composite $J' \circ J^{-1}$ is an automorphism of S fixing $0, 1, \infty$, and so is the identity map. For the second statement, note that J defines an isomorphism of $X_0(1)$ onto the Riemann sphere under which the meromorphic function z on the Riemann sphere corresponds to J on the $X_0(1)$. □

We wish to identify the function J. Recall from Chapter III that, for a lattice Λ in $\mathbb{C}$,

$$G_{2k}(\Lambda) \stackrel{\text{def}}{=} \sum_{\omega \in \Lambda, \omega \neq 0} \frac{1}{\omega^{2k}},$$

and that $G_{2k}(z) \stackrel{\text{def}}{=} G_{2k}(\mathbb{Z}z + \mathbb{Z})$, $z \in \mathbb{H}$. Moreover (III, 2.6), the map $z \mapsto (\wp(z) : \wp'(z) : 1)$ is an isomorphism from $\mathbb{C}/\mathbb{Z}z + \mathbb{Z}$ onto the elliptic curve

$$Y^2 Z = 4X^3 - g_4(z) X Z^2 - g_6(z) Z^3, \quad \Delta = g_4(z)^3 - 27 g_6(z)^2 \neq 0,$$

where $g_4(z) = 60 G_4(z)$ and $g_6(z) = 140 G_6(z)$. The j-invariant of this curve is

$$j(z) = \frac{1728 g_4(z)^3}{\Delta}.$$

From their definitions, it is clear that $G_{2k}(z)$, $\Delta(z)$, and $j(z)$ are invariant under $T : z \mapsto z + 1$, and so can be expressed in terms of the variable $q = e^{2\pi i z}$. In Serre 1973, VII, Equations (23,33,42), we can find the following expansions:

$$G_{2k}(z) = 2\zeta(2k) + \frac{2(2\pi i)^{2k}}{(2k-1)!} \sum_{n=1}^{\infty} \sigma_{2k-1}(n) q^n, \quad \sigma_k(n) = \sum_{d \mid n} d^k, \tag{42}$$

$$\Delta = (2\pi)^{12} (q - 24q^2 + 252q^3 - 1472q^4 + \cdots), \tag{43}$$

$$j = \frac{1}{q} + 744 + 196884q + 21493760q^2 + \sum_{n=3}^{\infty} c(n) q^n, \quad c(n) \in \mathbb{Z}. \tag{44}$$

The proof of the formula for $G_{2k}(z)$ is elementary, and the others follow from it together with elementary results on $\zeta(2k)$. The factor 1728 was traditionally included in the formula for j so that it has residue 1 at infinity.

The function j is invariant under $\Gamma_0(1) = \mathrm{SL}_2(\mathbb{Z})$, because $j(z)$ depends only on the lattice $\mathbb{Z}z + \mathbb{Z}$. Moreover,

- ◇ $j(\rho) = 0$, because $\mathbb{C}/\mathbb{Z}\rho + \mathbb{Z}$ has complex multiplication by $\rho^2 = \sqrt[3]{1}$, and therefore is of the form $Y^2 = X^3 + b$, which has j-invariant 0;
- ◇ $j(i) = 1728$, because $\mathbb{C}/\mathbb{Z}i + \mathbb{Z}$ has complex multiplication by i, and therefore is of the form $Y^2 = X^3 + aX$.

Consequently $j = 1728J$, and we can restate Proposition 2.1 as follows.

PROPOSITION 2.2 *The function j is the unique meromorphic function on $X_0(1)$ that is holomorphic except at ∞, where it has a simple pole, and takes the values*

$$j(i) = 1728, \quad j(\rho) = 0.$$

It defines an isomorphism from $X_0(1)$ onto the Riemann sphere, and so the field of meromorphic functions on $X_0(N)$ is $\mathbb{C}(j)$.

The meromorphic functions on $X_0(N)$

Define j_N to be the function on $\mathbb{H}$ such that $j_N(z) = j(Nz)$. For $\gamma \in \Gamma_0(1)$, one is tempted to say

$$j_N(\gamma z) \stackrel{\text{def}}{=} j(N\gamma z) = j(\gamma N z) = j(Nz) \stackrel{\text{def}}{=} j_N(z),$$

but, this is false in general, because $N\gamma z \neq \gamma N z$. However, it is true that $j_N(\gamma z) = j_N(z)$ if $\gamma \in \Gamma_0(N)$. To see this, let $\gamma = \left(\begin{smallmatrix} a & b \\ c & d \end{smallmatrix}\right) \in \Gamma_0(N)$, so that $c = Nc'$ with $c' \in \mathbb{Z}$. Then

$$j_N(\gamma z) = j\left(\frac{Naz + Nb}{cz + d}\right) = j\left(\frac{a(Nz) + Nb}{c'(Nz) + d}\right) = j(\gamma' N z),$$

where $\gamma' = \begin{pmatrix} a & Nb \\ c' & b \end{pmatrix} \in \Gamma_0(1)$, so

$$j(\gamma' N z) = j(Nz) = j_N(z).$$

Thus, we see that j_N is invariant under $\Gamma_0(N)$, and therefore defines a meromorphic function on $X_0(N)$.

THEOREM 2.3 *The field of meromorphic functions on $X_0(N)$ is $\mathbb{C}(j, j_N)$.*

PROOF. The curve $X_0(N)$ is a covering of $X_0(1)$ of degree m equal to $(\Gamma_0(1) : \Gamma_0(N))$. The general theory implies that the field of meromorphic functions on $X_0(N)$ has degree m over $\mathbb{C}(j)$, but we shall prove this again. Let $\{\gamma_1 = 1, ..., \gamma_m\}$ be a set of representatives for the right cosets of $\Gamma_0(N)$ in $\Gamma_0(1)$, so that,

$$\Gamma_0(1) = \bigsqcup_{i=1}^{m} \Gamma_0(N)\gamma_i \quad \text{(disjoint union)}.$$

Right multiplication by a $\gamma \in \Gamma_0(1)$ permutes the cosets $\Gamma_0(N)\gamma_i$, and so $\{\gamma_1\gamma, ..., \gamma_m\gamma\}$ is also a set of representatives for the right cosets of $\Gamma_0(N)$ in $\Gamma_0(1)$.

If $f(z)$ is a modular function for $\Gamma_0(N)$, then $f(\gamma_i z)$ depends only on the coset $\Gamma_0(N)\gamma_i$. Hence the functions $\{f(\gamma_i \gamma z)\}$ are a permutation of the functions $\{f(\gamma_i z)\}$, and every symmetric polynomial in the $f(\gamma_i z)$ is invariant under $\Gamma_0(1)$; since such a polynomial obviously satisfies the other conditions, it is a modular function for $\Gamma_0(1)$, and hence a rational function of j. Therefore $f(z)$ satisfies a polynomial of degree m with coefficients in $\mathbb{C}(j)$, namely, $\prod(Y - f(\gamma_i z))$. Since this holds for every meromorphic function on $X_0(N)$, we see that the field of such functions has degree at most m over $\mathbb{C}(j)$ (apply the primitive element theorem, FT, 5.1).

Next I claim that all the $f(\gamma_i z)$ are conjugate to $f(z)$ over $\mathbb{C}(j)$: for let $F(j, Y)$ be the minimum polynomial of $f(z)$ over $\mathbb{C}(j)$, so that $F(j, Y)$ is monic and irreducible when regarded as a polynomial in Y with coefficients in $\mathbb{C}(j)$; on replacing z with $\gamma_i z$ and remembering that $j(\gamma_i z) = j(z)$, we find that $F(j(z), f(\gamma_i z)) = 0$, which proves the claim.

If we can show that the functions $j(N\gamma_i z)$ are distinct, then it will follow that the minimum polynomial of j_N over $\mathbb{C}(j)$ has degree m, and that the field of meromorphic functions on $X_0(N)$ has degree m over $\mathbb{C}(j)$, and is generated by j_N.

Suppose that $j(N\gamma_i z) = j(N\gamma_j z)$ for some $i \neq j$. Recall (2.2) that j defines an isomorphism $\Gamma_0(1)\backslash\mathbb{H}^* \to S_2$ (Riemann sphere), and so

$$j(N\gamma_i z) = j(N\gamma_j z) \text{ all } z \Rightarrow \exists \gamma \in \Gamma_0(1) \text{ such that } N\gamma_i z = \gamma N\gamma_j z \text{ all } z,$$

and this implies that

$$\begin{pmatrix} N & 0 \\ 0 & 1 \end{pmatrix} \gamma_i = \pm\gamma \begin{pmatrix} N & 0 \\ 0 & 1 \end{pmatrix} \gamma_j.$$

Hence

$$\gamma_i\gamma_j^{-1} \in \Gamma_0(1) \cap \begin{pmatrix} N & 0 \\ 0 & 1 \end{pmatrix}^{-1} \Gamma_0(1) \begin{pmatrix} N & 0 \\ 0 & 1 \end{pmatrix} = \Gamma_0(N),$$

which contradicts the fact that γ_i and γ_j lie in different cosets. □

We saw in the proof that the minimum polynomial of j_N over $\mathbb{C}(j)$ is

$$F_N(j,Y) = \prod_{i=1}^{m}(Y - j(N\gamma_i z)).$$

The symmetric polynomials in the $j(N\gamma_i z)$ are holomorphic on $\mathbb{H}$. As they are rational functions of $j(z)$, they must in fact be polynomials in $j(z)$, and so $F_N(j,Y) \in \mathbb{C}[j,Y]$ (rather than $\mathbb{C}(j)[Y]$).

On replacing j with the variable X, we obtain a polynomial $F_N(X,Y) \in \mathbb{C}[X,Y]$,

$$F_N(X,Y) = \sum c_{r,s}X^rY^s, \quad c_{r,s} \in \mathbb{C}, \quad c_{0,m} = 1.$$

I claim that $F_N(X,Y)$ is the unique polynomial of degree $\leq m$ in Y, with $c_{0,m} = 1$, such that

$$F_N(j,j_N) = 0.$$

In fact, $F_N(X,Y)$ generates the ideal in $\mathbb{C}[X,Y]$ of all polynomials $G(X,Y)$ such that $G(j,j_N) = 0$, from which the claim follows.

PROPOSITION 2.4 *The polynomial $F_N(X,Y)$ has coefficients in $\mathbb{Q}$.*

PROOF. We know that

$$j(z) = q^{-1} + \sum_{n=0}^{\infty} c(n)q^n, \quad c(n) \in \mathbb{Z}.$$

When we substitute this into the equation

$$F_N(j(z), j(Nz)) = 0,$$

and equate coefficients of powers of q, we obtain a set of linear equations for the $c_{r,s}$ with coefficients in $\mathbb{Q}$, and when we adjoin the equation

$$c_{0,m} = 1,$$

then the system determines the $c_{r,s}$ uniquely. Because the system of linear equations has a solution in $\mathbb{C}$, it also has a solution in $\mathbb{Q}$ (look at ranks of matrices); because the solution is unique, the solution in $\mathbb{C}$ must in fact lie in $\mathbb{Q}$. Therefore $c_{r,s} \in \mathbb{Q}$. □

ASIDE 2.5 The polynomial $F_N(X,Y)$ was introduced by Kronecker more than 100 years ago. It is known to have integer coefficients and be symmetric in X and Y. For $N=2$, it is

$$X^3+Y^3-X^2Y^2+1488XY(X+Y)-162000(X^2+Y^2) \\ +40773375XY+8748000000(X+Y)-157464000000000.$$

It was computed for $N=2,3,5,7$ by Smith (1878, 1879), Berwick (1916), and Herrmann (1974) respectively. At this point the humans gave up, and employed computers, which found F_{11} in 1984. This last computation took about 20 hours on a VAX-780, and the result is a polynomial with coefficients up to 10^{60} that takes five pages to write out. Since then there have been improvements to the algorithms used to compute $F_N(X,Y)$, and $F_\ell(X,Y)$ has been computed for all primes ℓ up to at least 3,600 (Bröker et al. 2012). For us, it is important to know that the polynomial exists; it is not important to know what it is.

The curve $X_0(N)$ over $\mathbb{Q}$

Let C_N be the affine curve over $\mathbb{Q}$ with equation $F_N(X,Y)=0$, and let $\bar{C}_N$ be the projective curve defined by F_N made homogeneous. The map $z \mapsto (j(z), j(Nz))$ is a map $X_0(N) \smallsetminus \Xi \to C_N(\mathbb{C})$, where Ξ is the set where j or j_N has a pole. This map extends uniquely to a map $X_0(N) \to \bar{C}_N(\mathbb{C})$, which is an isomorphism except over the singular points of $\bar{C}_N$, and the pair $(X_0(N), X_0(N) \to \bar{C}_N(\mathbb{C}))$ is uniquely determined by $\bar{C}_N$ (up to a unique isomorphism): it is the canonical "desingularization" of $\bar{C}_N$ over $\mathbb{C}$.[2]

Now consider $\bar{C}_N$ over $\mathbb{Q}$. There is a canonical desingularization $X \to \bar{C}_N$ over $\mathbb{Q}$, i.e., a projective nonsingular curve X over $\mathbb{Q}$, and a regular map $X \to \bar{C}_N$ that is an isomorphism except over the singular points of $\bar{C}_N$, and the pair $(X, X \to \bar{C}_N)$ is uniquely determined by $\bar{C}_N$ (up to unique isomorphism). See I, Theorem 6.4. When we pass to the $\mathbb{C}$-points, we see that $(X(\mathbb{C}), X(\mathbb{C}) \to \bar{C}_N(\mathbb{C}))$ has the property characterizing $(X_0(N), X_0(N) \to \bar{C}_N(\mathbb{C}))$, and so there is a unique isomorphism of Riemann surfaces $X_0(N) \to X(\mathbb{C})$ compatible with the maps to $\bar{C}_N(\mathbb{C})$.

In summary, we have a well-defined curve X over $\mathbb{Q}$, a regular map $\gamma: X \to \bar{C}_N$ over $\mathbb{Q}$, and an isomorphism $X_0(N) \to X(\mathbb{C})$ whose composite with $\gamma(\mathbb{C})$ is (outside a finite set) $z \mapsto (j(z), j(Nz))$.

In future, we often use $X_0(N)$ to denote the curve X over $\mathbb{Q}$ — it should be clear from the context whether we mean the curve over $\mathbb{Q}$ or the Riemann

---

[2]The regular map $X_0(N) \smallsetminus \Xi \to C_N$ induces an isomorphism on the function fields, which realizes $X_0(N)$ as the normalization of $\bar{C}_N$.

surface. The affine curve $X_0(N) \smallsetminus \{\text{cusps}\} \subset X_0(N)$ is denoted by $Y_0(N)$; thus $Y_0(N)(\mathbb{C}) = \Gamma_0(1)\backslash\mathbb{H}$.

REMARK 2.6 The curve $F_N(X,Y) = 0$ is highly singular, because, without singularities, formula (6), p. 41, would predict much too high a genus.

The points on the curve $X_0(N)$

Since we cannot write down an equation for $X_0(N)$ as a nonsingular projective curve over $\mathbb{Q}$, we would at least like to know what its points are in any field containing $\mathbb{Q}$. This we can do.

We first look at the complex points of $X_0(N)$, i.e., at the Riemann surface $X_0(N)$. In this case, there is a diagram

$$\begin{array}{ccccccc}
\{(E,S)\}/\approx & \longleftrightarrow & \{(\Lambda,S)\}/\mathbb{C}^\times & \longleftrightarrow & \Gamma_0(N)\backslash M/\mathbb{C}^\times & \longleftrightarrow & \Gamma_0(N)\backslash\mathbb{H} \\
\downarrow & & \downarrow & & \downarrow & & \downarrow \\
\{E\}/\approx & \longleftrightarrow & \mathcal{L}/\mathbb{C}^\times & \longleftrightarrow & \Gamma_0(1)\backslash M/\mathbb{C}^\times & \longleftrightarrow & \Gamma_0(1)\backslash\mathbb{H}
\end{array}$$

whose terms we now explain. All the symbols $\leftrightarrow$ are natural bijections. The bottom row combines maps in Chapter III.

Recall that M is the set pairs $(\omega_1,\omega_2) \in \mathbb{C}\times\mathbb{C}$ such that $\Im(\omega_1/\omega_2) > 0$ (so $M/\mathbb{C}^\times \subset \mathbb{P}^1(\mathbb{C})$), and that the bijection $M/\mathbb{C}^\times \to \mathbb{H}$ sends (ω_1,ω_2) to ω_1/ω_2. The rest of the right hand square is now obvious.

Recall that $\mathcal{L}$ is the set of lattices in $\mathbb{C}$, and that the lattices defined by two pairs in M are equal if and only if the pairs lie in the same $\Gamma_0(1)$-orbit. Thus in passing from an element of M to its $\Gamma_0(1)$-orbit we are forgetting the basis and remembering only the lattice. In passing from an element of M to its $\Gamma_0(N)$-orbit, we remember a little of the basis, for suppose that

$$\begin{pmatrix}\omega_1'\\ \omega_2'\end{pmatrix} = \begin{pmatrix}a & b\\ c & d\end{pmatrix}\begin{pmatrix}\omega_1\\ \omega_2\end{pmatrix}, \quad \begin{pmatrix}a & b\\ c & d\end{pmatrix} \in \Gamma_0(N).$$

Then

$$\begin{aligned}
\omega_1' &= a\omega_1 + b\omega_2 \\
\omega_2' &= c\omega_1 + d\omega_2 \equiv d\omega_2 \mod N\Lambda.
\end{aligned}$$

Hence

$$\tfrac{1}{N}\omega_2' \equiv \tfrac{d}{N}\omega_2 \mod \Lambda.$$

Note that because $\begin{pmatrix} a & b \\ c & d \end{pmatrix}$ has determinant 1, $\gcd(d, N) = 1$, and so $\frac{1}{N}\omega_2'$ and $\frac{1}{N}\omega_2$ generate the same cyclic subgroup S of order N in $\mathbb{C}/\Lambda$. Therefore, the map

$$(\omega_1, \omega_2) \mapsto (\Lambda(\omega_1, \omega_2), \langle \tfrac{1}{N}\omega_2 \rangle)$$

defines a bijection from $\Gamma_0(N)\backslash M$ to the set of pairs consisting of a lattice Λ in $\mathbb{C}$ and a cyclic subgroup S of $\mathbb{C}/\Lambda$ of order N. Now $(\Lambda, S) \mapsto (\mathbb{C}/\Lambda, S)$ defines a one-to-one correspondence between this last set and the set of isomorphism classes of pairs (E, S) consisting of an elliptic curve over $\mathbb{C}$ and a cyclic subgroup S of $E(\mathbb{C})$ of order N. An isomorphism $(E, S) \to (E', S')$ is an isomorphism $E \to E'$ carrying S into S'.

Note that the quotient of E by S,

$$E/S \simeq \mathbb{C}/\Lambda(\omega_1, \tfrac{1}{N}\omega_2),$$

and that $\omega_1/(\frac{1}{N}\omega_2) = N\frac{\omega_1}{\omega_2}$. Thus, if $j(E) = j(z)$, then $j(E/S) = j(Nz)$.

Now, for any field $k \supset \mathbb{Q}$, define $\mathcal{E}_0(N)(k)$ to be the set of isomorphism classes of pairs E consisting of an elliptic curve E over k and a cyclic subgroup $S \subset E(k^{\mathrm{al}})$ of order N stable under $\mathrm{Gal}(k^{\mathrm{al}}/k)$ — thus the subgroup S is defined over k but not necessarily its individual elements. The above remarks show that there is a canonical bijection

$$\mathcal{E}_0(N)(\mathbb{C})/{\approx} \to Y_0(N)$$

whose composite with the map $Y_0(N) \to C_N(\mathbb{C})$ is

$$(E, S) \mapsto (j(E), j(E/S)).$$

Here $Y_0(N)$ denotes the Riemann surface $\Gamma_0(N)\backslash\mathbb{H}$.

THEOREM 2.7 *For any field $k \supset \mathbb{Q}$, there is a map*

$$\mathcal{E}_0(N)(k) \to Y_0(N)(k),$$

functorial in k, such that

(a) *the composite*

$$\mathcal{E}_0(N)(k) \to Y_0(N)(k) \to C_N(k)$$

is $(E, S) \mapsto (j(E), j(E/S))$;

(b) *the map* $\mathcal{E}_0(N)(k)/{\approx} \to Y_0(N)(k)$ *is surjective for all* k *and bijective for all algebraically closed* k.

The functoriality in k means that, for every homomorphism $\sigma\colon k \to k'$ of fields, the diagram

$$\begin{array}{ccc} \mathcal{E}_0(N)(k') & \longrightarrow & Y_0(N)(k') \\ \uparrow \sigma & & \uparrow \sigma \\ \mathcal{E}_0(N)(k) & \longrightarrow & Y_0(N)(k) \end{array}$$

commutes. In particular, $\mathcal{E}_0(N)(k^{\mathrm{al}}) \to Y_0(N)(k^{\mathrm{al}})$ commutes with the actions of $\mathrm{Gal}(k^{\mathrm{al}}/k)$. Since $Y_0(N)(k^{\mathrm{al}})^{\mathrm{Gal}(k^{\mathrm{al}}/k)} = Y_0(N)(k)$, this implies that

$$Y_0(N)(k) = (\mathcal{E}_0(N)(k^{\mathrm{al}}))/{\approx})^{\mathrm{Gal}(k^{\mathrm{al}}/k)}$$

for any field $k \supset \mathbb{Q}$.

This description of the points can be extended to $X_0(N)$ by adding to $\mathcal{E}_0(N)$ certain "degenerate" elliptic curves.

Here is a sketch of the proof of Theorem 2.7. From the bijection $\mathcal{E}_0(N)(\mathbb{C})/{\approx} \simeq Y_0(N)(\mathbb{C})$ we obtain an action of $\mathrm{Aut}(\Omega/\mathbb{Q})$ on $Y_0(N)(\mathbb{C})$, which one can show is continuous and regular, and so defines a model $E_0(N)$ of $Y_0(N)$ over $\mathbb{Q}$ (see I, 5.5). Over $\mathbb{C}$, we have regular maps

$$E_0(N) \to Y_0(N) \to C_N$$

whose composite (on points) is $(E, S) \mapsto (j(E), j(E/S))$. As this map commutes with the automorphisms of $\mathbb{C}$, the regular map $E_0(N) \to C_N$ is defined over $\mathbb{Q}$ (loc. cit.), and therefore so also is $E_0(N) \to Y_0(N)$. As it is an isomorphism on the $\mathbb{C}$-points, it is an isomorphism (I, 4.24). One can show that, for any field $k \subset \mathbb{C}$, the map $\mathcal{E}_0(N)(k) \to E_0(N)(k)$ is surjective with fibres equal to the geometric isomorphism classes of pairs (E, S), where two pairs are geometrically isomorphic if they become isomorphic over $\mathbb{C}$ (equivalently k^{al}).

Variants

2.8 For our applications to elliptic curves, we shall only need to use the quotients of $\mathbb{H}^*$ by the subgroups $\Gamma_0(N)$, but quotients by other subgroups are also of interest. For example, let

$$\begin{aligned}\Gamma_1(N) &= \left\{ \begin{pmatrix} a & b \\ c & d \end{pmatrix} \middle| a \equiv 1 \equiv d \mod N, \quad c \equiv 0 \mod N \right\} \\ &= \left\{ \begin{pmatrix} a & b \\ c & d \end{pmatrix} \middle| \begin{pmatrix} a & b \\ c & d \end{pmatrix} \equiv \begin{pmatrix} 1 & b \\ 0 & 1 \end{pmatrix} \mod N \right\}.\end{aligned}$$

The quotient $X_1(N) = \Gamma_1(N)\backslash\mathbb{H}^*$ again defines a curve, also denoted $X_1(N)$, over $\mathbb{Q}$, and there is a theorem similar to 2.7 but with $\mathcal{E}_1(N)(k)$ the set of pairs (E, P) consisting of an elliptic curve E over k and a point $P \in E(k)$ of order N. In this case, the map

$$\mathcal{E}_1(N)(k)/{\approx} \to Y_1(N)(k)$$

is a bijection for all fields whenever $4|N$. The curve $X_1(N)$ has genus 0 exactly for $N = 1, 2, \ldots, 10, 12$. Since $X_1(N)$ has a point with coordinates in $\mathbb{Q}$ for each of these N, i.e., there does exist an elliptic curve over $\mathbb{Q}$ with a point of that order (II, §5), $X_1(N)$ is isomorphic to $\mathbb{P}^1$ (see I, §2), and so $X_1(N)$ has infinitely many rational points. Therefore, for $N = 1, 2, \ldots, 10, 12$, there are infinitely many nonisomorphic elliptic curves over $\mathbb{Q}$ with a point of order N with coordinates in $\mathbb{Q}$. Mazur showed, that for all other N, $Y_1(N)$ is empty, and so these are the only possible orders for a point on an elliptic curve over $\mathbb{Q}$ (see II, 5.11).

3 Modular forms

It is difficult to construct functions on $\mathbb{H}$ invariant under a subgroup Γ of $\mathrm{SL}_2(\mathbb{Z})$ of finite index. One strategy is to construct functions, not invariant under Γ, but transforming in a certain fixed manner. A quotient of two functions transforming in the same manner will be invariant under Γ.[3] This idea suggests the notion of a modular form.

Definition of a modular form

DEFINITION 3.1 Let Γ be a subgroup of finite index in $\mathrm{SL}_2(\mathbb{Z})$. A ***modular form*** for Γ of weight[4] $2k$ is a function $f : \mathbb{H} \to \mathbb{C}$ satisfying the following conditions:

[3]This is similar to constructing rational functions on $\mathbb{P}^1$ as the quotient of two homogeneous polynomials of the same degree.

[4]k and $-k$ are also used.

(a) f is holomorphic on $\mathbb{H}$;

(b) for all $\gamma = \left(\begin{smallmatrix} a & b \\ c & d \end{smallmatrix}\right) \in \Gamma$, $f(\gamma z) = (cz+d)^{2k} f(z)$;

(c) f is holomorphic at the cusps.

Recall that the cusps are the points of $\mathbb{H}^*$ not in $\mathbb{H}$. Since Γ is of finite index in $\mathrm{SL}_2(\mathbb{Z})$, $T^h = \left(\begin{smallmatrix} 1 & h \\ 0 & 1 \end{smallmatrix}\right)$ is in Γ for some integer $h > 0$, which we may take to be as small as possible. Then condition (b) implies that $f(T^h z) = f(z)$, i.e., that $f(z+h) = f(z)$, and so

$$f(z) = f^*(q), \quad q = e^{2\pi i z/h},$$

and f^* is a function on a neighbourhood of $0 \in \mathbb{C}$, with 0 removed. To say that f is holomorphic at ∞ means that f^* is holomorphic at 0, and so

$$f(z) = \sum_{n \geq 0} c(n) q^n, \quad q = e^{2\pi i z/h}.$$

For a cusp $r \neq \infty$, choose a $\gamma \in \mathrm{SL}_2(\mathbb{Z})$ such that $\gamma(\infty) = r$, and then the requirement is that $f \circ \gamma$ be holomorphic at ∞. It suffices to check the condition for one cusp in each Γ-orbit.

A modular form f is called a ***cusp form*** if it is zero at the cusps. At the cusp ∞ this means that $c(0) = 0$, and so

$$f(z) = \sum_{n \geq 1} c(n) q^n.$$

REMARK 3.2 Note that, for $\gamma = \left(\begin{smallmatrix} a & b \\ c & d \end{smallmatrix}\right) \in \mathrm{SL}_2(\mathbb{Z})$,

$$d\gamma z = d\frac{az+b}{cz+d} = \frac{a(cz+d) - c(az+b)}{(cz+d)^2} dz = (cz+d)^{-2} dz.$$

Thus condition (3.1b) says that $f(z)(dz)^k$ is invariant under the action of the group Γ.

Write $\mathcal{M}_{2k}(\Gamma)$ for the vector space of modular forms of weight $2k$, and $\mathcal{S}_{2k}(\Gamma)$ for the subspace[5] of cusp forms. A modular form of weight 0 is a holomorphic modular function, i.e., a holomorphic function on the compact Riemann surface $X(\Gamma)$, and is therefore constant: $\mathcal{M}_0(\Gamma) = \mathbb{C}$. A

[5]The $\mathcal{S}$ is for "Spitzenform", the German name for cusp form. The French name is "forme parabolique".

product of modular forms of weight $2k$ and $2k'$ is a modular form of weight $2(k+k')$, which is a cusp form if one of the two forms is a cusp form. Therefore $\bigoplus_{k\geq 0}\mathcal{M}_{2k}(\Gamma)$ is a graded $\mathbb{C}$-algebra containing $\bigoplus_{k\geq 0}\mathcal{S}_{2k}(\Gamma)$ as an ideal.

PROPOSITION 3.3 *Let π denote the quotient map $\mathbb{H}^* \to \Gamma_0(N)\backslash\mathbb{H}^*$, and set $\pi^*\omega = f dz$ if ω is a holomorphic differential one-form ω on $\Gamma_0(N)\backslash\mathbb{H}^*$. Then $\omega \mapsto f$ is an isomorphism from the space of holomorphic differential one-forms on $\Gamma_0(N)\backslash\mathbb{H}^*$ to $\mathcal{S}_2(\Gamma_0(N))$.*

PROOF. The only surprise is that f is necessarily a cusp form rather than just a modular form. We examine this for the cusp ∞. Recall (p. 234) that there is a neighbourhood U of ∞ in $\Gamma_0(N)\backslash\mathbb{H}^*$ and an isomorphism $q\colon U \to D$ from U to a disk such that $q\circ\pi = e^{2\pi i z}$. Consider the differential $g(q)dq$ on U. Its inverse image on $\mathbb{H}$ is

$$g(e^{2\pi i z})d(e^{2\pi i z}) = 2\pi i\cdot g(e^{2\pi i z})\cdot e^{2\pi i z}dz = 2\pi i f dz,$$

where $f(z) = g(e^{2\pi i z})\cdot e^{2\pi i z}$. If g is holomorphic at 0, then

$$g(q) = \sum_{n\geq 0} c(n)q^n,$$

and so the q-expansion of f is $q\sum_{n\geq 0} c(n)q^n$, which is zero at ∞. □

COROLLARY 3.4 *The $\mathbb{C}$-vector space $\mathcal{S}_2(\Gamma_0(N))$ has dimension equal to the genus of $X_0(N)$.*

PROOF. When the Riemann–Roch theorem for a compact Riemann surface is applied to the divisor 0, it says that the dimension of the vector space of holomorphic differential one-forms is equal to the genus of the surface. □

Hence, there are explicit formulas for the dimension of $\mathcal{S}_2(\Gamma_0(N))$ — see p. 236. For example, it is zero for $N \leq 10$, and has dimension 1 for $N = 11$. In fact, the Riemann–Roch theorem gives formulas for the dimension of $\mathcal{S}_{2k}(\Gamma_0(N))$ for all N and k.

The modular forms for $\Gamma_0(1)$

In this section, we find the $\mathbb{C}$-algebra $\bigoplus_{k\geq 0}\mathcal{M}_{2k}(\Gamma_0(1))$.

We first explain a method of constructing functions satisfying (3.1b). As before, let $\mathcal{L}$ be the set of lattices in $\mathbb{C}$, and let $F:\mathcal{L} \to \mathbb{C}$ be a function such that

$$F(\lambda\Lambda) = \lambda^{-2k} F(\Lambda), \quad \lambda \in \mathbb{C}, \quad \Lambda \in \mathcal{L}.$$

Then

$$\omega_2^{2k} F(\Lambda(\omega_1, \omega_2))$$

depends only on the ratio $\omega_1 : \omega_2$, and so there is a function $f(z)$ defined on $\mathbb{H}$ such that

$$\omega_2^{2k} F(\Lambda(\omega_1, \omega_2)) = f(\omega_1/\omega_2) \quad \text{whenever } \Im(\omega_1/\omega_2) > 0.$$

For $\gamma = \left(\begin{smallmatrix} a & b \\ c & d \end{smallmatrix}\right) \in \mathrm{SL}_2(\mathbb{Z})$, $\Lambda(a\omega_1 + b\omega_2, c\omega_1 + d\omega_2) = \Lambda(\omega_1, \omega_2)$ and so

$$f(\tfrac{az+b}{cz+d}) = (cz+d)^{-2k} F(\Lambda(z,1)) = (cz+d)^{-2k} f(z).$$

When we apply this remark to the Eisenstein series

$$G_{2k}(\Lambda) = \sum_{\omega \in \Lambda, \omega \neq 0} \frac{1}{\omega^{2k}},$$

we find that the function $G_{2k}(z) \overset{\text{def}}{=} G_{2k}(\Lambda(z,1))$ satisfies (3.1b). In fact, more is true.

PROPOSITION 3.5 *For all $k > 1$, $G_{2k}(z)$ is a modular form of weight $2k$ for $\Gamma_0(1)$, and Δ is a cusp form of weight 12.*

PROOF. We know that $G_{2k}(z)$ is holomorphic on $\mathbb{H}$, and the formula (42), p. 238, shows that it is holomorphic at ∞, which is the only cusp for $\Gamma_0(1)$ (up to $\Gamma_0(1)$-equivalence). The statement for Δ is obvious from its definition $\Delta = g_4(z)^3 - 27g_4(z)^2$ and its q-expansion (43), p. 238. □

THEOREM 3.6 *The $\mathbb{C}$-algebra $\bigoplus_{k \geq 0} \mathcal{M}_{2k}(\Gamma_0(1))$ is generated by G_4 and G_6, and G_4 and G_6 are algebraically independent over $\mathbb{C}$. Therefore*

$$\mathbb{C}[G_4, G_6] \xrightarrow{\simeq} \bigoplus_{k \geq 0} \mathcal{M}_{2k}(\Gamma_0(1)),$$

$$\mathbb{C}[G_4, G_6] \simeq \mathbb{C}[X, Y]$$

(isomorphisms of graded $\mathbb{C}$-algebras if X and Y are given weights 4 and 6 respectively). Moreover,

$$f \mapsto f \cdot \Delta : \mathcal{M}_{2k-12}(\Gamma_0(1)) \to \mathcal{S}_{2k}(\Gamma_0(1))$$

is a bijection.

PROOF. We sketch the proof. Let f be a nonzero modular form of weight $2k$, and let $\omega = f(dz)^k$ be the corresponding k-fold differential form on $X_0(1) \stackrel{\text{def}}{=} \Gamma_0(1)\backslash\mathbb{H}^*$ (see 3.2). For $P \in \mathbb{H}^*$, let $\mathrm{ord}_P(f)$ denote the order of the zero of f at P and $\mathrm{ord}_P(\omega)$ the order of the zero of ω at the image of P in $X_0(1)$. Then

$$\begin{aligned}\mathrm{ord}_\infty(f) &= \mathrm{ord}_\infty(\omega) + k\\ \mathrm{ord}_P(f) &= e\,\mathrm{ord}_P(\omega) + k(e-1),\end{aligned}$$

where e is the order of the stabilizer of P in $\mathrm{PSL}_2(\mathbb{Z})$ (cf. the proof of 3.3). In particular,

$$\mathrm{ord}_P(f) = \mathrm{ord}_P(\omega)$$

for $P \in \bar{D}$, $P \neq i, \rho, \rho^2$. For a differential one-form ω on a curve of genus g, the divisor of ω is a canonical divisor, which has degree $2g-2$. For a k-fold differential, the degree is $k(2g-2)$. As $X_0(1)$ has genus 0, we deduce that

$$\mathrm{ord}_\infty(f) + \frac{1}{2}\mathrm{ord}_i(f) + \frac{1}{3}\mathrm{ord}_\rho(f) + \sum \mathrm{ord}_P(f) = -2k + \cdots = \frac{k}{6}, \quad (45)$$

where the sum is over a set of representatives for the remaining points in $\bar{D}$.

Let f be a nonzero modular form of weight $2k$. Then $\mathrm{ord}_P(f) \geq 0$ for $P = \infty$ and all $P \in \mathbb{H}$, and so (45) shows that $k \geq 0$. For $k = 0$, the only modular forms are the constants, and for $k = 1$, formula (45) shows that $\mathcal{M}_2 = 0$. For $k = 2$, f has zeros only on the points of the orbit of ρ, and there has a simple zero. It follows that $\dim \mathcal{M}_4 \leq 1$. Similarly, $\dim \mathcal{M}_6 \leq 1$. We conclude that

$$\dim \mathcal{M}_{2k}(\Gamma_0(1)) = \begin{cases} 0 & \text{if } k < 0 \text{ or } k = 1\\ \leq 1 & \text{if } k = 2, 3.\end{cases}$$

Next we prove that the sequence

$$0 \longrightarrow \mathcal{M}_{2k-12} \xrightarrow{f \mapsto f\cdot\Delta} \mathcal{M}_{2k} \xrightarrow{f \mapsto f(\infty)} \mathbb{C} \longrightarrow 0$$

is exact. As $G_{2k}(\infty) = 2\zeta(2k) \neq 0$, we see that the sequence is exact at $\mathbb{C}$. If $f \in \mathcal{M}_{2k}$ has $f(\infty) = 0$, then f/Δ is holomorphic on $\mathbb{H}$ (because Δ is never zero on $\mathbb{H}$), holomorphic at infinity (because $\mathrm{ord}_\infty(\Delta) = 1$), and of weight $2k-12$, and so lies in $\mathcal{M}_{2k-12}$. Now an induction argument starting from the results of the last paragraph show that, for $k \geq 2$,

$$\dim \mathcal{M}_{2k}(\Gamma_0(1)) = \begin{cases} [k/6] & \text{if } k \equiv 1 \mod 6\\ [k/6]+1 & \text{otherwise.}\end{cases}$$

Here $[x]$ is the largest integer $\leq x$.

Let $k \geq 2$. We next show by induction on k that every modular form of weight $2k$ is an isobaric polynomial in G_4 and G_6. Choose a pair of integers $m \geq 0$ and $n \geq 0$ such that $2m + 3n = k$ (this is always possible for $k \geq 2$). The modular form $g = G_4^m G_6^n$ is not zero at ∞. If $f \in \mathcal{M}_{2k}$, then $f - \frac{f(\infty)}{g(\infty)} g$ is zero at ∞, and so it can be written $\Delta \cdot h$ with $h \in \mathcal{M}_{2k-12}$. Now we apply the induction hypothesis.

Thus the map $\mathbb{C}[G_4, G_6] \to \bigoplus_k \mathcal{M}_{2k}$ is surjective, and it remains to show that it is injective. If not, the modular function G_2^3/G_6^2 satisfies an algebraic equation over $\mathbb{C}$, and so is constant. But $G_4(\rho) = 0 \neq G_6(\rho)$ whereas $G_4(i) \neq 0 = G_6(i)$. □

THEOREM 3.7 (JACOBI) *There is the following formula*

$$\Delta = (2\pi)^{12} q \prod_{n=1}^{\infty} (1 - q^n)^{24}, \quad q = e^{2\pi i z}.$$

PROOF. Let

$$F(z) = q \prod_{n=1}^{\infty} (1 - q^n)^{24}.$$

From the theorem, we know that the space of cusp forms of weight 12 has dimension 1, and therefore if we can show that $F(z)$ is such a form, then it is a multiple of Δ, and it will follow from the formula (43), p. 238, that the multiple is $(2\pi)^{12}$.

Because $\mathrm{SL}_2(\mathbb{Z})/\{\pm I\}$ is generated by $T = \begin{pmatrix} 1 & 1 \\ 0 & 1 \end{pmatrix}$ and $S = \begin{pmatrix} 0 & -1 \\ 1 & 0 \end{pmatrix}$, to check the conditions in (3.1), it suffices to check that F transforms correctly under T and S. For T this is obvious from the way we have defined F, and for S it amounts to checking that

$$F(-1/z) = z^{12} F(z).$$

This is trickier than it looks, but there are short (two-page) elementary proofs — see, for example, Serre 1973, VII, 4.4. □

EXERCISE 3.8 Define $\Delta(z) = \Delta(\mathbb{Z}z + \mathbb{Z})$ (see Chap. III), so that Δ is a basis for the $\mathbb{C}$-vector space of cusp forms of weight 12 for $\Gamma_0(1)$. Define $\Delta_{11}(z) = \Delta(11z)$, and show that it is a cusp form of weight 12 for $\Gamma_0(11)$. Deduce that $\Delta \cdot \Delta_{11}$ is a cusp form of weight 24 for $\Gamma_0(11)$.

EXERCISE 3.9 Assume Jacobi's formula,

$$\Delta(z) = (2\pi)^{12} q \prod_{n=1}^{\infty} (1-q^n)^{24},$$

($q = e^{2\pi i z}$), and that $S_2(\Gamma_0(11))$ has dimension 1. Show that

$$F(z) = q \prod_{n=1}^{\infty} (1-q^n)^2 (1-q^{11n})^2$$

is a cusp form of weight 2 for $\Gamma_0(11)$. [Hint: Let f be a nonzero element of $S_2(\Gamma_0(11))$, and let $g = \Delta \cdot \Delta_{11}$. Show that f^{12}/g is holomorphic on $\mathbb{H}^*$ and invariant under $\Gamma_0(1)$, and is therefore constant (because the only holomorphic functions on a compact Riemann surface are the constant functions). The only real difficulty is in handling the cusp 0, since I have more or less ignored cusps other than ∞.]

4 Modular forms and the L-functions of elliptic curves

In this section, we first see that elliptic curves over $\mathbb{Q}$ are classified up to isogeny by their L-functions, and then we see how the theory of modular forms leads to a list of candidates for the L-functions of such curves.

Dirichlet series

A ***Dirichlet series*** is a series of the form

$$f(s) = \sum_{n \geq 1} a(n) n^{-s}, \quad a(n) \in \mathbb{C}, \quad s \in \mathbb{C}.$$

The simplest example of such a series is, of course, the Riemann zeta function $\sum_{n \geq 1} n^{-s}$. If there exist positive constants A and b such that $|\sum_{n \leq x} a(n)| \leq Ax^b$ for all large x, then the series converges to an analytic function on the half-plane $\Re(s) > b$.

It is important to note that the function $f(s)$ determines the $a(n)$, i.e., if $\sum a(n)n^{-s}$ and $\sum b(n)n^{-s}$ are equal as functions of s on some half-plane, then $a(n) = b(n)$ for all n. In fact, by means of the Mellin transform and its inverse (see 4.3 below), f determines, and is determined by, the function

$g(q) = \sum a(n)q^n$, which is convergent on some disk about 0. Therefore, the claim follows from the similar statement for power series.

We shall be especially interested in Dirichlet series that can be expressed as Euler products, that is, as

$$f(s) = \prod_p \frac{1}{1 - P_p(p^{-s})},$$

where each P_p is a polynomial and the product is over the prime numbers.

Dirichlet series arise in two essentially different ways: from analysis and from arithmetic geometry and number theory. One of the *big* problems in mathematics is to show that the second set of Dirichlet series is a subset of the first, and to identify the subset. This is a major theme in the Langlands program, and the rest of the book will be concerned with explaining how work of Wiles and others succeeds in identifying the L-functions of elliptic curves over $\mathbb{Q}$ with L-functions arising from modular forms.

The L-function of an elliptic curve

Recall that for an elliptic curve E over $\mathbb{Q}$, we define

$$L(E,s) = \prod_{p \text{ good}} \frac{1}{1 - a_p p^{-s} + p^{1-s}} \prod_{p \text{ bad}} \frac{1}{1 - a_p p^{-s}},$$

where

$$a_p = \begin{cases} p + 1 - N_p & p \text{ good}, \\ 1 & p \text{ split nodal}, \\ -1 & p \text{ nonsplit nodal}, \\ 0 & p \text{ cuspidal}. \end{cases}$$

Recall also that the conductor $N = N_{E/\mathbb{Q}}$ of $\mathbb{Q}$ is $\prod_p p^{f_p}$, where $f_p = 0$ if E has good reduction at p, $f_p = 1$ if E has nodal reduction at p, and $f_p \geq 2$ otherwise (and $= 2$ unless $p = 2, 3$).

On expanding out the product, we obtain a Dirichlet series

$$L(E,s) = \sum_{n \geq 1} a_n n^{-s}.$$

This series has, among others, the following properties.

(a) (Rationality) Its coefficients a_n lie in $\mathbb{Q}$.

(b) (Euler product) It can be expressed as an Euler product; in fact, that is how we defined it.

(c) (Functional equation) Conjecturally it can be extended analytically to a meromorphic function on the whole complex plane, and $\Lambda(E,s) \stackrel{\text{def}}{=} N_{E/\mathbb{Q}}^{s/2}(2\pi)^{-s}\Gamma(s)L(E,s)$ satisfies the functional equation

$$\Lambda(E,s) = w_E\,\Lambda(E,2-s), \quad w_E = \pm 1.$$

Here $\Gamma(s)$ is the gamma function.

L-functions and isogeny classes

Recall (II, 1.8) that two elliptic curves (E,O) and (E',O') are said to be isogenous if there exists a nonconstant regular map $E \to E'$ sending O to O', and that isogeny is an equivalence relation.

An isogeny $E \to E'$ defines a homomorphism $E(\mathbb{Q}) \to E'(\mathbb{Q})$ which, in general, will be neither injective nor surjective, but which does have a finite kernel and cokernel.[6] Therefore, the ranks of $E(\mathbb{Q})$ and $E'(\mathbb{Q})$ are the same, but their torsion subgroups will, in general, be different. Surprisingly, isogenous curves over a finite field do have the same number of points.

THEOREM 4.1 *Let E and E' be elliptic curves over $\mathbb{Q}$. If E and E' are isogenous, then $N_p(E) = N_p(E')$ for all good p. Conversely, if $N_p(E) = N_p(E')$ for sufficiently many good p, then E is isogenous to E'.*

PROOF. Let p be a good prime for E and E'. An isogeny $\alpha: E \to E'$ induces an isogeny $\alpha_p: E_p \to E'_p$, which commutes with the Frobenius maps φ and φ' on E and E'. To see this, suppose that

$$\alpha_p(x:y:z) = (P(x,y,z):Q(x,y,z):R(x,y,z)), \quad P,Q,R \in \mathbb{F}_p[X,Y,Z].$$

Then

$$(\alpha_p \circ \varphi)(x:y:z) = (P(x^p,y^p,z^p):Q(x^p,y^p,z^p):R(x^p,y^p,z^p))$$
$$(\varphi' \circ \alpha_p)(x:y:z) = (P(x,y,z)^p:Q(x,y,z)^p:R(x,y,z)^p)$$

[6] For an isogeny φ of degree n, there exists an isogeny φ' such that $\varphi \circ \varphi' = n$, and $n: E(\mathbb{Q}) \to E(\mathbb{Q})$ has finite kernel and cokernel because $E(\mathbb{Q})$ is finitely generated.

which the characteristic p binomial theorem shows to be equal. Now the diagram

$$\begin{array}{ccc} E & \xrightarrow{1-\varphi} & E \\ \downarrow{\scriptstyle \alpha_p} & & \downarrow{\scriptstyle \alpha_p} \\ E' & \xrightarrow{1-\varphi} & E' \end{array}$$

commutes, and so

$$\deg\alpha \cdot \deg(1-\varphi) = \deg(1-\varphi') \cdot \deg\alpha.$$

As

$$\deg(1-\varphi) = N_p(E)$$
$$\deg(1-\varphi') = N_p(E')$$

(see the proof of Theorem 9.4), this proves that $N_p(E) = N_p(E')$.

The converse is much more difficult. It was conjectured by Tate about 1963, and proved under some hypotheses by Serre. It was proved in general (for all abelian varieties) by Faltings in his great article on Mordell's conjecture (Faltings 1983). □

The theorem of Faltings gives an effective procedure for deciding whether two elliptic curves over $\mathbb{Q}$ are isogenous: there is a constant P such that if $N_p(E) = N_p(E')$ for all good $p \leq P$, then E and E' are isogenous. This has been made into an effective algorithm. In practice, if your computer fails to find a p with $N_p(E) \neq N_p(E')$ in a few minutes you can be very confident that the curves are isogenous.

It is not obvious, but follows from the theory of Néron models, that isogenous elliptic curves have the same type of reduction at every prime. Therefore, isogenous curves have exactly the same L-functions and the same conductor. Because the L-functions is determined by, and determines the N_p, we have the following corollary.

COROLLARY 4.2 *Two elliptic curves E and E' are isogenous if and only if $L(E,s) = L(E',s)$.*

We therefore have a one-to-one correspondence:

$$\{\text{isogeny classes of elliptic curves over } \mathbb{Q}\} \leftrightarrow \{\text{ certain } L\text{-functions}\}.$$

In the remainder of this section we identify the L-functions corresponding to elliptic curves over $\mathbb{Q}$.

As noted, isogenous curves over $\mathbb{Q}$ have the same set of bad primes S. A theorem of Shafarevich says that, given a finite set of primes S of $\mathbb{Q}$, there are only finitely many elliptic curves over $\mathbb{Q}$ with good reduction outside S up to isomorphism. Hence, every isogeny class of elliptic curves over $\mathbb{Q}$ contains only finitely many isomorphism classes.

We sketch a proof of Shafarevich's theorem. We may enlarge S and so suppose that $2,3 \in S$. Let $\mathbb{Z}_S$ be the ring obtained from $\mathbb{Z}$ by inverting the primes in S. An elliptic curve with good reduction outside S can be written

$$E: Y^2 = X^3 + aX + b, \quad a,b \in \mathbb{Z}_S, \quad \Delta \in \mathbb{Z}_S^\times.$$

The group $\mathbb{Z}_S^\times$ is obviously finitely generated, and so $\mathbb{Z}_S^\times/\mathbb{Z}_S^{\times 12}$ is finite. It therefore suffices to show that there are only finitely many such curves E with Δ a fixed element modulo $\mathbb{Z}_S^{\times 12}$, say, with $\Delta = cd^{12}$, where c is fixed and $d \in \mathbb{Z}_S^\times$. As $\Delta = 4a^3 + 27b^2$, we see that $(a/d^4, b/d^6)$ is a $\mathbb{Z}_S$-point on the elliptic curve

$$4X^3 + 27Y^2 = c.$$

A theorem of Siegel says that such a curve has only finitely many points with coordinates in $\mathbb{Z}_S$.

The L-function of a modular form

Let f be a cusp form of weight $2k$ for $\Gamma_0(N)$. By definition, it is invariant under $z \mapsto z+1$ and is zero at the cusp ∞, and so can be expressed

$$f(z) = \sum_{n \geq 1} c(n) q^n, \quad q = e^{2\pi i z}, \quad c(n) \in \mathbb{C}.$$

The ***L-function*** of f is the Dirichlet series

$$L(f,s) = \sum c(n) n^{-s}, \quad s \in \mathbb{C}.$$

A rather rough estimate shows that $|c(n)| \leq Cn^k$ for some constant C, and so this Dirichlet series is convergent for $\Re(s) > k+1$.

REMARK 4.3 Let $f = \sum_{n\geq 1} c(n) q^n$ be a cusp form. The ***Mellin transform*** of f (more accurately, of the function $y \mapsto f(iy)\colon \mathbb{R}_{>0} \to \mathbb{C}$) is defined to be

$$g(s) = \int_0^\infty f(iy) y^s \frac{dy}{y}.$$

Ignoring questions of convergence, we find that

$$\begin{aligned}
g(s) &= \int_0^\infty \sum_{n=1}^\infty c(n) e^{-2\pi n y} y^s \frac{dy}{y}, \\
&= \sum_{n=1}^\infty c(n) \int_0^\infty e^{-t} (2\pi n)^{-s} t^s \frac{dt}{t}, \qquad t = 2\pi n y, \\
&= (2\pi)^{-s} \Gamma(s) \sum_{n=1}^\infty c(n) n^{-s}, \qquad \Gamma(s) \stackrel{\text{def}}{=} \int_0^\infty t^{s-1} e^{-t} dt, \\
&= (2\pi)^{-s} \Gamma(s) L(f,s).
\end{aligned}$$

For the experts, the Mellin transform is the version of the Fourier transform appropriate for the multiplicative group $\mathbb{R}_{>0}$.

Modular forms whose L-functions have a functional equations

Let $\alpha_N = \begin{pmatrix} 0 & -1 \\ N & 0 \end{pmatrix}$. Then

$$\alpha_N \begin{pmatrix} a & b \\ c & d \end{pmatrix} \alpha_N^{-1} = \begin{pmatrix} 0 & -1 \\ N & 0 \end{pmatrix} \begin{pmatrix} a & b \\ c & d \end{pmatrix} \begin{pmatrix} 0 & 1/N \\ -1 & 0 \end{pmatrix} = \begin{pmatrix} d & -c/N \\ -Nb & a \end{pmatrix},$$

and so conjugation by α_N preserves $\Gamma_0(N)$. Define

$$(w_N f)(z) = (\sqrt{N} z)^{2k} f(-1/z).$$

Then w_N preserves $S_{2k}(\Gamma_0(N))$ and $w_N^2 = 1$. Therefore the only possible eigenvalues for w_N are ± 1, and $S_{2k}(\Gamma_0(N))$ is a direct sum of the corresponding eigenspaces $S_{2k} = S_{2k}^{+1} \oplus S_{2k}^{-1}$.

THEOREM 4.4 (HECKE) *Let $f \in S_{2k}(\Gamma_0(N))$ be a cusp form in the ε-eigenspace for w_N, where $\varepsilon = \pm 1$. Then $L(f,s)$ extends analytically to a holomorphic function on the whole complex plane, and*

$$\Lambda(f,s) \stackrel{\text{def}}{=} N^{s/2} (2\pi)^{-s} \Gamma(s) L(f,s)$$

satisfies the functional equation

$$\Lambda(f,s) = \varepsilon(-1)^k \Lambda(f, 2k-s).$$

PROOF. We sketch the proof (following Knapp 1992, Theorem 9.8).

Let $f \in S_{2k}(\Gamma_0(N))$, and consider the function

$$\varphi(\tau) = |f(\tau)|\sigma^k.$$

By studying its behaviour near the cusps, one shows φ is bounded on a fundamental domain for $\Gamma_0(N)$. It is also invariant under $\Gamma_0(N)$, and so it is bounded on the whole of $\mathbb{H}$ (Knapp 1992, Lemma 9.6).

Let $f = \sum_{n=1}^{\infty} c(n)q^n$ be the q-expansion of f at the cusp ∞. Then

$$c(n) = \int_{-\frac{1}{2}}^{\frac{1}{2}} f(\tau)e^{-2\pi i n\tau} d\rho.$$

We just showed that $|f(\tau)| \leq B\sigma^{-k}$ for some constant B, and so $|c(n)| \leq B\sigma^{-k}e^{2\pi n\sigma}$ for all $\sigma > 0$. On taking $\sigma = 1/n$, we find that

$$|c(n)| \leq Cn^k \text{ with } C = Be^{2\pi}.$$

Therefore $\left|\sum_{n\leq x} c(n)\right| \leq Cx^{k+1}$, and so $L(f,s)$ converges for $\Re(s) > k+1$.

Now assume that f is in the ε-eigenspace for w_N, i.e., that $w_N f = \varepsilon f$. When we write this equality out, we obtain the following inversion law

$$f(i/N\sigma) = \varepsilon N^k i^{2k} \sigma^{2k} f(i\sigma).$$

Recall (4.3) that $(2\pi)^{-s}\Gamma(s)L(s,f) = \int_0^\infty f(i\sigma)\sigma^s \frac{d\sigma}{\sigma}$, and so

$$\Lambda(f,s) = N^{s/2}\int_0^\infty f(i\sigma)\sigma^{s-1}d\sigma.$$

Because $\varphi(\tau)$ is bounded on $\mathbb{H}$, the integral

$$\int_{1/\sqrt{N}}^\infty f(i\sigma)\sigma^{s-1}d\sigma$$

converges for all $s \in \mathbb{C}$ and defines an entire function. We rewrite $\Lambda(f,s)$ for $\Re(s) > k+1$ as

$$\Lambda(f,s) = N^{s/2}\int_0^{1/\sqrt{N}} f(i\sigma)\sigma^{s-1}d\sigma + N^{s/2}\int_{1/\sqrt{N}}^\infty f(i\sigma)\sigma^{s-1}d\sigma.$$

When we replace σ by $(N\sigma)^{-1}$ in the first term and apply the inversion law for f, we obtain the formula

$$\Lambda(f,s) = \varepsilon N^{(2k-s)/2} i^{2k} \int_{1/\sqrt{N}}^{\infty} f(i\sigma)\sigma^{2k-s-1} d\sigma + \\ N^{s/2} \int_{1/\sqrt{N}}^{\infty} f(i\sigma)\sigma^{s-1} d\sigma. \quad (*)$$

This shows that $\Lambda(f,s)$ is entire. As $\Gamma(s)$ is nowhere zero, it follows that $L(f,s)$ is entire. When we replace s with $2k-s$ in (*) and multiply by $\varepsilon i^{2k} = \varepsilon(-1)^k$, we obtain the formula

$$\varepsilon(-1)^k \Lambda(f,2k-s) = N^{\frac{s}{2}} \int_{1/\sqrt{N}}^{\infty} f(i\sigma)\sigma^{s-1} d\sigma + \\ \varepsilon N^{(2k-s)/2} i^{2k} \int_{1/\sqrt{N}}^{\infty} f(i\sigma)\sigma^{2k-s-1} d\sigma.$$

On comparing this (*), we obtain the required functional equation. □

Thus we see that, for $k = 2$, $L(f,s)$ has exactly the functional equation we hope for the L-function $L(E,s)$ of an elliptic curve E.

Modular forms whose L-functions are Euler products

Write

$$q\prod_{1}^{\infty}(1-q^n)^{24} = \sum \tau(n)q^n.$$

The function $n \mapsto \tau(n)$ is called the ***Ramanujan τ-function.*** In 1916, Ramanujan made two remarkable conjectures,[7] namely,

(a) $\tau(n) \le n^{11/2}\sigma(n)$, where $\sigma(n)$ is the number of positive divisors of n, and

(b) $$\begin{cases} \tau(mn) = \tau(m)\tau(n) & \text{if } \gcd(m,n) = 1; \\ \tau(p)\cdot\tau(p^n) = \tau(p^{n+1}) + p^{11}\tau(p^{n-1}) & \text{if } p \text{ is prime and } n \ge 1. \end{cases}$$

[7] Sarnak (lecture 2018) states that "the two conjectures changed the theory of modular forms and of number theory in the 20th century in a very major way", and he calls the conjecture that $\tau(n) \le n^{11/2}\sigma(n)$ "one of the most remarkable insights anyone ever had".

For a prime p, (a) says that $|\tau(p)| \leq 2p^{11/2}$. Readers will recognize that this will be true if $\tau(p)$ is the sum of two reciprocal roots of a polynomial "$P_{11}(T)$" occurring in the zeta function of an algebraic variety (see p. 209). Deligne's proof of the Riemann hypothesis for algebraic varieties over finite fields was inspired by a theorem of Rankin that $|\tau(n)| = O(n^{5.8})$. In turn, Deligne deduced Ramanujan's conjecture from his theorem.

Conjecture (b) was proved by Mordell in 1917 in a paper in which he introduced the first examples of Hecke operators, which we now explain.

Consider a modular form f of weight $2k$ for $\Gamma_0(N)$ (for example, $\Delta = (2\pi)^{12} q \prod(1-q^n)^{24}$, which is a modular form of weight 12 for $\Gamma_0(1)$), and write

$$L(f,s) = \sum_{n \geq 0} c(n) n^{-s}.$$

PROPOSITION 4.5 *The Dirichlet series $L(f,s)$ has an Euler product expansion of the form*

$$L(f,s) = \prod_{p|N} \frac{1}{1-c(p)p^{-s}} \prod_{p \nmid N} \frac{1}{1-c(p)p^{-s}+p^{2k-1-s}}$$

if (and only if)

$$(*) \begin{cases} c(mn) = c(m)c(n) \quad \text{if } \gcd(m,n) = 1; \\ c(p) \cdot c(p^r) = c(p^{r+1}) + p^{2k-1} c(p^{r-1}),\ r \geq 1, \text{ if } p \nmid N; \\ c(p^r) = c(p)^r,\ r \geq 1, \quad \text{if } p|N. \end{cases}$$

PROOF. For a prime p not dividing N, let

$$L_p(s) = \sum c(p^m) p^{-ms} = 1 + c(p)p^{-s} + \cdots + c(p^m)(p^{-s})^m + \cdots.$$

By inspection, the coefficient of $(p^{-s})^m$ in the product

$$(1 - c(p)p^{-s} + p^{2k-1}p^{-s}) \cdot L_p(s)$$

is 1 if $m = 0$, 0 if $m = 1$, and

$$c(p^{r+1}) - c(p)c(p^r) + p^{2k-1}c(p^{r-1})$$

if $m = r+1 \geq 2$. Therefore

$$L_p(s) = \frac{1}{1-c(p)p^{-s}+p^{2k-1-s}}$$

if and only if the second equation in (*) holds.

For a prime p dividing N, let

$$L_p(s) = \sum c(p^r)p^{-rs}.$$

A similar argument shows that

$$L_p(s) = \frac{1}{1 - c(p)p^{-s}}$$

if and only if the third equation in (*) holds.

Let $n \in \mathbb{N}$ factor as $n = \prod p_i^{r_i}$. Then the coefficient of $(p^{-s})^n$ in $\prod L_p(s)$ is $\prod c(p_i^{r_i})$, which equals $c(n)$ if (*) holds. □

REMARK 4.6 The proposition says that $L(f,s)$ is equal to an Euler product of the above form if and only if $n \mapsto c(n)$ is weakly multiplicative and if the $c(p^m)$ satisfy a suitable recurrence relation. Note that $(*)$, together with the normalization $c(1) = 1$, shows that the $c(n)$ are determined by the $c(p)$ for p prime.

Hecke defined linear maps (the ***Hecke operators***)

$$T(n): S_{2k}(\Gamma_0(N)) \to S_{2k}(\Gamma_0(N)), \quad n \geq 1,$$

and proved the following theorems.

THEOREM 4.7 *The maps $T(n)$ have the following properties:*

(a) $T(mn) = T(m)T(n)$ *if* $\gcd(m,n) = 1$;

(b) $T(p) \cdot T(p^r) = T(p^{r+1}) + p^{2k-1}T(p^{r-1})$ *if p does not divide N;*

(c) $T(p^r) = T(p)^r$, $r \geq 1$, $p|N$;

(d) *all* $T(n)$ *commute.*

We prove this in the next subsection.

THEOREM 4.8 *Let f be a cusp form of weight $2k$ for $\Gamma_0(N)$ that is simultaneously an eigenvector for all $T(n)$, say, $T(n)f = \lambda(n)f$, and let*

$$f(s) = \sum_{n=1}^{\infty} c(n)q^n, \quad q = e^{2\pi i s}.$$

Then

$$c(n) = \lambda(n)c(1).$$

We prove this in the next subsection. Note that $c(1) \neq 0$, because otherwise $c(n) = 0$ for all n, and $f = 0$.

COROLLARY 4.9 *Let f be as in Theorem 4.8, and normalize f so that $c(1) = 1$. Then*

$$L(f,s) = \prod_{p|N} \frac{1}{1-c(p)p^{-s}} \prod_{p\nmid N} \frac{1}{1-c(p)p^{-s}+p^{2k-1-s}}.$$

PROOF. According to Theorems 4.7 and 4.8, the $c(n)$ satisfy the condition (*) in Proposition 4.5. □

EXAMPLE 4.10 Since $\mathcal{S}_{12}(\Gamma_0(1))$ has dimension 1, Δ must be an eigenform for all $T(n)$, which implies (b) of Ramanujan's conjecture.

Definition of the Hecke operators; proof of Theorems 4.7 and 4.8

We first define the Hecke operators for the full group $\Gamma_0(1) = \mathrm{SL}_2(\mathbb{Z})$. Recall (p. 110) that there are canonical bijections

$$\mathcal{L}/\mathbb{C}^\times \leftrightarrow \Gamma_0(1)\backslash M/\mathbb{C}^\times \leftrightarrow \Gamma_0(1)\backslash\mathbb{H}.$$

Moreover, the equation

$$f(z) = F(\Lambda(z,1))$$

defines a one-to-one correspondence between

(a) functions $F:\mathcal{L} \to \mathbb{C}$ such that $F(\lambda\Lambda) = \lambda^{-2k} F(\Lambda), \quad \lambda \in \mathbb{C}^\times$;

(b) functions $f:\mathbb{H} \to \mathbb{C}$ such that $f(\gamma z) = (cz+d)^{2k} f(z)$, $\gamma = \left(\begin{smallmatrix} a & b \\ c & d \end{smallmatrix}\right)$.

We work first with $\mathcal{L}$.

Let $\mathcal{D}$ be the free abelian group generated by the $\Lambda \in \mathcal{L}$; thus an element of $\mathcal{D}$ is a finite sum

$$\sum n_\Lambda[\Lambda], \quad n_\Lambda \in \mathbb{Z}, \quad \Lambda \in \mathcal{L},$$

and two such sums $\sum n_\Lambda[\Lambda]$ and $\sum n'_\Lambda[\Lambda]$ are equal if and only if $n_\Lambda = n'_\Lambda$ for all Λ.

For $n \geq 1$, define maps

$$T(n):\mathcal{D} \to \mathcal{D}, \quad [\Lambda] \mapsto \sum_{(\Lambda:\Lambda')=n} [\Lambda']$$

$$R(n):\mathcal{D} \to \mathcal{D}, \quad [\Lambda] \mapsto [n\Lambda].$$

PROPOSITION 4.11 (a) $T(mn) = T(m) \circ T(n)$ if $\gcd(m,n) = 1$.
(b) $T(p^r) \circ T(p) = T(p^{r+1}) + pR(p) \circ T(p^{r-1})$.

PROOF. (a) For a lattice Λ,

$$T(mn)[\Lambda] = \sum [\Lambda''] \quad \text{(sum over } \Lambda'' \text{, with } (\Lambda : \Lambda'') = mn\text{),}$$
$$T(m) \circ T(n)[\Lambda] = \sum [\Lambda''] \quad \text{(sum over pairs } (\Lambda', \Lambda'') \text{ with } (\Lambda : \Lambda') = n,\ (\Lambda' : \Lambda'') = m\text{).}$$

But if Λ'' is a lattice of index mn, then Λ/Λ'' is a commutative group of order mn with $\gcd(m,n) = 1$, and so has a unique subgroup of order m. The inverse image of this subgroup in Λ will be the unique lattice $\Lambda' \supset \Lambda''$ such that $(\Lambda' : \Lambda'') = m$. Thus the two sums are the same.

(b) For a lattice Λ,

$$T(p^r) \circ T(p)[\Lambda] = \sum [\Lambda''] \quad \text{(sum over pairs } (\Lambda', \Lambda'') \text{ with } (\Lambda : \Lambda') = p,\ (\Lambda' : \Lambda'') = p^r\text{),}$$
$$T(p^{r+1})[\Lambda] = \sum [\Lambda''] \quad \text{(sum over } \Lambda'' \text{ with } (\Lambda : \Lambda'') = p^{r+1}\text{),}$$
$$pR(p) \circ T(p^{n-1})[\Lambda] = p \cdot \sum R(p)[\Lambda'] \quad \text{(sum over } \Lambda' \text{ with } (\Lambda : \Lambda') = p^{r-1}\text{),}$$
$$= p \cdot \sum [\Lambda''] \quad \text{(sum over } \Lambda'' \subset p\Lambda \text{ with } (p\Lambda : \Lambda'') = p^{r-1}\text{).}$$

Each of these is a sum of lattices Λ'' of index p^{r+1} in Λ. Fix such a lattice Λ'', and let a be the number of times that $[\Lambda'']$ occurs in the first sum, and b the number of times it occurs in the third sum. It occurs exactly once in the second sum, and so we have to prove that

$$a = 1 + pb.$$

There are two cases to consider.

The lattice Λ'' is not contained in $p\Lambda$. In this case, $b = 0$, and a is the number of lattices Λ' such that $(\Lambda : \Lambda') = p$ and $\Lambda' \supset \Lambda''$. Such lattices are in one-to-one correspondence with the subgroups of $\Lambda/p\Lambda$ of index p containing the image $\bar{\Lambda}''$ of Λ'' in $\Lambda/p\Lambda$. But $(\Lambda : p\Lambda) = p^2$ and

$\Lambda/p\Lambda \neq \bar{\Lambda}'' \neq 0$, and so there is only one such subgroup, namely $\bar{\Lambda}''$ itself. Therefore there is only one possible Λ', namely $p\Lambda + \Lambda''$, and so $a = 1$.

The lattice $\Lambda'' \supset p\Lambda$. Here $b = 1$. Every lattice Λ' of index p in Λ contains $p\Lambda$, hence also Λ'', and the number of such Λ' is the number of lines through the origin in $\Lambda/p\Lambda \approx \mathbb{F}_p^2$, i.e., the number of points in $\mathbb{P}^1(\mathbb{F}_p)$, which is $p+1$ as required. □

COROLLARY 4.12 *For all* m *and* n,

$$T(m) \circ T(n) = \sum d \cdot R(d) \circ T(mn/d^2)$$

(the sum is over the positive divisors d *of* $\gcd(m,n)$*).*

PROOF. Prove by induction on s that

$$T(p^r)T(p^s) = \sum_{i \leq r,s} p^i \cdot R(p^i) \circ T(p^{r+s-2i}),$$

and then apply (a) of the proposition. □

COROLLARY 4.13 *Let* $\mathcal{H}$ *be the* $\mathbb{Z}$*-subalgebra of* $\mathrm{End}(\mathcal{D})$ *generated by the operators* $T(p)$ *and* $R(p)$ *with* p *prime; then* $\mathcal{H}$ *is commutative, and it contains* $T(n)$ *for all* n.

PROOF. Obvious from the proposition. □

Let F be a function $\mathcal{L} \to \mathbb{C}$. We can extend F by linearity to a function $F: \mathcal{D} \to \mathbb{C}$,

$$F\left(\sum n_\Lambda[\Lambda]\right) = \sum n_\Lambda F(\Lambda).$$

For any linear map $T: \mathcal{D} \to \mathcal{D}$, we define $T \cdot F$ to be the function $\mathcal{L} \to \mathbb{C}$ such that $T \cdot F(\Lambda) = F(T\Lambda)$. For example,

$$(T(n) \cdot F)(\Lambda) = \sum_{(\Lambda:\Lambda')=n} F(\Lambda'),$$

and if $F(\lambda\Lambda) = \lambda^{-2k} F(\Lambda)$, then

$$R(n) \cdot F = n^{-2k} F.$$

PROPOSITION 4.14 *If* $F: \mathcal{L} \to \mathbb{C}$ *has the property that* $F(\lambda\Lambda) = \lambda^{-2k} F(\Lambda)$ *for all* λ, Λ*, then so also does* $T(n) \cdot F$*, and*

(a) $T(mn) \cdot F = T(m) \cdot T(n) \cdot F \quad$ *if* $\gcd(m,n) = 1$;

(b) $T(p)\cdot T(p^r)\cdot F = T(p^{r+1})\cdot F + p^{1-2k}T(p^{r-1})\cdot F.$

PROOF. Immediate consequence of Proposition 4.11. □

Now let $f(z)$ be a modular form of weight $2k$, and let F be the associated function on $\mathcal{L}$. We define $T(n)\cdot f$ to be the function on $\mathbb{H}$ associated with $n^{2k-1}\cdot T(n)\cdot F$. Thus

$$(T(n)\cdot f)(z) = n^{2k-1}(T(n)\cdot F)(\Lambda(z,1)).$$

Theorem 4.7 in the case $N = 1$ follows easily from the Proposition. To prove Theorem 4.8 we need an explicit description of the lattices of index n in a fixed lattice.

Write $M_2(\mathbb{Z})$ for the ring of 2×2 matrices with coefficients in $\mathbb{Z}$.

LEMMA 4.15 *Let $A \in M_2(\mathbb{Z})$ have determinant n. Then there exists a $U \in M_2(\mathbb{Z})^\times$ such that*

$$UA = \begin{pmatrix} a & b \\ 0 & d \end{pmatrix}, \quad ad = n, \quad a \geq 1, \quad 0 \leq b < d.$$

Moreover, the integers a, b, d are uniquely determined.

PROOF. Let $A = \left(\begin{smallmatrix} a & b \\ c & d \end{smallmatrix}\right)$, and let $ra + sc = a'$, where $a' = \gcd(a,c) \geq 1$. Then $\gcd(r,s) = 1$, and so there exist e, f such that $re + sf = 1$. Now

$$\begin{pmatrix} r & s \\ -f & e \end{pmatrix}\begin{pmatrix} a & b \\ c & d \end{pmatrix} = \begin{pmatrix} a' & b' \\ c' & d' \end{pmatrix}$$

and $\det\left(\begin{smallmatrix} r & s \\ -f & e \end{smallmatrix}\right) = 1$. Now apply the appropriate elementary row operations to get UA into upper triangular form. For the uniqueness, note that multiplication by such a U does not change the greatest common divisor of the entries in any column, and so a is uniquely determined. Now d is uniquely determined by the equation $ad = n$, and b is obviously uniquely determined modulo d. □

For the lattice $\Lambda(z,1)$, the sublattices of index n are exactly the lattices $\Lambda(az+b,d)$, where (a,b,d) runs through the triples in the lemma. Therefore

$$(T(n)\cdot f)(z) = n^{2k-1}\sum_{a,b,d} d^{-2k} f\left(\frac{az+b}{d}\right),$$

where the sum is over the same triples. On substituting this into the q-expansion

$$f = \sum_{m \geq 1} c(m) q^m$$

one finds (after a little work) that

$$T(n) \cdot f = c(n) q + \cdots.$$

Therefore, if $T(n) \cdot f = \lambda(n) f$, then

$$\lambda(n) c(1) = c(n).$$

This proves Theorem 4.8 in the case $N = 1$.

When $N \neq 1$, the theory of the Hecke operators is much the same, only a little more complicated. For example, instead of $\mathcal{L}$, one must work with the set of pairs (Λ, S), where $\Lambda \in \mathcal{L}$ and S is a cyclic subgroup of order N in $\mathbb{C}/\Lambda$. This is no problem for the $T(n)$ with $\gcd(n, N) = 1$, but the $T(p)$ with $p|N$ have to be treated differently.[8] For example, Proposition 4.14b holds only for the p that do not divide N; if p divides N, then $T(p^r) \cdot F = T(p)^r \cdot F$, $r \geq 1$.

It follows from Corollary 4.9 that the problem of finding cusp forms f whose L-functions have Euler product expansions becomes a problem of finding simultaneous eigenforms for the linear maps $T(n): \mathcal{S}_{2k}(\Gamma_0(N)) \to \mathcal{S}_{2k}(\Gamma_0(N))$. Hecke was unable to do this because he lacked the spectral theorem.

Linear algebra: the spectral theorem

Recall that a ***hermitian form*** on a vector space V is a mapping $\langle\, ,\rangle: V \times V \to \mathbb{C}$ such that $\langle v, w\rangle = \overline{\langle w, v\rangle}$ and $\langle\, ,\rangle$ is linear in one variable and conjugate-linear in the other. Such a form is said to be ***positive definite*** if $\langle v, v\rangle > 0$ whenever $v \neq 0$. A linear map $\alpha: V \to V$ is ***self-adjoint*** (or ***hermitian***) relative to $\langle\, ,\rangle$ if

$$\langle \alpha v, w\rangle = \langle v, \alpha w\rangle, \quad \text{all } v, w.$$

THEOREM 4.16 (SPECTRAL THEOREM) *Let V be a finite-dimensional complex vector space with a positive definite hermitian form $\langle\, ,\rangle$.*

[8] In the literature, $T(p)$ with $p|N$ is sometimes denoted by $U(p)$.

(a) *Every self-adjoint linear map* $\alpha: V \to V$ *is diagonalizable, i.e.,* V *is a direct sum of eigenspaces for* α.

(b) *Let* $\alpha_1, \alpha_2, \ldots$ *be a sequence of commuting self-adjoint linear maps* $V \to V$*; then* V *has a basis of consisting of vectors that are eigenvectors for all* α_i.

PROOF. (a) Because $\mathbb{C}$ is algebraically closed, α has an eigenvector e_1. Let V_1 be $(\mathbb{C}e_1)^{\perp}$. Then V_1 is stable under α, and so contains an eigenvector e_2. Let $V_2 = (\mathbb{C}e_1 \oplus \mathbb{C}e_2)^{\perp}$, and so on.

(b) We know that $V = \bigoplus V(\lambda_i)$, where the λ_i are the distinct eigenvalues of α_1. Because α_2 commutes with α_1, it stabilizes each $V(\lambda_i)$, and so each $V(\lambda_i)$ can be decomposed into a direct sum of eigenspaces for α_2. Continuing in this fashion, we arrive at a decomposition $V = \bigoplus V_j$ such that each α_i acts as a scalar on each V_j. Choose a basis for each space V_j, and take their union. □

This suggests that we should look for a hermitian form on $\mathcal{S}_{2k}(\Gamma_0(N))$ for which the $T(n)$ are self-adjoint.

The Petersson inner product

As Beltrami pointed out, the unit disk $\{(x,y) \in \mathbb{R}^2 \mid x^2 + y^2 < 1\}$ forms a model for hyperbolic geometry: if one defines a "line" to be a segment of a circle orthogonal to the circumference of the disk, angles to be the usual angles, and distances in terms of cross-ratios, one obtains a geometry that satisfies all the axioms for Euclidean geometry except that, given a point P and a line ℓ, there exist *more* than one line through P not meeting ℓ. The map $z \mapsto \frac{z-i}{z+i}$ sends the upper half-plane $\{(x,y) \in \mathbb{R}^2 \mid y > 0\}$ isomorphically onto the unit disk, and, being fractional-linear, maps circles and lines to circles and lines (collectively, not separately) and preserves angles. Therefore the upper half-plane is also a model for hyperbolic geometry. The group of transformations of the upper half-plane preserving distances and orientation is $\mathrm{PSL}_2(\mathbb{R}) \stackrel{\text{def}}{=} \mathrm{SL}_2(\mathbb{R})/\{\pm I\}$, which therefore plays the same role as the group of orientation preserving affine transformations of the Euclidean plane. The next proposition shows that the measure $\mu(U) = \iint_U \frac{dxdy}{y^2}$ is invariant under transformations in $\mathrm{PSL}_2(\mathbb{R})$, and therefore plays the same role as the measure $\iint_U dxdy$ on sets in the Euclidean plane.[9]

[9]Poincaré studied the geometry of the upper half plane equipped with its metric $\frac{dxdy}{y^2}$; for this reason, both the half plane and the metric are often named after him.

PROPOSITION 4.17 *For a subset U of the upper half-plane, let*

$$\mu(U) = \iint_U \frac{dxdy}{y^2};$$

then $\mu(\gamma U) = \mu(U)$ for all $\gamma \in \mathrm{SL}_2(\mathbb{R})$.

PROOF. If $\gamma = \begin{pmatrix} a & b \\ c & d \end{pmatrix}$, then

$$\frac{d\gamma}{dz} = \frac{1}{(cz+d)^2}, \quad \Im(\gamma z) = \frac{\Im(z)}{|cz+d|^2}.$$

The next lemma shows that

$$\gamma^*(dxdy) = \left|\frac{d\gamma}{dz}\right|^2 dxdy \quad (z = x+iy),$$

and so

$$\gamma^*\left(\frac{dxdy}{y^2}\right) = \frac{dxdy}{y^2}.$$

□

LEMMA 4.18 *For a holomorphic function $w(z)$, the map $z \mapsto w(z)$ multiplies areas by $|w'(z)|^2$.*

PROOF. Write $w(z) = u(x,y) + iv(x,y)$, so that $z \mapsto w(z)$ is the map

$$(x,y) \mapsto (u(x,y), v(x,y)),$$

whose jacobian is

$$\begin{vmatrix} u_x & v_x \\ u_y & v_y \end{vmatrix} = u_x v_y - v_x u_y.$$

On the other hand, $w'(z) = u_x + iv_x$, so that

$$|w'(z)|^2 = u_x^2 + v_x^2.$$

The Cauchy–Riemann equations state that $u_x = v_y$ and $v_x = -u_y$, and so the two expressions agree. □

If f and g are modular forms of weight $2k$ for $\Gamma_0(N)$, then

$$f(z)\cdot\overline{g(z)}y^{2k}$$

is invariant under $\mathrm{SL}_2(\mathbb{R})$, which suggests defining

$$\langle f,g\rangle = \iint_D f\bar{g}y^{2k}\frac{dxdy}{y^2}$$

for D a fundamental domain for $\Gamma_0(N)$ — the above discussion shows that, assuming the integral converges, $\langle f,g\rangle$ will be independent of the choice of D.

THEOREM 4.19 (PETERSSON) *The above integral converges provided at least one of f or g is a cusp form. It therefore defines a positive definite hermitian form on the vector space $\mathcal{S}_{2k}(\Gamma_0(N))$ of cusp forms. The Hecke operators $T(n)$ are self-adjoint for all n relatively prime to N.*

PROOF. That the integral converges if f is a cusp form follows from the fact that $|f(\tau)|\sigma^k$ is bounded on the whole of $\mathbb{H}$ (see the proof of 4.4). The self-adjointness of the Hecke operators follows from a lengthy direct calculation (see Knapp 1992, 8.22, 9.18). □

On putting the theorems of Hecke and Petersson together, we find that there exists a decomposition

$$\mathcal{S}_{2k}(\Gamma_0(N)) = \bigoplus V_i$$

of $\mathcal{S}_{2k}$ into a direct sum of orthogonal subspaces V_i, each of which is a simultaneous eigenspace for all $T(n)$ with $\gcd(n,N)=1$. The $T(p)$ for $p|N$ stabilize each V_i and commute, and so there does exist at least one f in each V_i that is also an eigenform for the $T(p)$ with $p|N$. If we scale f so that $c(1)=1$, and write $f = q+\sum_{n\geq 2}c(n)q^n$, then

$$L(f,s) = \prod_{p|N}\frac{1}{1-c_p p^{-s}}\prod_{p\nmid N}\frac{1}{1-c(p)p^{-s}+p^{2k-1-2s}}.$$

The operator w_N is self-adjoint for the Petersson product, and commutes with $T(n)$ if $\gcd(n,N)=1$, and so each space V_i decomposes into orthogonal eigenspaces

$$V_i = V_i^{+1}\oplus V_i^{-1}$$

for w_N. Unfortunately, w_N does not commute with the $T(p)$ if $p|N$, and so the decomposition is not necessarily stable under these $T(p)$. Thus, the above results do not imply that there is even one f that is simultaneously an eigenvector for w_N (and hence has a functional equation by 4.4) and for *all* $T(n)$ (and hence is equal to an Euler product 4.9).

New forms: the theorem of Atkin and Lehner

The problem left in the last section has a simple remedy. If $M|N$, then $\Gamma_0(M) \supset \Gamma_0(N)$, and so $\mathcal{S}_{2k}(\Gamma_0(M)) \subset \mathcal{S}_{2k}(\Gamma_0(N))$. Recall that N appears in the functional equation for $L(f,s)$, and so it is not surprising that we run into trouble when we mix f of "level" N with f that are really of level $M|N$, $M < N$.

The way out of the problem is to say that a cusp form lying in some subspace $\mathcal{S}_{2k}(\Gamma_0(M))$, $M|N$, $M < N$, is ***old***. The old forms form a subspace $\mathcal{S}_{2k}^{\text{old}}(\Gamma_0(N))$ of $\mathcal{S}_{2k}(\Gamma_0(N))$ whose orthogonal complement $\mathcal{S}_{2k}^{\text{new}}(\Gamma_0(N))$ is called the space of ***new forms***. It is stable under all the operators $T(n)$ and w_N, and so $\mathcal{S}_{2k}^{\text{new}}$ decomposes into a direct sum of orthogonal subspaces W_i,

$$\mathcal{S}_{2k}^{\text{new}}(\Gamma_0(N)) = \bigoplus W_i,$$

each of which is a simultaneous eigenspace for all $T(n)$ with $\gcd(n,N) = 1$. Since the $T(p)$ for $p|N$ and w_N each commute with the $T(n)$ for $\gcd(n,N) = 1$, each stabilizes each W_i.

THEOREM 4.20 (ATKIN–LEHNER 1970) *The spaces W_i in the above decomposition all have dimension* 1.

It follows that each W_i is also an eigenspace for w_N and $T(p)$, $p|N$. Each W_i contains (exactly) one cusp form f whose q-expansion is of the form $q + \sum_{n\geq 2} c(n)q^n$. For this form, $L(f,s)$ has an Euler product expansion, and $\Lambda(f,s)$ satisfies a functional equation

$$\Lambda(f,s) = \varepsilon\Lambda(f,2-s),$$

where $\varepsilon = \pm 1$ is the eigenvalue of w_N acting on W_i. If the $c(n) \in \mathbb{Z}$, then $\Lambda(f,s)$ is a candidate for being the L-function of an elliptic curve E over $\mathbb{Q}$.

REMARK 4.21 Once the Hecke–Petersson theory had been developed, Ramanujan's conjecture had an obvious generalization. Let $f \in \mathcal{S}_{2k}^{\text{new}}(\Gamma_0(N))$

be a simultaneous eigenform for the operators $T(n)$ with n prime to N, and write $f = \sum_{n\geq 1} c(n)q^n$ with $q = e^{2\pi is}$. Normalize f so that $c(1) = 1$. The Ramanujan–Petersson conjecture states that

$$|a_p| \leq 2 \cdot p^{(2k-1)/2}.$$

In other words, the roots of the polynomial

$$T^2 - a_p T + p^{2k-1}$$

have absolute value $p^{(2k-1)/2}$.

Deligne (1971) showed that the roots of the above polynomial occur as the eigenvalues of the Frobenius operator acting on the H^{2k-1} of a nonsingular projective variety over $\mathbb{F}_p$.[10] Therefore, the Ramanujan–Petersson conjecture follows from the Riemann hypothesis for the variety, proved in Deligne 1974.

5 Statement of the main theorems

Recall that to an elliptic curve E over $\mathbb{Q}$, we have attached an L-function $L(E,s) = \sum a_n n^{-s}$ that (a) has coefficients $a_n \in \mathbb{Z}$, (b) can be expressed as an Euler product, and (c) conjecturally satisfies a functional equation (involving the conductor N of E). Moreover, isogenous elliptic curves have the same L-function (4.1). We therefore have a map

$$\{\text{elliptic curves over } \mathbb{Q} \text{ up to isogeny}\} \xrightarrow{E \mapsto L(E,s)} \{\text{Dirichlet series}\}.$$

As noted earlier, Faltings (1983) shows that this map is injective: two elliptic curves are isogenous if they have the same L-function.

PROBLEM 5.1 Describe the collection of Dirichlet functions arising from elliptic curves over $\mathbb{Q}$.

The theory of Hecke and Petersson, together with the theorem of Atkin and Lehner, shows that the subspace $\mathcal{S}_2^{\text{new}}(\Gamma_0(N)) \subset \mathcal{S}_2(\Gamma_0(N))$ of new forms decomposes into a direct sum

$$\mathcal{S}_2^{\text{new}}(\Gamma_0(N)) = \bigoplus W_i$$

[10] Under some restrictive hypotheses, but "the general case is not much more difficult," according to Deligne 1974, p. 302.

of one-dimensional subspaces W_i that are simultaneous eigenspaces for all the Hecke operators $T(n)$ with $\gcd(n,N) = 1$. Because they have dimension 1, each W_i is also an eigenspace for w_N and for the $T(p)$ with $p|N$. An element of one of the subspaces W_i, i.e., a simultaneous eigenforms in $\mathcal{S}_2^{\text{new}}(\Gamma_0(N))$, is traditionally called a ***newform***, and we shall adopt this terminology.

In each W_i there is exactly one form $f_i = \sum c(n)q^n$ with $c(1) = 1$ (said to be ***normalized***). Because f_i is an eigenform for all the Hecke operators, it has an Euler product, and because it is an eigenform for w_N, it satisfies a functional equation (involving N). If the $c(n)$ are[11] in $\mathbb{Z}$, then $L(f_i,s)$ is a candidate for being the L-function of an elliptic curve over $\mathbb{Q}$.

THEOREM 5.2 (EICHLER–SHIMURA) *Let $f = \sum c(n)q^n$ be a normalized newform for $\Gamma_0(N)$ of weight 2. If all $c(n) \in \mathbb{Z}$, then there exists an elliptic curve E_f of conductor N such that $L(E_f,s) = L(f,s)$.*

The early forms of the theorem (in the articles of Eichler and Shimura), were less precise — in particular, they predate the work of Atkin and Lehner in which newforms were defined.

The theorem of Eichler–Shimura has two parts: given f, construct the curve E_f (up to isogeny); having constructed E_f, prove that $L(E_f,s) = L(f,s)$. We discuss the proofs of the two parts in Sections 6 and 7 below.

DEFINITION 5.3 An elliptic curve E over $\mathbb{Q}$ is said to be ***modular of level*** N if its L-function $L(E,s)$ is the L-function $L(f,s)$ of a normalized newform f of weight 2 for some $\Gamma_0(N)$.[12]

To complete the correspondence between elliptic curves over $\mathbb{Q}$ and normalized newforms of weight 2, we need to determine which elliptic curves are modular. Weil (1967) surprised everyone by proving that they all are if their L-functions have sufficiently good functional equations, i.e., satisfy a strong form of the Hasse–Weil conjecture.

[11] In the next section, we shall see that the $c(n)$ automatically lie in some finite extension of $\mathbb{Q}$, and that if they lie in $\mathbb{Q}$ then they lie in $\mathbb{Z}$

[12] This is equivalent to the earlier defnition p. 215. Given a nonconstant map $X_0(N) \to E$ over $\mathbb{C}$, the canonical differential one-form on E pulls back to a differential one-form on $X_0(N)$, which corresponds to a modular form f. The Eichler–Shimura theorem shows that the zeta function of E is $L(f,s)$. Conversely, suppose that the zeta function of E equals $L(f,s)$ with f as above. From f, we get an elliptic curve E_f and a nonconstant map $X_0(N) \to E_f$ over $\mathbb{Q}$ (see §6). The elliptic curve E_f has the same zeta function as E, and so is isogenous to it by the theorem of Faltings.

Recall that a ***Dirichlet character with modulus*** m is a homomorphism

$$\chi : (\mathbb{Z}/m\mathbb{Z})^\times \to \mathbb{C}^\times.$$

We extend χ to a map $\mathbb{Z} \to \mathbb{C}$ by setting

$$\chi(n) = \begin{cases} \chi(n \bmod m) & \text{if } n \text{ is relatively prime to } m, \\ 0 & \text{otherwise.} \end{cases}$$

Let $L(s) = \sum a_n n^{-s}$ be an L-function, and let N and k be positive integers. For a Dirichlet character χ with modulus m prime to N, let

$$L_\chi(s) = \sum \chi(n) a_n n^{-s}, \quad \Lambda_\chi(s) = \left(\frac{m}{2\pi}\right)^{-s} \Gamma(s) L_\chi(s).$$

Weil proved, under some mild hypotheses on $L(s)$, that if the functions $\Lambda_\chi(s)$, where χ runs over the Dirichlet characters χ with modulus prime to N, satisfy certain functional equations relating $\Lambda_\chi(s)$ and $\Lambda_\chi(2k - s)$, then $L(s) = L(f, s)$ for some newform f of weight $2k$ for $\Gamma_0(N)$.

Now consider an elliptic curve E over $\mathbb{Q}$ with conductor N. Recall that the Hasse–Weil conjecture (10.3) asserts that the L-function of E satisfies a functional equation relating $\Lambda(E, s)$ and $\Lambda(E, 2 - s)$. Weil made the more precise conjecture "on certain theoretical grounds" that this is also true for the modified L-functions $\Lambda_\chi(s)$. Together with his theorem, this conjecture implies the following conjecture.

CONJECTURE 5.4 *Let E be an elliptic curve over $\mathbb{Q}$ with conductor N. Then the L-function $L(E, s)$ equals $L(f, s)$ for some newform f of weight 2 for $\Gamma_0(N)$; in particular, E is modular.*

Weil did not state 5.4 as a conjecture, but rather "recommended it to the interested reader as an exercise". Ogg, in his review of the article, referred to it as a "rather startling" conjecture.

For several years, Conjecture 5.4 was referred to as ***Weil's conjecture***. Then it was noted that, in a set of problems circulated in Japanese at a 1955 conference,[13] Taniyama had asked in rather vague form,[14] whether every elliptic curve over a number field is modular, and so his name was

[13]The conference in question was the famous 1955 Tokyo-Nikko conference on algebraic number theory. It was attended by most of the leading Western algebraic number theorists (Artin, Chevalley, Deuring, Néron, Serre, Weil, ...) as well as the new generation of young Japanese mathematicians (Iwasawa, Kubota, Nakayama, Satake, Shimura, Taniyama, ...).

[14]"Let C be an elliptic curve defined over an algebraic number field k, and $L_C(s)$ the

added to the conjecture. It then transpired that Shimura, in conversation with various mathematicians, had conjectured that all elliptic curves over $\mathbb{Q}$ are modular, and so his name was also added. After a campaign by Serge Lang, some authors dropped Weil's name, and it became known as the ***Shimura–Taniyama conjecture***. More recently, mathematicians have referred to it as the ***modularity conjecture*** (now theorem).

Whatever the history, it was Weil's article that brought the conjecture to the attention of the wider mathematical community, provided the first persuasive evidence for it, and included the coincidence of the conductor of E with the level of f in the statement. The last assertion made it possible to test the conjecture numerically: for a given N, list the (finitely many) f for $\Gamma_0(N)$, list the (finitely many) isogeny classes of elliptic curves over $\mathbb{Q}$ with conductor N, and check that the lists agree.

In lectures in 1985, Frey suggested that the elliptic curve in II, 3.4, defined by a counterexample to Fermat's Last Theorem, has properties that are "so excellent that one suspects that such a curve cannot exist"; in particular, it should not be modular (Frey 1986). This encouraged Serre to rethink some old conjectures of his, and formulate two conjectures, one of which implies that Frey's curve is indeed not modular. In 1986, Ribet proved sufficient of Serre's conjectures to be able to show that Frey's curve cannot be modular. We shall discuss this work in Section 9.

Thus, at this stage (1986) it was known that Conjecture 5.4 for elliptic curves over $\mathbb{Q}$ implies Fermat's Last Theorem, which inspired Wiles to attempt to prove the conjecture. After a premature announcement in 1993, Wiles proved in 1994 (with the help of R. Taylor) that all semistable elliptic curves over $\mathbb{Q}$, in particular, the Frey curve, are modular. Recall that semistable just means that the curve does not have cuspidal reduction at any prime. In a series of articles, Breuil, Conrad, Diamond, and Taylor improved the result so that it now says that all elliptic curves E over $\mathbb{Q}$ is modular. In other words, the map

$$f \mapsto E_f : \{f\} \to \{E \text{ over } \mathbb{Q}\}/\sim$$

L-function of C over k in the sense that $\zeta_C(s) = \zeta_k(s)\zeta_k(s-1)/L_C(s)$ is the zeta function of C over k. If Hasse's conjecture is true for $\zeta_C(s)$, then the Fourier series obtained from $L_C(s)$ by the inverse Mellin transformation must be an automorphic form of dimension -2 of a special type (see Hecke). If so, it is very plausible that this form is an elliptic differential of the field of associated automorphic functions. Now, going through these observations backwards, is it possible to prove Hasse's conjecture by finding a suitable automorphic form from which $L_C(s)$ can be obtained?"

is surjective. This is the ***modularity theorem***. We shall discuss the strategy of Wiles's proof in Section 8.

6 How to get an elliptic curve from a cusp form

Not long after Leibniz and Newton had developed calculus, mathematicians[15] discovered that they could not evaluate integrals of the form

$$\int \frac{dx}{\sqrt{f(x)}},$$

where $f(x) \in \mathbb{R}[x]$ is a cubic or quartic polynomial without a repeated factor. In fact, such an integral cannot be evaluated in terms of elementary functions. Thus, they were forced to treat them as new functions and to study their properties. For example, Euler showed that

$$\int_a^{t_1} \frac{dx}{\sqrt{f(x)}} + \int_a^{t_2} \frac{dx}{\sqrt{f(x)}} = \int_a^{t_3} \frac{dx}{\sqrt{f(x)}},$$

where t_3 is a rational function of t_1, t_2. The explanation for this lies with elliptic curves.

Consider the elliptic curve $Y^2 = f(X)$ over $\mathbb{R}$, and the differential one-form $\omega = \frac{1}{y}dx + 0dy$ on $\mathbb{R}^2$. As we learn in calculus, to integrate ω over a segment of the elliptic curve, we should parameterize the curve. Assume that the segment $\gamma(a,t)$ of the elliptic curve over the closed interval $[a,t]$ in the x-axis can be smoothly parameterized by x. Thus $x \mapsto (x, \sqrt{f(x)})$ maps the interval $[a,t]$ smoothly onto the segment $\gamma(a,t)$, and

$$\int_{\gamma(a,t)} \frac{dx}{y} = \int_a^t \frac{dx}{\sqrt{f(x)}}.$$

Hence, the elliptic integral can be regarded as an integral over a segment of an elliptic curve.

A key point, which will be explained shortly, is that the restriction of ω to E is translation invariant, i.e., if t_Q denotes the map $P \mapsto P + Q$ on E, then $t_Q^*\omega = \omega$ (on E). Hence

$$\int_{\gamma(a,t)} \omega = \int_{\gamma(a+x(Q),t+x(Q))} \omega$$

[15]In 1655, Wallis attempted to compute the length of an arc of an elliptic curve, and found such integrals, which became known as elliptic integrals. The literature on elliptic integrals and their history is vast.

for all $Q \in E(\mathbb{R})$ (here $x(Q)$ is the x-coordinate of Q). Now Euler's theorem becomes the statement

$$\int_{\gamma(a,t_1)} \omega + \int_{\gamma(a,t_2)} \omega = \int_{\gamma(a,t_1)} \omega + \int_{\gamma(t_1,t_3)} \omega = \int_{\gamma(a,t_3)} \omega$$

where t_3 is determined by

$$(t_2, \sqrt{f(t_2)}) - (a, \sqrt{f(a)}) + (t_1, \sqrt{f(t_1)}) = (t_3, \sqrt{f(t_3)})$$

(difference and sum for the group structure on $E(\mathbb{R})$).

In this way, the study of elliptic integrals leads to the study of elliptic curves.

Differential one-forms on elliptic curves

Consider an elliptic curve

$$E : Y^2 Z = X^3 + aXZ^2 + bZ^3, \quad a, b \in \mathbb{C}, \quad \Delta \neq 0.$$

As E has genus 1, the holomorphic differential one-forms on it form a vector space of dimension 1. This vector space is generated by $\omega = \frac{dx}{2y}$ (more accurately, by the restriction to $E^{\mathrm{aff}}(\mathbb{C})$ of $\frac{1}{2y} dx + 0 dy$ on $\mathbb{C}^2$). Note that, on E^{aff},

$$2y dy = (3x^2 + a) dx,$$

and so

$$\frac{dx}{2y} = \frac{dy}{3x^2 + a}$$

where both are defined. As $\Delta \neq 0$, the functions $2y$ and $3x^2 + a$ have no common zero, and so ω is holomorphic on E^{aff}. One can check that it also holomorphic at the point at infinity. For any $Q \in E(\mathbb{C})$, the translate $t_Q^* \omega$ of ω is also holomorphic, and so $t_Q^* \omega = c\omega$ for some $c \in \mathbb{C}$. Now $Q \mapsto c : E(\mathbb{C}) \to \mathbb{C}$ is a holomorphic function on $\mathbb{C}$, and all such functions are constant (see III, 2.2). Since the function takes the value 1 when $Q = 0$, it is 1 for all Q, and so ω is invariant under translation. Alternatively, one can simply note that the inverse image of ω under the map (III, 3.7)

$$(x, y) \mapsto (\wp(z), \wp'(z)), \quad \mathbb{C} \smallsetminus \Lambda \to E^{\mathrm{aff}}(\mathbb{C})$$

is

$$\frac{d\wp(z)}{2\wp'(z)} = \frac{dz}{2},$$

which is clearly translation invariant on $\mathbb{C}$ because $d(z + c) = dz$.

The jacobian variety of a Riemann surface

Consider an elliptic curve E over $\mathbb{C}$ and a nonzero holomorphic differential one-form ω. We choose a point $P_0 \in E(\mathbb{C})$ and try to define a map

$$P \mapsto \int_{P_0}^{P} \omega \colon E(\mathbb{C}) \to \mathbb{C}.$$

This is not well-defined because the value of the integral depends on the path we choose from P_0 to P — nonhomotopic paths may give different answers. However, if we choose a basis $\{\gamma_1, \gamma_2\}$ for $H_1(E(\mathbb{C}), \mathbb{Z})$ (equivalently, a basis for $\pi_1(E(\mathbb{C}), P_0)$), then the integral is well-defined modulo the lattice Λ in $\mathbb{C}$ generated by

$$\int_{\gamma_1} \omega, \quad \int_{\gamma_2} \omega.$$

In this way, we obtain an isomorphism

$$P \mapsto \int_{P_0}^{P} \omega \colon E(\mathbb{C}) \to \mathbb{C}/\Lambda.$$

Note that this construction is inverse to that in III, §3.

Jacobi and Abel made a similar construction for any compact Riemann surface X. Suppose that X has genus g, and let $\omega_1, \ldots, \omega_g$ be a basis for the vector space $\Omega^1(X)$ of holomorphic one-forms on X. Choose a point $P_0 \in X$. Then there is a smallest lattice Λ in $\mathbb{C}^g$ such that the map

$$P \mapsto \left(\int_{P_0}^{P} \omega_1, \ldots, \int_{P_0}^{P} \omega_g \right) : X \to \mathbb{C}^g/\Lambda$$

is well-defined. By a ***lattice*** in $\mathbb{C}^g$, we mean the free $\mathbb{Z}$-submodule of rank $2g$ generated by a basis for $\mathbb{C}^g$ regarded as a real vector space. The quotient $\mathbb{C}^g/\Lambda$ is a complex manifold, called the ***jacobian variety*** $\mathrm{Jac}(X)$ of X, which can be regarded as a higher-dimensional analogue of $\mathbb{C}/\Lambda$. Note that it is has a commutative group structure.

We can make the definition of $\mathrm{Jac}(X)$ more canonical. Let $\Omega^1(X)^\vee$ be the dual of $\Omega^1(X)$ as a complex vector space. For each $\gamma \in H_1(X, \mathbb{Z})$,

$$\omega \mapsto \int_{\gamma} \omega$$

is an element of $\Omega^1(X)^\vee$, and in this way we obtain an injective homomorphism

$$H_1(X,\mathbb{Z}) \hookrightarrow \Omega^1(X)^\vee,$$

which one can prove identifies $H_1(X,\mathbb{Z})$ with a lattice in $\Omega^1(X)^\vee$. Define

$$\mathrm{Jac}(X) = \Omega^1(X)^\vee / H_1(X,\mathbb{Z}).$$

When we fix a $P_0 \in X$, every $P \in X$ defines an element

$$\omega \mapsto \int_{P_0}^{P} \omega \mod H_1(X,\mathbb{Z})$$

of $\mathrm{Jac}(X)$, and so we get a map $X \to \mathrm{Jac}(X)$. The choice of a different P_0 changes the map by a translation.

Construction of the elliptic curve over $\mathbb{C}$

We apply the above theory to the Riemann surface $X_0(N)$. Let π be the map $\pi\colon \mathbb{H} \to X_0(N)$ (not quite onto). For every $\omega \in \Omega^1(X_0(N))$, $\pi^*\omega$ equals $f dz$ with $f \in \mathcal{S}_2(\Gamma_0(N))$, and the map $\omega \mapsto f$ is a bijection

$$\Omega^1(X_0(N)) \to \mathcal{S}_2(\Gamma_0(N))$$

(see 3.3). The Hecke operator $T(n)$ acts on $\mathcal{S}_2(\Gamma_0(N))$, and hence on the vector space $\Omega^1(X_0(N))$ and its dual.

PROPOSITION 6.1 *There is a canonical action of $T(n)$ on $H_1(X_0(N),\mathbb{Z})$, which is compatible with the map $H_1(X_0(N),\mathbb{Z}) \to \Omega^1(X_0(N))^\vee$. In other words, the action of $T(n)$ on $\Omega^1(X_0(N))^\vee$ stabilizes its sublattice $H_1(X_0(N),\mathbb{Z})$, and therefore induces an action on the quotient* $\mathrm{Jac}(X_0(N))$.

PROOF. One can give an explicit set of generators for $H_1(X_0(N),\mathbb{Z})$, explicitly describe an action of $T(n)$ on them, and then explicitly verify that this action is compatible with the map $H_1(X_0(N),\mathbb{Z}) \to \Omega^1(X_0(N))^\vee$. Alternatively, as we discuss in the next section, there are more geometric reasons why the $T(n)$ should act on $\mathrm{Jac}(X_0(N))$. □

REMARK 6.2 From the action of $T(n)$ on $H_1(X_0(N),\mathbb{Z}) \approx \mathbb{Z}^{2g}$ we get a characteristic polynomial $P(Y) \in \mathbb{Z}[Y]$ of degree $2g$. What is its relation to the characteristic polynomial $Q(Y) \in \mathbb{C}[Y]$ of $T(n)$ acting on $\Omega^1(X)^\vee \approx \mathbb{C}^g$? The obvious guess is that $P(Y)$ is the product of $Q(Y)$ with its complex conjugate $\overline{Q(Y)}$. The proof that this is so is an exercise in linear algebra. See the next section (7.9).

Now let $f = \sum c(n)q^n$ be a normalized newform for $\Gamma_0(N)$ with $c(n) \in \mathbb{Z}$. The map

$$\alpha \mapsto \alpha(f): \Omega^1(X_0(N))^\vee \to \mathbb{C}$$

identifies $\mathbb{C}$ with the largest quotient of $\Omega^1(X_0(N))^\vee$ on which each $T(n)$ acts as multiplication by $c(n)$. The image of $H_1(X_0(N), \mathbb{Z})$ is a lattice Λ_f, and we set $E_f = \mathbb{C}/\Lambda_f$ — it is an elliptic curve over $\mathbb{C}$. Note that we have constructed maps

$$X_0(N) \to \mathrm{Jac}(X_0(N)) \to E_f$$

such that the inverse image of the differential one-form on E_f represented by dz is the differential one-form on $X_0(N)$ represented by $f dz$.

Construction of the elliptic curve over $\mathbb{Q}$

We briefly explain why the above construction in fact gives an elliptic curve over $\mathbb{Q}$. There will be a few more details in the next section.

For a compact Riemann surface X, we defined

$$\mathrm{Jac}(X) = \Omega^1(X)^\vee / H_1(X, \mathbb{Z}) \approx \mathbb{C}^g/\Lambda, \quad g = \mathrm{genus}\, X.$$

This is a complex manifold, but as in the case of an elliptic curve, it is possible to construct enough functions on it to embed it into projective space, and so realize it as a projective algebraic variety.

Now suppose that X is a nonsingular projective curve over an field k. Weil showed (as part of the work mentioned on p. 207) that it is possible to attach to X a projective algebraic variety $\mathrm{Jac}(X)$ over k, which, in the case $k = \mathbb{C}$ becomes the variety defined in the last paragraph. There is again a map $X \to \mathrm{Jac}(X)$, well-defined up to translation by the choice of a point $P_0 \in X(k)$. The variety $\mathrm{Jac}(X)$ is an abelian variety, i.e., not only is it projective, but it is equipped with a group structure given by regular maps.

In particular, there is such a variety attached to the curve $X_0(N)$ defined in Section 2. Moreover (see the next section), the Hecke operators $T(n)$ define endomorphisms of $\mathrm{Jac}(X_0(N))$. Because it has an abelian group structure, every integer m defines an endomorphism of $\mathrm{Jac}(X_0(N))$, and we define E_f to be the largest "quotient" of $\mathrm{Jac}(X_0(N))$ on which $T(n)$ and $c(n)$ agree for all n relatively prime to N. One can prove that this operation of "passing to the quotient" commutes with change of the ground field, and so in this way we obtain an elliptic curve over $\mathbb{Q}$ that becomes equal over

$\mathbb{C}$ to the curve defined in the last subsection. On composing $X_0(N) \to \mathrm{Jac}(X_0(N))$ with $\mathrm{Jac}(X_0(N)) \to E_f$ we obtain a map $X_0(N) \to E_f$. More precisely, we have the following statement.

THEOREM 6.3 *Let $f = \sum c(n)q^n$ be a newform in $S_2(\Gamma_0(N))$, normalized to have $c(1) = 1$, and assume that all $c(n) \in \mathbb{Z}$. Then there exists an elliptic curve E_f and a map $\alpha: X_0(N) \to E_f$ with the following properties:*

(a) *α factors uniquely through $\mathrm{Jac}(X_0(N))$,*

$$X_0(N) \to \mathrm{Jac}(X_0(N)) \to E_f,$$

and the second map realizes E_f as the largest quotient of $\mathrm{Jac}(X_0(N))$ on which the endomorphisms $T(n)$ and $c(n)$ of $\mathrm{Jac}(X_0(N))$ agree;

(b) *the inverse image of any invariant differential one-form ω on E_f under $\mathbb{H} \to X_0(N) \to E_f$ is a nonzero rational multiple of $f\,dz$.*

7 Why the L-function of E_f agrees with the L-function of f

In this section we sketch a proof of the identity of Eichler and Shimura relating the Hecke correspondence $T(p)$ to the Frobenius map at p, and hence the L-function of f to the L-function of E_f.

The ring of correspondences of a curve

Let X and X' be projective nonsingular curves over a perfect field k.

A ***correspondence*** T between X and X', written $T: X \vdash X'$, is a pair of finite surjective regular maps

$$X \overset{\alpha}{\leftarrow} Y \overset{\beta}{\rightarrow} X'.$$

It can be thought of as a many-valued map $X \to X'$ sending a point $P \in X(k)$ to the set $\{\beta(Q_i)\}$ where the Q_i run through the elements of $\alpha^{-1}(P)$ (the Q_i need not be distinct). Better, recall that $\mathrm{Div}(X)$ is the free abelian group on the set of points of X, so that an element of $\mathrm{Div}(X)$ is a finite formal sum

$$D = \sum n_P[P], \quad n_P \in \mathbb{Z}, \quad P \in X(k).$$

A correspondence T then defines a map

$$\mathrm{Div}(X) \to \mathrm{Div}(X'), \quad [P] \mapsto \sum_i [\beta(Q_i)]$$

(notation as above). This map multiplies the degree of a divisor by $\deg(\alpha)$. It therefore sends the divisors of degree zero on X into the divisors of degree zero on X', and one can show that it sends principal divisors to principal divisors. Hence it defines a map $T: J(X) \to J(X')$, where

$$J(X) \stackrel{\text{def}}{=} \left(\mathrm{Div}^0(X_{k^{\mathrm{al}}})/\{\text{ principal divisors}\}\right)^{\mathrm{Gal}(k^{\mathrm{al}}/k)}.$$

We define the ***ring of correspondences*** $\mathcal{A}(X)$ on X to be the subring of $\mathrm{End}(J(X))$ generated by the maps defined by correspondences.

If T is the correspondence

$$X \stackrel{\beta}{\leftarrow} Y \stackrel{\alpha}{\rightarrow} X,$$

then the transpose T^{tr} of T is the correspondence

$$X \stackrel{\alpha}{\leftarrow} Y \stackrel{\beta}{\rightarrow} X.$$

A morphism $\alpha: X \to X'$ can be thought of as a correspondence

$$X \leftarrow \Gamma \rightarrow X'$$

where $\Gamma \subset X \times X'$ is the graph of α and the maps are the projections. The transpose of a morphism α is the many valued map $P \mapsto \alpha^{-1}(P)$.

REMARK 7.1 Let U and U' be the curves obtained from X and X' by removing a finite number of points. Then a regular map $\alpha: U \to U'$ extends *uniquely* to a regular map $\bar{\alpha}: X \to X'$: take $\bar{\alpha}$ to be the regular map whose graph is the closure of the graph of α (see 4.19). On applying this remark twice, we see that a correspondence $U \vdash U'$ extends uniquely to a correspondence $X \vdash X'$. It follows that two correspondences $X \vdash X'$ are equal if they are equal as correspondences of sets $U(k^{\mathrm{al}}) \vdash U'(k^{\mathrm{al}})$.

REMARK 7.2 Let

$$X \stackrel{\alpha}{\leftarrow} Y \stackrel{\beta}{\rightarrow} X'.$$

be a correspondence $T: X \vdash X'$. For any regular function f on X', we define $T(f)$ to be the regular function $P \mapsto \sum f(\beta Q_i)$ on X (notation as above). Similarly, T will define a homomorphism $\Omega^1(X') \to \Omega^1(X)$.

The Hecke correspondence

For $p \nmid N$, the Hecke correspondence $T(p): Y_0(N) \to Y_0(N)$ is defined to be

$$Y_0(N) \stackrel{\alpha}{\leftarrow} Y_0(pN) \stackrel{\beta}{\to} Y_0(N),$$

where α is the obvious projection map and β is the map induced by $z \mapsto pz: \mathbb{H} \to \mathbb{H}$.

On points, it has the following description. Recall that a point of $Y_0(pN)$ is represented by a pair (E, S), where E is an elliptic curve and S is a cyclic subgroup of E of order pN. Because $p \nmid N$, every such subgroup decomposes uniquely into subgroups of order N and p, $S = S_N \times S_p$. The map α sends the point represented by (E, S) to the point represented by (E, S_N), and β sends it to the point represented by $(E/S_p, S/S_p)$. Since E_p has $p+1$ cyclic subgroups, the correspondence is $1 : p+1$.

The unique extension of $T(p)$ to a correspondence $X_0(N) \to X_0(N)$ acts on $\Omega^1(X_0(N)) = \mathcal{S}_2(\Gamma_0(N))$ as the Hecke correspondence defined in Section 4. This description of $T(p)$, $p \nmid N$, makes sense, and is defined on, the curve $X_0(N)$ over $\mathbb{Q}$. Similar remarks apply to the $T(p)$ for $p|N$.

The Frobenius map

Let C be a curve defined over the algebraic closure $\mathbb{F}$ of $\mathbb{F}_p$. If C is defined by equations

$$\sum a_{i_0 i_1 \ldots} X_0^{i_0} X_1^{i_1} \cdots = 0,$$

then we let $C^{(p)}$ be the curve defined by the equations

$$\sum a_{i_0 i_1 \ldots}^p X_0^{i_0} X_1^{i_1} \cdots = 0,$$

and we let the ***Frobenius map*** $\varphi_p: C \to C^{(p)}$ send the point $(b_0 : b_1 : b_2 : \ldots)$ to $(b_0^p : b_1^p : b_2^p : \ldots)$. If C is defined over $\mathbb{F}_p$, then $C = C^{(p)}$ and φ_p is the Frobenius map defined earlier.

Recall that a nonconstant morphism $\alpha: C \to C'$ of curves defines an inclusion $\alpha^*: k(C') \hookrightarrow k(C)$ of function fields, and that the degree of α is defined to be $[k(C) : \alpha^* k(C')]$. The map α is said to be ***separable*** or ***purely inseparable*** according as $k(C)$ is a separable or purely inseparable extension of $\alpha^* k(C')$. If the separable degree of $k(C)$ over $\alpha^* k(C')$ is m, then the map $C(k^{\mathrm{al}}) \to C'(k^{\mathrm{al}})$ is $m : 1$ except over the finite set where it is ramified.

PROPOSITION 7.3 *The Frobenius map $\varphi_p : C \to C^{(p)}$ is purely inseparable of degree p, and every purely inseparable map $\varphi : C \to C'$ of degree p (of complete nonsingular curves) factors as*

$$C \xrightarrow{\varphi_p} C^{(p)} \xrightarrow{\simeq} C'.$$

PROOF. For $C = \mathbb{P}^1$, this is obvious, and the general case follows because $\mathbb{F}(C)$ is a separable extension of $\mathbb{F}(T)$. See Silverman 2009, II, 2.12, for the details. □

Brief review of the points of order p on elliptic curves

Let E be an elliptic curve over an algebraically closed field k. The map $p: E \to E$ (multiplication by p) is of degree p^2. If k has characteristic zero, then the map is separable, which implies that its kernel has order p^2. If k has characteristic p, the map is never separable: either it is purely inseparable (and so E has no points of order p) or its separable and inseparable degrees are both p (and so E has p points of order dividing p). The first case occurs for only finitely many values of j.

The Eichler–Shimura relation

The curve $X_0(N)$ and the Hecke correspondence $T(p)$ are defined over $\mathbb{Q}$. For almost all primes $p \nmid N$, $X_0(N)$ will reduce to a nonsingular curve $\tilde{X}_0(N)$.[16] For such a good prime p, the correspondence $T(p)$ defines a correspondence $\tilde{T}(p)$ on $\tilde{X}_0(N)$.

THEOREM 7.4 *For a prime p where $X_0(N)$ has good reduction,*

$$\tilde{T}(p) = \varphi_p + \varphi_p^{tr}.$$

(Equality in the ring $\mathcal{A}(\tilde{X}_0(N))$ of correspondences on $\tilde{X}_0(N)$ over the algebraic closure $\mathbb{F}$ of $\mathbb{F}_p$.)

PROOF. According to 7.1, it suffices to show that they agree as correspondences of sets $U(\mathbb{Q}_p^{\mathrm{al}}) \vdash U(\mathbb{Q}_p^{\mathrm{al}})$ for some open subset U of $X_0(N)$.

Over $\mathbb{Q}_p^{\mathrm{al}}$ we have the following description of $T(p)$ (see above): a point P on $Y_0(N)$ is represented by a homomorphism of elliptic curves

[16] In fact, it is known that $X_0(N)$ has good reduction for all primes $p \nmid N$, but this is hard to prove. It is easy to see that $X_0(N)$ has bad reduction at the primes dividing N.

$\alpha: E \to E'$ with cyclic kernel of order N; let $S_0, \ldots, S_p$ be the subgroups of order p in E; then $T_p(P) = \{Q_0, \ldots, Q_p\}$, where Q_i is represented by $E/S_i \to E'/\alpha(S_i)$.

Consider a point $\tilde{P}$ on $\tilde{X}_0(N)$ with coordinates in $\mathbb{F}$ — by Hensel's lemma it will lift to a point on $X_0(N)$ with coordinates in $\mathbb{Q}_p^{\rm al}$. Ignoring a finite number of points of $\tilde{X}_0(N)$, we can suppose $\tilde{P} \in \tilde{Y}_0(N)$ and hence is represented by a map $\tilde{\alpha}: \tilde{E} \to \tilde{E}'$, where $\alpha: E \to E'$ has cyclic kernel of order N. By ignoring a further finite number of points, we may suppose that $\tilde{E}$ has p points of order dividing p.

Let $\alpha: E \to E'$ be a lifting of $\tilde{\alpha}$ to $\mathbb{Q}_p^{\rm al}$. The reduction map $E_p(\mathbb{Q}_p^{\rm al}) \to \tilde{E}_p(\mathbb{F}_p^{\rm al})$ has a kernel of order p. Number the subgroups of order p in E so that S_0 is the kernel of this map. Then each S_i, $i \neq 0$, maps to a subgroup of order p in $\tilde{E}$.

The map $p: \tilde{E} \to \tilde{E}$ has factorizations

$$\tilde{E} \xrightarrow{\varphi} \tilde{E}/S_i \xrightarrow{\psi} \tilde{E}, \quad i = 0, 1, \ldots, p.$$

When $i = 0$, φ is a purely inseparable map of degree p (it is the reduction of the map $E \to E/S_0$ — it therefore has degree p and has zero kernel), and so ψ must be separable of degree p (we are assuming $\tilde{E}$ has p points of order dividing p). Proposition 7.3 shows that there is an isomorphism $\tilde{E}^{(p)} \to \tilde{E}/S_0$. Similarly $\tilde{E}'^{(p)} \approx \tilde{E}'/S_0$. Therefore Q_0 is represented by $\tilde{E}^{(p)} \to \tilde{E}'^{(p)}$, which also represents $\varphi_p(P)$.

When $i \neq 0$, φ is separable (its kernel is the reduction of S_i), and so ψ is purely inseparable. Therefore $\tilde{E} \approx \tilde{E}_i^{(p)}$, and similarly $\tilde{E}' \approx \tilde{E}_i'^{(p)}$, where $\tilde{E}_i/\tilde{E}/S_i$ and $\tilde{E}_i' = \tilde{E}'/S_i$. It follows that $\{Q_1, \ldots, Q_p\} = \varphi_p^{-1}(P) = \varphi_p^{\rm tr}(P)$. □

The zeta function of an elliptic curve revisited

Recall (III, 3.21) that, for an elliptic curve $E = \mathbb{C}/\Lambda$ over $\mathbb{C}$, the degree of a nonzero endomorphism of E is the determinant of α acting on Λ. More generally (III, 6.4), for an elliptic curve E over an algebraically closed field k and ℓ a prime not equal to the characteristic of k,

$$\deg \alpha = \det(\alpha | T_\ell E) \tag{46}$$

where $T_\ell E$ is the Tate module $T_\ell E = \varprojlim E(k)_{\ell^n}$ of E.

When Λ is a free module over some ring R and $\alpha: \Lambda \to \Lambda$ is R-linear, we let $\mathrm{Tr}(\alpha|\Lambda)$ denote the trace of the matrix of α relative to some basis for Λ — it is independent of the choice of basis.

PROPOSITION 7.5 *Let E be an elliptic curve over $\mathbb{F}_p$. Then the trace of the Frobenius endomorphism φ_p on $T_\ell E$ satisfies*

$$\mathrm{Tr}(\varphi_p|T_\ell E) = a_p \overset{\mathrm{def}}{=} p+1-N_p.$$

PROOF. For any 2×2 matrix A, $\det(A-I_2) = \det A - \mathrm{Tr}\, A + 1$. On applying this to the matrix of φ_p acting on $T_\ell E$, and using (46), we find that

$$\deg(\varphi_p - 1) = \deg(\varphi_p) - \mathrm{Tr}(\varphi_p|T_\ell E) + 1.$$

As we saw in the proof of IV, 9.4, $\deg(\varphi_p - 1) = N_p$ and $\deg(\varphi_p) = p$. □

As we noted above, a correspondence $T\colon X \vdash X$ defines a map $J(X) \to J(X)$. When E is an elliptic curve, $E(k) = J(E)$, and so T acts on $E(k)$, and hence also on $T_\ell(E)$.

COROLLARY 7.6 *Let E be an elliptic curve over $\mathbb{F}_p$. Then*

$$\mathrm{Tr}(\varphi_p^{\mathrm{tr}}|T_\ell E) = \mathrm{Tr}(\varphi_p|T_\ell E).$$

PROOF. Because φ_p has degree p, $\varphi_p \circ \varphi_p^{\mathrm{tr}} = p$. Therefore, if α,β are the eigenvalues of φ_p, so that in particular $\alpha\beta = \deg\varphi = p$, then

$$\mathrm{Tr}(\varphi_p^{\mathrm{tr}}|T_\ell E) = p/\alpha + p/\beta = \beta + \alpha.$$ □

The action of the Hecke operators on $H_1(E,\mathbb{Z})$

Again, we first need an elementary result from linear algebra.

Let V be a real vector space and suppose that we are given the structure of a complex vector space on V. This means that we are given an $\mathbb{R}$-linear map $J\colon V \to V$ such that $J^2 = -1$. The map J extends by linearity to $V \otimes_{\mathbb{R}} \mathbb{C}$, which splits into a direct sum

$$V \otimes_{\mathbb{R}} \mathbb{C} = V^+ \oplus V^-,$$

where $V^\pm$ are the ± 1 eigenspaces of J.

PROPOSITION 7.7 *(a) The map*

$$V \xrightarrow{v \mapsto v\otimes 1} V \otimes_{\mathbb{R}} \mathbb{C} \xrightarrow{\text{project}} V^+$$

is an isomorphism of complex *vector spaces.*

(b) The map $v\otimes z \mapsto v \otimes \bar{z}\colon V\otimes_{\mathbb{R}}\mathbb{C} \to V\otimes_{\mathbb{R}}\mathbb{C}$ is an $\mathbb{R}$-linear involution of $V\otimes_{\mathbb{R}}\mathbb{C}$ interchanging V^+ and V^-.

PROOF. Easy exercise. □

COROLLARY 7.8 *Let α be an endomorphism of V which is $\mathbb{C}$-linear. Write A for the matrix of α regarded as an $\mathbb{R}$-linear endomorphism of V, and A_1 for the matrix of α as a $\mathbb{C}$-linear endomorphism of V. Then*

$$A \sim A_1 \oplus \bar{A}_1.$$

(By this I mean that the matrix A is equivalent to the matrix $\begin{pmatrix} A_1 & 0 \\ 0 & \bar{A}_1 \end{pmatrix}$.)

PROOF. Follows immediately from the above Proposition.[17] □

COROLLARY 7.9 *For a prime p not dividing N,*

$$\mathrm{Tr}(T(p) \mid H_1(X_0(N),\mathbb{Z})) = \mathrm{Tr}(T(p) \mid \Omega^1(X_0(N))) + \overline{\mathrm{Tr}(T(p) \mid \Omega^1(X_0(N)))}.$$

PROOF. To say that $H_1(X_0(N),\mathbb{Z})$ is a lattice in $\Omega^1(X_0(N))^\vee$ means that

$$H_1(X_0(N),\mathbb{Z}) \otimes_{\mathbb{Z}} \mathbb{R} \simeq \Omega^1(X_0(N))^\vee$$

as real vector spaces. Clearly

$$\mathrm{Tr}(T(p) \mid H_1(X_0(N),\mathbb{Z})) = \mathrm{Tr}(T(p) \mid H_1(X_0(N),\mathbb{Z}) \otimes_{\mathbb{Z}} \mathbb{R}),$$

and so we can apply the preceding corollary. □

The proof that $c(p) = a_p$

THEOREM 7.10 *Let $f = \sum c(n)q^n$ be a newform in $S_2(\Gamma_0(N))$, normalized to have $c(1) = 1$, and assume that all $c(n) \in \mathbb{Z}$. Let $X_0(N) \to E$ be the map given by Theorem 6.3. Then, for all $p \nmid N$,*

$$c(p) = a_p \stackrel{\text{def}}{=} p + 1 - N_p(E).$$

[17] When V has dimension 2, which is the only case we are interested in, we can identify V (as a real or complex vector space) with $\mathbb{C}$. For the map "multiplication by $\alpha = a + ib$" the statement becomes,

$$\begin{pmatrix} a & -b \\ b & a \end{pmatrix} \sim \begin{pmatrix} a+ib & 0 \\ 0 & a-ib \end{pmatrix},$$

which is true because the two matrices are semisimple and have the same trace and determinant.

PROOF. We assume initially that $X_0(N)$ has genus 1. Then $X_0(N) \to E$ is an isogeny, and we can take

$$E = X_0(N).$$

Let p be a prime not dividing N. Then $X_0(N) = E$ has good reduction at p, and for any $\ell \neq p$, the reduction map $T_\ell E \to T_\ell \tilde{E}$ is an isomorphism. The Eichler–Shimura relation states that

$$\tilde{T}(p) = \varphi_p + \varphi_p^{\mathrm{tr}}.$$

On taking traces on $T_\ell \tilde{E}$, we find (using 7.5, 7.6, 7.9) that

$$2c(p) = a_p + a_p.$$

The proof of the general case is very similar except that, at various places in the argument, an elliptic curve has to be replace either by a curve or the jacobian variety of a curve. Ultimately, one uses that $T_\ell E$ is the largest quotient of $T_\ell \operatorname{Jac}(X_0(N))$ on which $T(p)$ acts as multiplication by $c(p)$ for all $p \nmid N$ (perhaps after tensoring with $\mathbb{Q}_\ell$). □

ASIDE 7.11 The original papers on the Eichler–Shimura theorem are Eichler 1954 and Shimura 1958. In his paper, Eichler proved the Ramanujan–Petersson conjecture for some modular forms of weight 2 by relating the conjecture to the Riemann hypothesis for curves over finite fields — he was the first to make this connection. According to Sarnak (lecture 2018): "All of modern automorphic forms in the connection with Frobenius, with algebraic geometry, with Grothendieck's work, starts with Eichler."

As noted in 4.21, the Ramanujan–Petersson conjecture has been proved by Deligne as a consequence of his proof of the Riemann hypothesis for algebraic varieties over finite fields. However, the conjecture has been generalized further to automorphic representations, and in that context there remain many open questions. See Sarnak 2005.

ASIDE 7.12 Let X be a Riemann surface. The map $[P]-[P_0] \mapsto \int_{P_0}^{P} \omega$ extends by linearity to map $\operatorname{Div}^0(X) \to \operatorname{Jac}(X)$. The famous theorem of Abel–Jacobi says that this induces an isomorphism $J(X) \to \operatorname{Jac}(X)$ (Fulton 1995, 21d). The jacobian variety $\operatorname{Jac}(X)$ of a curve X over a field k (constructed in general by Weil) has the property that $\operatorname{Jac}(X)(k) = J(X)$, at least when $J(k) \neq \emptyset$. For more on jacobian and abelian varieties over arbitrary fields, see Milne 1986a,b.

8 Wiles's proof

Somebody with an average or even good mathematical background might feel that all he ends up with after reading [...]'s paper is what he suspected before anyway: The proof of Fermat's Last Theorem is indeed very complicated.

M. Flach

In this section, I explain the strategy of Wiles's proof of the modularity conjecture for semistable elliptic curves over $\mathbb{Q}$ (i.e., curves with at worst nodal reduction).

Recall that if 0, $P_1, \ldots, P_s$, are points on the 2-sphere S, then $\pi \stackrel{\text{def}}{=} \pi_1(S \smallsetminus \{P_1, \ldots, P_s\}, O)$ is generated by loops $\gamma_1, \ldots, \gamma_s$ around each of the points $P_1, \ldots, P_s$, and that π classifies the coverings of S unramified except over $P_1, \ldots, P_s$.

Something similar is true for $\mathbb{Q}$. Let K be a finite extension of $\mathbb{Q}$, and let $\mathcal{O}_K$ be the ring of integers in K. In $\mathcal{O}_K$, the ideal $p\mathcal{O}_K$ factors into a product of powers of prime ideals: $p\mathcal{O}_K = \prod \mathfrak{p}^{e_\mathfrak{p}}$. The prime p is said to be ***unramified*** in K if no $e_\mathfrak{p} > 1$.

Now assume that $K/\mathbb{Q}$ is Galois with Galois group G. Let p be prime, and choose a prime ideal $\mathfrak{p}$ dividing $p\mathcal{O}_K$ (so that $\mathfrak{p} \cap \mathbb{Z} = (p)$). Let $G(\mathfrak{p})$ be the subgroup of G of σ such that $\sigma\mathfrak{p} = \mathfrak{p}$. The action of $G(\mathfrak{p})$ on $\mathcal{O}_K/\mathfrak{p} = k(\mathfrak{p})$ defines a surjection $G(\mathfrak{p}) \to \mathrm{Gal}(k(\mathfrak{p})/\mathbb{F}_p)$, which is an isomorphism if and only if p is unramified in K. When p is unramified, the element $F_\mathfrak{p} \in G(\mathfrak{p}) \subset G$ mapping to the Frobenius element $x \mapsto x^p$ in $\mathrm{Gal}(k(\mathfrak{p})/\mathbb{F}_p)$ is called the ***Frobenius element*** at $\mathfrak{p}$. Thus $F_\mathfrak{p} \in G$ is characterised by the conditions:

$$\begin{cases} F_\mathfrak{p}\mathfrak{p} &= \mathfrak{p}, \\ F_\mathfrak{p}x &\equiv x^p \mod \mathfrak{p}, \text{ for all } x \in \mathcal{O}_K. \end{cases}$$

If $\mathfrak{p}'$ also divides $p\mathcal{O}_K$, then there exists a $\sigma \in G$ such that $\sigma\mathfrak{p} = \mathfrak{p}'$, and so $F_{\mathfrak{p}'} = \sigma F_\mathfrak{p} \sigma^{-1}$. Therefore, the conjugacy class of $F_\mathfrak{p}$ depends only on p — we often write F_p for any one of the $F_\mathfrak{p}$. The analogue of π being generated by the loops γ_i is that G is generated by the $F_\mathfrak{p}$ (varying p).

The above discussion extends to infinite extensions. Fix a finite nonempty set S of prime numbers, and let K_S be the union of all $K \subset \mathbb{C}$ that are of finite degree over $\mathbb{Q}$ and unramified outside S — it is an infinite Galois extension of $\mathbb{Q}$. For each $p \in S$, there is an element $F_p \in \mathrm{Gal}(K_S/\mathbb{Q})$, well-defined up to conjugation, called the ***Frobenius element*** at p.

PROPOSITION 8.1 *Let E be an elliptic curve over $\mathbb{Q}$. Let ℓ be a prime, and let*

$$S = \{p \mid E \text{ has bad reduction at } p\} \cup \{\ell\}.$$

Then all points of order ℓ^n on E have coordinates in K_S, i.e., $E(K_S)_{\ell^n} = E(\mathbb{Q}^{\mathrm{al}})_{\ell^n}$ for all n.

PROOF. Let $P \in E(\mathbb{Q}^{\mathrm{al}})$ be a point of ℓ-power order, and let K be a finite Galois extension of $\mathbb{Q}$ such that $P \in E(K)$. Let H be the subgroup of $G \overset{\mathrm{def}}{=} \mathrm{Gal}(K/\mathbb{Q})$ of elements fixing P. Then H is the kernel of $G \to \mathrm{Aut}(\langle P \rangle)$, and so is normal. After replacing K with K^H, we may suppose that G acts faithfully on $\langle P \rangle$. Let $p \in S$, and let $\mathfrak{p}$ be a prime ideal of $\mathcal{O}_K$ dividing (p). The reduction map $E(K)_{\ell^n} \to E(k(\mathfrak{p}))_{\ell^n}$ is injective, and so if σ lies in the kernel of $G(\mathfrak{p}) \to \mathrm{Gal}(k(\mathfrak{p})/k)$, it must fix P, and so be trivial. This shows that K is unramified at p. Since this is true for all $p \in S$, we can deduce that $K \subset K_S$. □

EXAMPLE 8.2 The smallest field containing the coordinates of the points of order 2 on the curve $E: Y^2Z = X^3 + aXZ^2 + bZ^3$ is the splitting field of $X^3 + aX + b$. From algebraic number theory we know that this field is unramified at the primes not dividing the discriminant Δ of $X^3 + aX + b$, i.e., at the primes where E has good reduction (ignoring 2).

For an elliptic curve over a field k, we define the Tate module $T_\ell E$ to be the Tate module of $E_{k^{\mathrm{al}}}$. Thus, for E over $\mathbb{Q}$ and S as in the proposition, $T_\ell E$ is the free $\mathbb{Z}_\ell$-module of rank 2 such that

$$T_\ell E/\ell^n T_\ell E = E(K_S)_{\ell^n} = E(\mathbb{Q}^{\mathrm{al}})_{\ell^n}$$

for all n. The action of G_S on the quotients defines a continuous action of G_S on $T_\ell E$, i.e., a continuous homomorphism (also referred to as a representation)

$$\rho_\ell \colon G_S \to \mathrm{Aut}_{\mathbb{Z}_\ell}(T_\ell E) \approx \mathrm{GL}_2(\mathbb{Z}_\ell).$$

PROPOSITION 8.3 *Let E, ℓ, S be as in the previous proposition. For all $p \notin S$,*

$$\mathrm{Tr}(\rho_\ell(F_p) \mid T_\ell E) = a_p \overset{\mathrm{def}}{=} p + 1 - N_p(E).$$

PROOF. Because $p \notin S$, E has good reduction to an elliptic curve E_p over $\mathbb{F}_p$, and the reduction map $P \mapsto \bar{P}$ induces an isomorphism $T_\ell E \to T_\ell E_p$. By definition F_p maps to the Frobenius element in $\mathrm{Gal}(\mathbb{F}/\mathbb{F}_p)$, and the two have the same action on $T_\ell E$. Therefore the proposition follows from (7.5). □

DEFINITION 8.4 A continuous homomorphism $\rho\colon G_S \to \mathrm{GL}_2(\mathbb{Z}_\ell)$ is said to be ***modular*** if $\mathrm{Tr}(\rho(F_p)) \in \mathbb{Z}$ for all $p \notin S$ and there exists a cusp form $f = \sum c(n)q^n$ in $S_{2k}(\Gamma_0(N))$ for some k and N such that

$$\mathrm{Tr}(\rho(F_p)) = c(p)$$

for all $p \notin S$.

Thus, to prove that E is modular we must prove that $\rho_\ell\colon G_S \to \mathrm{Aut}(T_\ell E)$ is modular for some ℓ. Note that then ρ_ℓ will be modular for all ℓ.

Similarly, one says that a continuous homomorphism $\rho\colon G_S \to \mathrm{GL}_2(\mathbb{F}_\ell)$ is modular if there exists a cusp form $f = \sum c(n)q^n$ in $S_{2k}(\Gamma_0(N))$ for some k and N such that

$$\mathrm{Tr}(\rho(F_p)) \equiv c(p) \mod \ell$$

for all $p \notin S$. There is the following remarkable conjecture.

CONJECTURE 8.5 (SERRE) *Every odd irreducible representation $\rho : G_S \to \mathrm{GL}_2(\mathbb{F}_\ell)$ is modular.*

"Odd" means that $\det \rho(c) = -1$, where c is complex conjugation. "Irreducible" means that there is no one-dimensional subspace of $\mathbb{F}_\ell^2$ stable under the action of G_S. Let $E_\ell = E(\mathbb{Q}^{\mathrm{al}})_\ell$. The Weil pairing (Silverman 2009, III, 8), shows that $\bigwedge^2 E_\ell \simeq \mu_\ell$ (the group of ℓ-roots of 1 in $\mathbb{Q}^{\mathrm{al}}$). Since $c\zeta = \zeta^{-1}$, this shows that the representation of G_S on E_ℓ is odd. It need not be irreducible; for example, if E has a point of order ℓ with coordinates in $\mathbb{Q}$, then it will not be.

As we shall discuss in the next section, Serre in fact gave a recipe for defining the level N and weight $2k$ of modular form.

By the early 1990s, there was much numerical evidence supporting Serre's conjecture, but few theorems. The most important of these was the following.

THEOREM 8.6 (LANGLANDS, TUNNELL) *If $\rho : G_S \to \mathrm{GL}_2(\mathbb{F}_3)$ is odd and irreducible, then it is modular.*

Note that $\mathrm{GL}_2(\mathbb{F}_3)$ has order $8 \cdot 6 = 48$. The action of $\mathrm{PGL}_2(\mathbb{F}_3)$ on the projective plane over $\mathbb{F}_3$ identifies it with S_4, and so $\mathrm{GL}_2(\mathbb{F}_3)$ is a double cover $\tilde{S}_4$ of S_4.

The theorem of Langlands and Tunnell in fact concerned representations $G_S \to \mathrm{GL}_2(\mathbb{C})$. In the nineteenth century, Klein classified the finite subgroups of $\mathrm{GL}_2(\mathbb{C})$: their images in $\mathrm{PGL}_2(\mathbb{C})$ are cyclic, dihedral, A_4, S_4, or A_5. Langlands constructed candidates for the modular forms, and verified they had the correct property in the A_4 case. Tunnell extended this to the S_4 case, and, since $\mathrm{GL}_2(\mathbb{F}_3)$ embeds into $\mathrm{GL}_2(\mathbb{C})$, this verifies Serre's conjecture for $\mathbb{F}_3$.

Fix a representation $\rho_0 \colon G_S \to \mathrm{GL}_2(\mathbb{F}_\ell)$. In future, R will always denote a complete local Noetherian ring with residue field $\mathbb{F}_\ell$, for example, $\mathbb{F}_\ell$, $\mathbb{Z}_\ell$, or $\mathbb{Z}_\ell[[X]]$. Two homomorphism $\rho_1, \rho_2 \colon G_S \to \mathrm{GL}_2(R)$ will be said to be ***strictly equivalent*** if

$$\rho_1 = M\rho_2 M^{-1}, \quad M \in \mathrm{Ker}(\mathrm{GL}_2(R) \to \mathrm{GL}_2(k)).$$

A ***deformation*** of ρ_0 is a strict equivalence class of homomorphisms $\rho \colon G_S \to \mathrm{GL}_2(R)$ whose composite with $\mathrm{GL}_2(R) \to \mathrm{GL}_2(\mathbb{F}_p)$ is ρ_0.

Let $*$ be a set of conditions on representations $\rho \colon G_S \to \mathrm{GL}(R)$. Mazur showed that, for certain $*$, there is a universal $*$-deformation of ρ_0, i.e., a ring $\tilde{R}$ and a deformation $\tilde{\rho} \colon G_S \to \mathrm{GL}_2(\tilde{R})$ satisfying $*$ such that for any other deformation $\rho \colon G_S \to \mathrm{GL}_2(R)$ satisfying $*$, there is a unique homomorphism $\tilde{R} \to R$ for which the composite $G_S \xrightarrow{\tilde{\rho}} \mathrm{GL}_2(\tilde{R}) \to \mathrm{GL}_2(R)$ is ρ.

Now assume that ρ_0 is modular. Work of Hida and others show that, for certain $*$, there exists a deformation $\rho_{\mathbb{T}} \colon G_S \to \mathrm{GL}_2(\mathbb{T})$ that is universal for *modular* deformations satisfying $*$. Because $\tilde{\rho}$ is universal for *all* $*$-representations, there exists a unique homomorphism $\delta \colon \tilde{R} \to \mathbb{T}$ carrying $\tilde{\rho}$ into $\rho_{\mathbb{T}}$. It is onto, and it is injective if and only if *every* $*$-representation is modular.

It is now possible to explain Wiles's strategy. First, state conditions $*$ as strong as possible but which are satisfied by the representation of G_S on $T_\ell E$ for E a semistable elliptic curve over $\mathbb{Q}$. Fixing a modular ρ_0 we get a homomorphism $\delta \colon \tilde{R} \to \mathbb{T}$.

THEOREM 8.7 (WILES) *The homomorphism $\delta \colon \tilde{R} \to \mathbb{T}$ is an isomorphism (and so every $*$-representation lifting ρ_0 is modular).*

Now let E be an elliptic curve over $\mathbb{Q}$, and assume initially that the representation of G_S on E_3 is irreducible. By the theorem of Langlands and Tunnell, the representation $\rho_0 \colon G_S \to \mathrm{Aut}(E(K_S)_3)$ is modular, and so, by the theorem of Wiles, $\rho_3 \colon G_S \to \mathrm{Aut}(T_3 E)$ is modular, which implies that E is modular.

What if the representation of G_S on $E(K_S)_3$ is not irreducible, for example, if $E(\mathbb{Q})$ contains a point of order three? The representations of G_S on $E(K_S)_3$ and $E(K_S)_5$ cannot both be reducible, because otherwise either E or a curve isogenous to E will have rational points of order 3 and 5, hence a point of order 15, which is impossible by Mazur's theorem (II, 5.11). Unfortunately, there is no Langlands–Tunnell theorem for 5. Instead, Wiles uses the following elegant argument.

He shows that, given E, there exists a semistable elliptic curve E' over $\mathbb{Q}$ such that:

(a) $E'(K_S)_3$ is irreducible;

(b) $E'(K_S)_5 \approx E(K_S)_5$ as G_S-modules.

Because of (a), the preceding argument applies to E' and shows it to be modular. Hence the representation $\rho_5 \colon G_S \to \mathrm{Aut}(T_5 E')$ is modular, which implies that $\rho_0 \colon G_S \to \mathrm{Aut}(E'(K_S)_5) \approx \mathrm{Aut}(E(K_S)_5)$ is modular. Now, Wiles can apply his original argument with 3 replaced by 5.

9 Fermat, at last

Fix a prime number ℓ, and let E be an elliptic curve over $\mathbb{Q}$. As we now explain, it is possible to decide whether or not E has good reduction at a prime p purely by considering the action of $G = \mathrm{Gal}(\mathbb{Q}^{\mathrm{al}}/\mathbb{Q})$ on the modules $E(\mathbb{Q}^{\mathrm{al}})_{\ell^n}$ for $n \geq 1$.

Let M be a finite abelian group, and let $\rho \colon G \to \mathrm{Aut}(M)$ be a continuous homomorphism (discrete topology on $\mathrm{Aut}(M)$). The kernel H of ρ is an open subgroup of G, and therefore its fixed field $\mathbb{Q}^{\mathrm{al}H}$ is a finite extension of $\mathbb{Q}$. We say that ρ is ***unramified*** at p if p is unramified in $\mathbb{Q}^{\mathrm{al}H}$. With this terminology, we have the following criterion.

THEOREM 9.1 *The elliptic curve E has good reduction at a prime $p \neq \ell$ if and only if the representation of G on $E(\mathbb{Q}^{\mathrm{al}})_{\ell^n}$ is unramified for all n.*

The necessity follows from Proposition 8.1. For the sufficiency, one can show from the Néron model of E that, if E has bad reduction, then not all of its ℓ^n-torsion points can have coordinates in $\mathbb{Q}_\ell^{\mathrm{un}}$.

There is a similar criterion for $p = \ell$.

THEOREM 9.2 *Let ℓ be a prime. The elliptic curve E has good reduction at ℓ if and only if the representation of G on E_{ℓ^n} is flat for all n.*

For the experts, the representation of G on $E(\mathbb{Q}^{\mathrm{al}})_{\ell^n}$ is flat if there is a finite flat group scheme H over $\mathbb{Z}_\ell$ such that $H(\mathbb{Q}_\ell^{\mathrm{al}}) \approx E(\mathbb{Q}_\ell^{\mathrm{al}})_{\ell^n}$ as G-modules. Some authors say "finite" or "crystalline" instead of flat.

These criteria show that it is possible to detect whether E has bad reduction at p, and hence whether p divides the conductor of E, from knowing how G acts on $E(\mathbb{Q}^{\mathrm{al}})_{\ell^n}$ for *all* n — it may not be possible to detect bad reduction simply by looking at $E(\mathbb{Q}^{\mathrm{al}})_\ell$ for example.

Recall that Serre conjectured that every odd irreducible representation $\rho: G \to \mathrm{GL}_2(\mathbb{F}_\ell)$ is modular, i.e., that there exists an $f = \sum c(n)q^n \in S_{2k}(\Gamma_0(N))$, some k and N, such that

$$\mathrm{Tr}(\rho(F_p)) = c(n) \mod \ell$$

whenever ρ is unramified at p.

CONJECTURE 9.3 (REFINED SERRE) *Every odd irreducible representation $\rho: G \to \mathrm{GL}_2(\mathbb{F}_\ell)$ is modular for a specific k and N. For example, a prime $p \neq \ell$ divides N if and only if ρ is ramified at p, and p divides N if and only if ρ is not flat.*

THEOREM 9.4 (RIBET AND OTHERS) *If $\rho: G \to \mathrm{GL}_2(\mathbb{F}_\ell)$ is modular, then it is possible to choose the cusp form to have the weight $2k$ and level N predicted by Serre.*

We omit the proof, which is difficult.

Now let E be the curve defined in II, Exercise 3.4, corresponding to a solution to $X^\ell + Y^\ell = Z^\ell$, $\ell > 3$. It is not hard to verify, using nontrivial facts about elliptic curves, that the representation ρ_0 of G on $E(\mathbb{Q}^{\mathrm{al}})_\ell$ is irreducible; moreover, that it is unramified for $p \neq 2, \ell$, and that it is flat for $p = \ell$. The last statement follows from the facts that E has at worst nodal reduction at p, and if it does have bad reduction at p, then $p^\ell | \Delta$.

Now

$$\begin{aligned} E \text{ modular} &\implies \rho_0 \text{ modular} \\ &\overset{\text{Ribet}}{\implies} \rho_0 \text{ modular for a cusp form of weight 2, level 2.} \end{aligned}$$

Such a cusp form would correspond to a holomorphic differential on $X_0(2)$, but $X_0(N)$ has genus 0 when $N = 2$ (see p. 236), and so there is no such cusp form. Now Wiles's theorem proves that E does not exist.

ASIDE 9.5 Great problems are important because, like the Riemann hypothesis, they have important applications, or, like Fermat's Last Theorem, they reveal our ignorance and inspire great mathematics. Fermat's Last Theorem has certainly inspired great mathematics but it needs to be said that, even after its solution, our ignorance of the rational solutions of polynomial equations over $\mathbb{Q}$ remains almost as profound as before because the method applies only to Fermat's equation (or very similar equations). We do not even know, for example, whether it is possible for there to exist an algorithm for deciding whether a polynomial equation with coefficients in $\mathbb{Q}$ has a solution with coordinates in $\mathbb{Q}$.

NOTES Remarkably, Serre's conjecture 9.3 has been proved. The level 1 case was proved by Khare (2006), and the full conjecture by Khare and Wintenberger (2009). As described by Darmon 2017, the proof uses a glorious extension of Wiles's 3 – 5 switching technique in which essentially all the primes are used. For an engaging account of the collaboration that led to the proof of the conjecture, see Khare 2020.

NOTES Among the many works inspired by the proof of Fermat's Last Theorem, I mention only

- ◇ the book Mozzochi 2000, which gives an engaging eyewitness account of the events surrounding the proof,
- ◇ the book Diamond and Shurman 2005, which gives a much more detailed description of the modularity theorem and its background than was possuble in this chapter,
- ◇ the book of the 1995 instructional conference (Cornell et al. 1997) devoted to explaining the work of Ribet and others on Serre's conjecture and of Wiles and others on the modularity conjecture, and
- ◇ the article Darmon 2017, which includes a brief summary of some of the work inspired by "Wiles's marvelous proof".

10 Epilogue

In a lecture in 2017, Serre roughly divided the history of number theory into two parts: the abelian period from the time of Gauss and his quadratic reciprocity law through Artin and his general reciprocity law to the mid 1960s, and the nonabelian period from the mid 1960s to the present. According to Serre, the change was marked by Grothendieck's introduction of motives (about 1964) and Langlands's announcement of his program (about 1966). We briefly explain what these are, and their relation to elliptic curves.

Motives

We have seen that the zeta function of an elliptic curve over $\mathbb{Q}$ decomposes as

$$\zeta(E,s) = \frac{\zeta(s)\cdot\zeta(s-1)}{L(E,s)},$$

where $\zeta(s)$ is the usual Riemann zeta function. The fundamental idea of motives is that underlying this, and all similar decompositions, there is a decomposition of the "motive" of E,

$$h(E) = h^0(E) \oplus h^1(E) \oplus h^2(E).$$

Here $h^1(E)$ has zeta function $L(E,s)$, and $h^0(E)$ has zeta function $\zeta(s)$, and so should equal the motive $h(\mathbb{P}^0)$ of a point over $\mathbb{Q}$. The zeta function of $h^2(E)$ is a shift of $\zeta(s)$, which means that $h^2(E)$ should be a "Tate twist" of $h(\mathbb{P}^0)$.

Consider smooth projective varieties X and Y over a field k. An irreducible subvariety Z in $X \times Y$ of codimension $\dim(X)$ defines a correspondence from Y to X — it can be thought of as a many-valued map. The formal sums with $\mathbb{Q}$-coefficients of such subvarieties, modulo a suitable equivalence relation, form the $\mathbb{Q}$-vector space $C(Y,X)$ of correspondences from Y to X. The smooth projective varieties over k form a category with the correspondences as morphisms . The category of motives $\mathsf{Mot}(k)$ over k is obtained from this category by adjoining the kernels of idempotents and inverting the motive $h^2(\mathbb{P}^1)$. Every smooth projective variety X over k defines a motive $h(X)$, and motives over number fields define zeta functions. For example, the motive $h(E)$ of an elliptic curve over $\mathbb{Q}$ does decompose as described above. The conjecture of Birch and Swinnerton-Dyer can be stated for an arbitrary motive over a number field (Bloch, Kato, ...).

It is possible to form tensor products of motives. When we take the equivalence relation defining the correspondences to be numerical equivalence, $\mathsf{Mot}(k)$ is an abelian category whose Hom-groups are finite-dimensional vector spaces. When k has characteristic zero, the subcategory generated by a finite collection of motives is equivalent to the category of representations of a reductive algebraic group, called the ***motivic Galois group.***

For a brief elementary introduction to motives, see Milne 2013.

The Langlands program

What is the Langlands program? As Casselman says, it is all about L-functions.

In number theory, many phenomena cannot be understood algebraically. For example, to understand how the prime numbers are distributed, we need the Riemann zeta function. As another example, Artin's reciprocity law gives an algebraic description of the decomposition of primes in an abelian extension of number fields, but there is no such algebraic description for a nonabelian extension. Finally, we have seen that to understand the arithmetic of an elliptic curve over $\mathbb{Q}$, even its rational points, we need its L-function.

In brief, since the time of Euler, number theorists have been introducing functions in order to try to understand the phenomena they are studying. Usually, these are L-functions that have Euler product expansions and (conjecturally) a functional equation of a very specific type. The fundamental question is: what are the L-functions that arise in this way?

In 1966 Langlands provided a conjectural answer. Given a reductive algebraic group over $\mathbb{Q}$, he described a way of constructing L-functions that have Euler product expansions, and for each L-function, he constructed the various terms (N, the Γ-factors, the constant term, ...) that should go into its functional equation. These L-functions are said to be ***automorphic***. When the reductive group is GL_2, they include the Mellin transforms of modular forms. Every L-function arising naturally in number theory should be automorphic. In particular, all zeta functions of motives should be automorphic. The modularity theorem can now be interpreted as saying that the zeta function of the motive $h^1(E)$ of an elliptic curve E over $\mathbb{Q}$ is automorphic.

References

ARTIN, M. AND SWINNERTON-DYER, H. P. F. 1973. The Shafarevich–Tate conjecture for pencils of elliptic curves on $K3$ surfaces. *Invent. Math.* 20:249–266.

ATKIN, A. O. L. AND LEHNER, J. 1970. Hecke operators on $\Gamma_0(m)$. *Math. Ann.* 185:134–160.

BARNET-LAMB, T., GERAGHTY, D., HARRIS, M., AND TAYLOR, R. 2011. A family of Calabi-Yau varieties and potential automorphy II. *Publ. Res. Inst. Math. Sci.* 47:29–98.

BEST, A. J., BOBER, J., BOOKER, A. R., COSTA, E., CREMONA, J., DERICKX, M., LOWRY-DUDA, D., LEE, M., ROE, D., SUTHERLAND, A. V., AND VOIGHT, J. 2020. Computing classical modular forms. arXiv:2002.04717.

BHARGAVA, M., SKINNER, C., AND ZHANG, W. 2019. A majority of elliptic curves over $\mathbb{Q}$ satisfy the Birch and Swinnerton-Dyer conjecture (June 22, 2019). arXiv:1407.1826.

BIRCH, B. J. 1969a. Diophantine analysis and modular functions, pp. 35–42. *In* Algebraic Geometry (Internat. Colloq., Tata Inst. Fund. Res., Bombay, 1968). Oxford Univ. Press, London.

BIRCH, B. J. 1969b. Weber's class invariants. *Mathematika* 16:283–294.

BIRCH, B. J. 1970. Elliptic curves and modular functions, pp. 27–32. *In* Symposia Mathematica, Vol. IV (INDAM, Rome, 1968/69). Academic Press, London.

BIRCH, B. J. 1975. Heegner points of elliptic curves, pp. 441–445. *In* Symposia Mathematica, Vol. XV (Convegno di Strutture in Corpi Algebrici, INDAM, Rome, 1973). Academic Press, London.

BIRCH, B. J. 2004. Heegner points: the beginnings, pp. 1–10. *In* Heegner points and Rankin L-series, volume 49 of *Math. Sci. Res. Inst. Publ.* Cambridge Univ. Press, Cambridge.

BIRCH, B. J. AND SWINNERTON-DYER, H. P. F. 1963. Notes on elliptic curves. I. *J. Reine Angew. Math.* 212:7–25.

BIRCH, B. J. AND SWINNERTON-DYER, H. P. F. 1965. Notes on elliptic curves. II. *J. Reine Angew. Math.* 218:79–108.

BRIESKORN, E. AND KNÖRRER, H. 1986. Plane algebraic curves. Birkhäuser Verlag, Basel. Translated from the German by John Stillwell.

BRÖKER, R., LAUTER, K., AND SUTHERLAND, A. V. 2012. Modular polynomials via isogeny volcanoes. *Math. Comp.* 81:1201–1231.

CARTAN, H. 1963. Elementary theory of analytic functions of one or several complex variables. Éditions Scientifiques Hermann, Paris.

CASSELS, J. W. S. 1962a. Arithmetic on curves of genus 1. III. The Tate–Šafarevič and Selmer groups. *Proc. London Math. Soc. (3)* 12:259–296.

CASSELS, J. W. S. 1962b. Arithmetic on curves of genus 1. IV. Proof of the Hauptvermutung. *J. Reine Angew. Math.* 211:95–112.

CASSELS, J. W. S. 1964. Arithmetic on curves of genus 1. VII. The dual exact sequence. *J. Reine Angew. Math.* 216:150–158.

CASSELS, J. W. S. 1965. Arithmetic on curves of genus 1. VIII. On conjectures of Birch and Swinnerton-Dyer. *J. Reine Angew. Math.* 217:180–199.

CASSELS, J. W. S. 1966. Diophantine equations with special reference to elliptic curves. *J. London Math. Soc.* 41:193–291.

CASSELS, J. W. S. 1986. Mordell's finite basis theorem revisited. *Math. Proc. Cambridge Philos. Soc.* 100:31–41.

CASSELS, J. W. S. 1991. Lectures on elliptic curves, volume 24 of *London Mathematical Society Student Texts*. Cambridge University Press, Cambridge.

COATES, J. 2013. Lectures on the Birch–Swinnerton-Dyer conjecture. *ICCM Not.* 1:29–46.

COATES, J. 2016. The conjecture of Birch and Swinnerton-Dyer, pp. 207–223. *In* Open problems in mathematics. Springer-Verlag, New York.

COATES, J. 2017. The oldest problem. *ICCM Not.* 5:8–13.

COATES, J. AND WILES, A. 1977. On the conjecture of Birch and Swinnerton-Dyer. *Invent. Math.* 39:223–251.

COHEN, H. 1993. A course in computational algebraic number theory, volume 138 of *Graduate Texts in Mathematics*. Springer-Verlag, New York.

COHEN, H. 2000. Advanced topics in computational number theory, volume 193 of *Graduate Texts in Mathematics*. Springer-Verlag, New York.

CONWAY, J. H. AND SLOANE, N. J. A. 1999. Sphere packings, lattices and groups, volume 290 of *Grundlehren der Mathematischen Wissenschaften*. Springer-Verlag, New York, third edition.

CORNELL, G., SILVERMAN, J. H., AND STEVENS, G. 1997. Modular forms and Fermat's last theorem. Springer-Verlag, New York. Papers from the Instructional Conference on Number Theory and Arithmetic Geometry held at Boston University, Boston, MA, August 9–18, 1995.

COX, D., LITTLE, J., AND O'SHEA, D. 2007. Ideals, varieties, and algorithms. Undergraduate Texts in Mathematics. Springer-Verlag, New York, third edition. An introduction to computational algebraic geometry and commutative algebra.

CREMONA, J. E. 1997. Algorithms for modular elliptic curves. Cambridge University Press, Cambridge, second edition.

DARMON, H. 2017. Andrew Wiles's marvelous proof. *Notices Amer. Math. Soc.* 64:209–216.

DEDEKIND, R. AND WEBER, H. 1882. Theorie der algebraischen Functionen einer Veränderlichen. *J. Reine Angew. Math.* 92:181–290.

DELIGNE, P. 1971. Formes modulaires et représentations l-adiques, pp. Exp. No. 355, 139–172. *In* Séminaire Bourbaki. Vol. 1968/69: Exposés 347–363, volume 175 of *Lecture Notes in Math.* Springer-Verlag, New York.

DELIGNE, P. 1974. La conjecture de Weil. I. *Inst. Hautes Études Sci. Publ. Math.* pp. 273–307.

DEURING, M. 1941. Die Typen der Multiplikatorenringe elliptischer Funktionenkörper. *Abh. Math. Sem. Hansischen Univ.* 14:197–272.

DIAMOND, F. AND SHURMAN, J. 2005. A first course in modular forms, volume 228 of *Graduate Texts in Mathematics*. Springer-Verlag, New York.

DICKSON, L. E. 1966. History of the theory of numbers. Vol. II: Diophantine analysis. Chelsea Publishing Co., New York.

DUMMIGAN, N. 1995. The determinants of certain Mordell-Weil lattices. *Amer. J. Math.* 117:1409–1429.

EICHLER, M. 1954. Quaternäre quadratische Formen und die Riemannsche Vermutung für die Kongruenzzetafunktion. *Arch. Math.* 5:355–366.

FALTINGS, G. 1983. Endlichkeitssätze für abelsche Varietäten über Zahlkörpern. *Invent. Math.* 73:349–366.

FISHER, T. 2018. A formula for the Jacobian of a genus one curve of arbitrary degree. *Algebra Number Theory* 12:2123–2150.

FREY, G. 1986. Links between stable elliptic curves and certain Diophantine equations. *Ann. Univ. Sarav. Ser. Math.* 1:iv+40.

FULTON, W. 1969. Algebraic curves. An introduction to algebraic geometry. W. A. Benjamin, Inc., New York-Amsterdam. Second edition 1989. Third edition 2008 freely available at `http://www.math.lsa.umich.edu/~wfulton/`.

FULTON, W. 1995. Algebraic topology, volume 153 of *Graduate Texts in Mathematics*. Springer-Verlag, New York.

GOLDFELD, D. 1982. Sur les produits partiels eulériens attachés aux courbes elliptiques. *C. R. Acad. Sci. Paris Sér. I Math.* 294:471–474.

GROSS, B. H. AND ZAGIER, D. B. 1986. Heegner points and derivatives of L-series. *Invent. Math.* 84:225–320.

HARTSHORNE, R. 1977. Algebraic geometry. Springer-Verlag, New York.

HASSE, H. 1936. Zur Theorie der abstrakten elliptischen Funktionenkörper. *J. Reine Angew. Math.* 175:55–62, 69–82, 193–208.

HEEGNER, K. 1952. Diophantische Analysis und Modulfunktionen. *Math. Z.* 56:227–253.

HINDRY, M. AND SILVERMAN, J. H. 2000. Diophantine geometry, volume 201 of *Graduate Texts in Mathematics*. Springer-Verlag, New York.

KHARE, C. 2006. Serre's modularity conjecture: the level one case. *Duke Math. J.* 134:557–589.

KHARE, C. 2020. Jean-Pierre Wintenberger. *Notices Amer. Math. Soc.* 67:61–63.

KHARE, C. AND WINTENBERGER, J.-P. 2009. Serre's modularity conjecture. I,II. *Invent. Math.* 178:485–586.

KNAPP, A. W. 1992. Elliptic curves, volume 40 of *Mathematical Notes*. Princeton University Press, Princeton, NJ.

KOBLITZ, N. 1977. p-adic numbers, p-adic analysis, and zeta-functions. Springer-Verlag, New York.

KOBLITZ, N. 1994. A course in number theory and cryptography, volume 114 of *Graduate Texts in Mathematics*. Springer-Verlag, New York, second edition.

KODAIRA, K. 1960. On compact analytic surfaces, pp. 121–135. *In* Analytic functions. Princeton Univ. Press, Princeton, N.J. (= Collected Works, Vol. III, [51].).

KOLYVAGIN, V. A. 1988a. Finiteness of $E(\mathbb{Q})$ and $Ш(E,\mathbb{Q})$ for a subclass of Weil curves. *Izv. Akad. Nauk SSSR Ser. Mat.* 52:522–540, 670–671.

KOLYVAGIN, V. A. 1988b. The Mordell-Weil and Shafarevich-Tate groups for Weil elliptic curves. *Izv. Akad. Nauk SSSR Ser. Mat.* 52:1154–1180, 1327.

KRIZ, D. 2020. Supersingular main conjectures, Sylvester's conjecture and Goldfeld's conjecture. arXiv:2002.04767.

LANG, S. AND TATE, J. 1958. Principal homogeneous spaces over abelian varieties. *Amer. J. Math.* 80:659–684.

LEE, J. M. 2003. Introduction to smooth manifolds, volume 218 of *Graduate Texts in Mathematics*. Springer-Verlag, New York.

LIND, C.-E. 1940. Untersuchungen über die rationalen Punkte der ebenen kubischen Kurven vom Geschlecht Eins. Thesis, University of Uppsala.

LIU, J. 2019. Legendre's theorem, Hasse invariant and Jacobi symbol. arXiv:1912.11750.

LORENZINI, D. 1996. An invitation to arithmetic geometry, volume 9 of *Graduate Studies in Mathematics*. American Mathematical Society, Providence, RI.

MILNE, J. S. 1968. The Tate-Šafarevič group of a constant abelian variety. *Invent. Math.* 6:91–105.

MILNE, J. S. 1975. On a conjecture of Artin and Tate. *Ann. of Math. (2)* 102:517–533.

MILNE, J. S. 1981. Comparison of the Brauer group with the Tate-šafarevič group. *J. Fac. Sci. Univ. Tokyo Sect. IA Math.* 28:735–743 (1982).

MILNE, J. S. 1986a. Abelian varieties, pp. 103–150. *In* Arithmetic geometry (Storrs, Conn., 1984). Springer, New York.

MILNE, J. S. 1986b. Jacobian varieties, pp. 167–212. *In* Arithmetic geometry (Storrs, Conn., 1984). Springer, New York.

MILNE, J. S. 1999. Descent for Shimura varieties. *Michigan Math. J.* 46:203–208.

MILNE, J. S. 2013. Motives — Grothendieck's dream, pp. 325–342. *In* Open problems and surveys of contemporary mathematics, volume 6 of *Surv. Mod. Math.* Int. Press, Somerville, MA.

MILNE, J. S. 2016. The Riemann hypothesis over finite fields from Weil to the present day, pp. 487–565. *In* The legacy of Bernhard Riemann after one hundred and fifty years. Vol. II, volume 35 of *Adv. Lect. Math. (ALM)*. Int. Press, Somerville, MA.

MILNE, J. S. 2017. Algebraic groups, volume 170 of *Cambridge Studies in Advanced Mathematics*. Cambridge University Press, Cambridge.

MORDELL, L. 1922. On the rational solutions of the indeterminate equations of the third and fourth degrees. *Proc. Cambridge Philos. Soc.* 21.

MOZZOCHI, C. J. 2000. The Fermat diary. American Mathematical Society, Providence, RI.

OESTERLÉ, J. 1990. Empilements de sphères. *Astérisque* pp. Exp. No. 727, 375–397. Séminaire Bourbaki, Vol. 1989/90.

OORT, F. AND SCHAPPACHER, N. 2016. Early history of the Riemann hypothesis in positive characteristic, pp. 595–631. *In* The legacy of Bernhard Riemann after one hundred and fifty years. Vol. II, volume 35 of *Adv. Lect. Math. (ALM)*. Int. Press, Somerville, MA.

PARK, J., POONEN, B., VOIGHT, J., AND WOOD, M. M. 2019. A heuristic for boundedness of ranks of elliptic curves. *J. Eur. Math. Soc. (JEMS)* 21:2859–2903.

POONEN, B. 2018. Heuristics for the arithmetic of elliptic curves. *In* Proceedings of the International Congress of Mathematicians—Rio de Janeiro 2018. Vol. II. Invited lectures, pp. 399–414. World Sci. Publ., Hackensack, NJ.

RICE, A. AND BROWN, E. 2012. Why ellipses are not elliptic curves. *Math. Mag.* 85:163–176.

ROQUETTE, P. 2018. The Riemann hypothesis in characteristic p in historical perspective, volume 2222 of *Lecture Notes in Mathematics*. Springer-Verlag, New York. History of Mathematics Subseries.

RUBIN, K. 1987. Tate–Shafarevich groups and L-functions of elliptic curves with complex multiplication. *Invent. Math.* 89:527–559.

SAMUEL, P. 2008. Algebraic theory of numbers. Dover, Mineola, New York.